Shock-Dominated Flow

Shock-Dominated Flow

Editors

Hexia Huang
Ye Tian
Huijun Tan

MDPI

Basel • Beijing • Wuhan • Barcelona • Belgrade • Novi Sad • Cluj • Manchester

Editors
Hexia Huang
Nanjing University of
Aeronautics and Astronautics
Nanjing 210016
China

Ye Tian
China Aerodynamic Research
and Development Center
Mianyang
China

Huijun Tan
Nanjing University of
Aeronautics and Astronautics
Nanjing
China

Editorial Office
MDPI AG
Grosspeteranlage 5
4052 Basel, Switzerland

This is a reprint of articles from the Special Issue published online in the open access journal *Aerospace* (ISSN 2226-4310) (available at: https://www.mdpi.com/journal/aerospace/special_issues/9211TG3Z74).

For citation purposes, cite each article independently as indicated on the article page online and as indicated below:

Lastname, A.A.; Lastname, B.B. Article Title. *Journal Name* **Year**, *Volume Number*, Page Range.

ISBN 978-3-7258-2011-5 (Hbk)
ISBN 978-3-7258-2012-2 (PDF)
doi.org/10.3390/books978-3-7258-2012-2

Contents

Contents

About the Editors

Hexia Huang

Dr. Hexia Huang is an associate professor who received Bachelor's and doctorate degrees from the Nanjing University of Aeronautics and Astronautics in 2011 and 2017, respectively. Huang is a postdoctoral researcher at Tsinghua University and a member of the American Institute of Aeronautics and Astronautics (AIAA), the Chinese Society of Aeronautics and Astronautics, and the Chinese Society of Astronautics. Huang was selected for the Young Talented Lift Scientists Sponsorship Program and "Changkong Star" Scholar Program of Nanjing University of Aeronautics and Astronautics. Huang has been engaged in the research of inlets of aerospace propulsion systems for a long time and has conducted a large amount of research in the fields of internal flow aerodynamics of inlets, shock wave–boundary layer interactions, the aerodynamic design and flow mechanism of high-speed inlets, flow control technologies, and inlet/engine/nozzle integration. He has undertaken over 30 scientific research projects and has published over 40 academic papers in the Journal of Fluid Mechanics, Physics of Fluids, AIAA Journal, etc.

Ye Tian

Dr. Ye Tian obtained his PhD degree in 2016 from the China Aerodynamics Research and Development Center (CARDC) and won the excellent doctoral thesis of the Chinese Society of Mechanics. Currently, he is a researcher at CARDC. He is mainly engaged in scramjet combustion organization design, including design technology based on artificial intelligence, and serves as the secretary of the Combustion Aerodynamics Committee and a member of the Scramjet Committee. He has published more than 100 papers and has served on the editorial board of several journals, such as the Journal of Experiments in Fluid Mechanics.

Huijun Tan

Professor Huijun Tan received his Bachelor's and doctorate degrees from Nanjing University of Aeronautics and Astronautics in 1999 and 2003, respectively. He is the leader of the "Advanced Engine Inlet and Exhaust Technology", which is an innovation team awarded by the National Defense Science and Technology Institute. He has been elected as a recipient of the National Science Fund for Distinguished Young Scholars, National Defense Outstanding Young Scholars, etc. He is mainly engaged in research on engine inlet aerodynamics, inlet/aircraft integration technology, complex internal flow mechanisms, flow control, etc. He has won the first prize from the National Defense Technology Institute and the Ministry of Education. He has published over 100 papers and granted over 100 patents. One doctoral thesis under his guidance was awarded the National Excellent Doctoral Dissertation in Aerospace Science and Technology.

aerospace

MDPI

Editorial

Shock-Dominated Flow

He-Xia Huang

College of Energy and Power Engineering, Nanjing University of Aeronautics and Astronautics, Nanjing 210016, China; huanghexia@nuaa.edu.cn

This 2024 Special Issue of *Aerospace*, an open-access journal from MDPI, is entitled "Shock-Dominated Flow" and was guest-edited by Dr. He-xia Huang, Professor Hui-jun Tan, and Professor Ye Tian. It comprises 10 articles, primarily focusing on the fluid mechanics and associated flow control methods of shock-dominated flows. These contributions offer new insights into this academic research field.

For high-speed aircraft [1,2], engines [3,4], or missiles [5], the external and internal flow is characterized by shock-dominated flow [6], which determines the aerodynamic performance, practical envelope, and flight range. Specifically utilizing the compression effect of shocks, high-speed engines can decelerate the incoming supersonic/hypersonic flow to a suitable range to match the ramjet or scramjet combustion [7,8]. However, the shock also induces unfavorable drag [9], pressure/thermal load [10], flameout, and even structural failure [11]. Therefore, an efficient flow control method for shock-dominated flow is required [12]. With the development of the aerospace field towards higher speeds, better performance, and more intelligent control [13,14], there is an urgent need to propose related theories, reveal the flow mechanism of shock-dominated flow, and develop some flow control methods to eliminate the accompanying hazard.

This Special Issue publishes recent advances in shock-dominated flow features and flow control methods related to aerospace. The hypersonic inlet is an aerodynamic interface between the aircraft and the engine, which utilizes a series of shocks to compress the incoming flow [8]. Due to the existence of a surface boundary layer, it faces severe shock–boundary layer interactions [15]. Gao et al. [16] numerically studied the transient flow evolution in a hypersonic inlet/isolator under incoming wind shear and found that the cowl-shock-induced separation bubble moved downstream and upstream, with the total pressure recovery coefficients increasing by approximately 10%. Moreover, the wind shear had substantial impacts on the downstream shock train. While the shock train was located near the throat initially, the wind shear may force it to move upstream, resulting in an inlet unstart. Therefore, the operation of a hypersonic inlet/isolator should consider the wind shear effect. When the hypersonic nozzle operated at a low nozzle pressure ratio state, the flow was over-expanded, which induced a shock within the nozzle [17]. Yu et al. [18] numerically analyzed the effect of external flow on the shock-induced separation in a single-expansion ramp nozzle. As the external flow Mach number increased, the internal flow separation experienced a transition from restricted shock separation (RSS) to free shock separation (FSS) and was finally converted to a fully unseparated state. As the separation induced by the shock behaved with some unsteadiness, Wang et al. [19] proposed a dynamic mode decomposition criterion for the spiked-blunt body flow at Ma = 2.2. The results showed that using the energy sorting criterion, the dynamic mode decomposition (DMD) method had an advantage in identifying the dominated flow structures of such an unsteady flow. Moreover, they observed that the spiked-blunt flow appeared with multiple dominated frequencies, among which, the primary frequency was 3.3 kHz, originating from the periodic motion of the aftershock.

To control the shock–boundary layer interaction well, several flow control methods have been put forward [12]. Wang et al. [20] proposed a vortex generator with high-frequency oscillation to control the shock wave–boundary layer interaction (SWBLI) in

Citation: Huang, H.-X. Shock-Dominated Flow. *Aerospace* **2024**, *11*, 686. https://doi.org/10.3390/aerospace11080686

Received: 19 August 2024
Accepted: 20 August 2024
Published: 21 August 2024

the inlet. The "extrusion" and "suction" effects during the oscillation process charged the airflow, which enhanced the momentum exchange. The unsteady numerical results demonstrated that this method effectively suppressed the shock-induced separation, with the separation bubble reduced by 31.76% and the total pressure recovery coefficient increased by 6.4%. Yu et al. [21] proposed a passive flow control method based on micro-serrations; the height of the hair of serration was lower than the thickness of the boundary layer δ. The separation length was able to be shortened by 9.13%, with a leading stair of 0.1δ, a depth of the subsequent serrations of 0.2δ, and a width of 0.05δ. Compared to a vortex generator or a micro-serration, the plasma-based flow control method, as an active flow control method, had a shorter response time, an ability to regulate the control intensity, high levels of injected momentum, and no additional mass loss [22], which has become a hot research topic in recent years [23,24]. Yang et al. [25] utilized surface arc plasma actuators to control the two-stage compression corner shocks–boundary layer interaction. A wind tunnel experiment under Mach number 6.0 was conducted. The experimental results showed that this method could weaken the shock intensity to a certain extent. As the discharging voltage increased from 0.5 kV to 1 kV, the influence range of the hot plume was able to extend from 65 mm to 85 mm. Furthermore, Yang et al. [26] used a 30-channel discharge array to control the shock–boundary layer interaction and shock–shock interaction in a hypersonic double-wedge to reduce the wave drag, thermal load, and pressure load. The Edney V-type shock–shock interaction was effectively controlled, and such an interaction disappeared or was intermittent when the jet plume emerged. Cai et al. [27] used a dielectric barrier discharge plasma actuator to lower the noise level in a Mach 4.0 cavity flow. The delayed detached eddy simulation (DDES) and plasma phenomenological model were built as described in the paper. The results demonstrated that the dielectric barrier discharge (DBD) plasma actuator effectively suppressed the supersonic cavity flow noise by 2.27 dB. The movement of a dominating vortex was changed to affect the maximum noise level.

For the muzzle launch system, owing to the high levels of kinetic energy at the muzzle, a significant shock diffraction phenomenon occurs when the muzzle expels out of the tubes [28,29]. Li et al. [30,31] experimentally and numerically studied transient shock evolution during muzzle jets and their interaction with the confined boundaries. The initial shock–shock collisions were formed, which delayed the evolution of the shocks and multiple reflected shocks. As the adjacent boundaries confined the expansion of the jet, the jet exhibited a circumferential asymmetric shape and induced transverse flow, forming a complex vortical flow. Once the jet approached the ground, the shock and the vortices were intensified, yielding a reflected shock, which increased the flight Mach number of the moving body from 1.4 to 1.6.

Acknowledgments: The Editors of this Special Issue would like to thank each of the authors for their contributions and for making this Special Issue a success. Additionally, the Guest Editors would like to thank the reviewers and the *Aerospace* Editorial Office.

Conflicts of Interest: The author declares no conflict of interest.

References

1. Küchemann, D. Hypersonic aircraft and their aerodynamic problems. *Prog. Aerosp. Sci.* **1965**, *6*, 271–353.
2. Van Wie, D.M. Hypersonics: Past, present, and potential future. *Johns Hopkins APL Tech. Dig.* **2021**, *35*, 335–341.
3. Zhao, D. Ramjets/Scramjets aerodynamics: A progress review. *Prog. Aerosp. Sci.* **2023**, *143*, 100958.
4. Urzay, J. Supersonic combustion in air-breathing propulsion systems for hypersonic flight. *Annu. Rev. Fluid Mech.* **2018**, *50*, 593–627.
5. Tracy, C.L.; Wright, D. Modeling the performance of hypersonic boost-glide missiles. *Sci. Glob. Secur.* **2020**, *28*, 135–170.
6. Matsuo, K.; Miyazato, Y.; Kim, H.D. Shock train and pseudo-shock phenomena in internal gas flows. *Prog. Aerosp. Sci.* **1999**, *35*, 33–100.
7. Gnani, F.; Zare-Behtash, H.; Kontis, K. Pseudo-shock waves and their interactions in high-speed intakes. *Prog. Aerosp. Sci.* **2016**, *82*, 36–56.
8. Huang, H.; Tan, H.; Li, F.; Tang, X.; Qin, Y.; Xie, L.; Xu, Y.; Li, C.; Gao, S.; Zhang, Y.; et al. A review of the shock-dominated flow in a hypersonic inlet/isolator. *Prog. Aerosp. Sci.* **2023**, *143*, 100952.

9. Ahmed, M.Y.M.; Qin, N. Forebody shock control devices for drag and aero-heating reduction: A comprehensive survey with a practical perspective. *Prog. Aerosp. Sci.* **2020**, *112*, 100585.
10. McQuellin, L.P.; Neely, A.; Currao, G. Considerations for a hypersonic flight test investigating fluid-thermal-structural interactions. In Proceedings of the 23rd AIAA International Space Planes and Hypersonic Systems and Technologies Conference, Montréal, QC, Canada, 10–12 March 2020.
11. Chang, J.; Li, N.; Xu, K.; Bao, W.; Yu, D. Recent research progress on unstart mechanism, detection and control of hypersonic inlet. *Prog. Aerosp. Sci.* **2017**, *89*, 1–22.
12. Huang, W.; Wu, H.; Yang, Y.; Yan, L.; Li, S.-B. Recent advances in the shock wave/boundary layer interaction and its control in internal and external flows. *Acta Astronaut.* **2020**, *174*, 103–122.
13. Rabault, J.; Kuchta, M.; Jensen, A.; Réglade, U.; Cerardi, N. Artificial neural networks trained through deep reinforcement learning discover control strategies for active flow control. *J. Fluid Mech.* **2019**, *865*, 281–302.
14. Ma, Y.; Guo, M.; Tian, Y.; Le, J. Recent advances and prospects in hypersonic inlet design and intelligent optimization. *Aerosp. Sci. Technol.* **2024**, *146*, 108953.
15. Holger, B.; Harvey, J.K. (Eds.) *Shock Wave-Boundary-Layer Interactions*; Cambridge University Press: Cambridge, UK, 2011; Volume 32.
16. Gao, S.; Huang, H.; Meng, Y.; Tan, H.; Liu, M.; Guo, K. Transient Flow Evolution of a Hypersonic Inlet/Isolator with Incoming Windshear. *Aerospace* **2023**, *10*, 1021. [CrossRef]
17. Lv, Z.; Xu, J.; Song, G.; Li, R.; Ge, J. Review on the aerodynamic issues of the exhaust system for scramjet and turbine based combined cycle engine. *Prog. Aerosp. Sci.* **2023**, *143*, 100956.
18. Yu, Y.; Mao, Y.; Yu, T.; Yang, Y.; Xu, S.; Liang, S. The Influence of External Flow Field on the Flow Separation of Overexpanded Single-Expansion Ramp Nozzle. *Aerospace* **2023**, *10*, 958. [CrossRef]
19. Wang, Y.; Xu, J.; Qin, Q.; Guan, R.; Cai, L. Experimental Investigation of the Shock-Related Unsteadiness around a Spiked-Blunt Body Based on a Novel DMD Energy Sorting Criterion. *Aerospace* **2024**, *11*, 188. [CrossRef]
20. Wang, M.; Wang, Z.; Zhang, Y.; Cheng, D.; Tan, H.; Wang, K.; Gao, S. Control of Cowl Shock/Boundary Layer Interaction in Supersonic Inlet Based on Dynamic Vortex Generator. *Aerospace* **2023**, *10*, 729. [CrossRef]
21. Yu, F.; Gao, Z.; Zhang, Q.; Yue, L.; Chen, H. Mitigation of Shock-Induced Separation Using Square-Shaped Micro-Serrations—A Preliminary Study. *Aerospace* **2024**, *11*, 148. [CrossRef]
22. Wu, Y.; Li, Y. Progress and outlook of plasma flow control. *Acta Aeronaut. Astronaut. Sin.* **2015**, *36*, 381–405.
23. Huang, H.-X.; Tan, H.-J.; Sun, S.; Zhang, Y.-C.; Cheng, L. Transient interaction between plasma jet and supersonic compression ramp flow. *Phys. Fluids* **2018**, *30*, 041703.
24. Yang, H.; Liang, H.; Yang, B.; Zong, H.; Li, J.; Zhang, D.; Xiong, Y.; Wu, Y.; Li, Y. Experimental study on the hypersonic double incident shock wave/boundary layer interaction regulated by plasma actuation array. *Phys. Fluids* **2024**, *36*, 066116.
25. Yang, B.; Yang, H.; Zhang, C.; Zhao, N.; Liang, H.; Zhang, D. Experimental Investigation on the Control of Hypersonic Shock Wave/Boundary Layer Interaction Using Surface Arc Plasma Actuators at Double Compression Corner. *Aerospace* **2023**, *10*, 1016. [CrossRef]
26. Yang, B.; Yang, H.; Zhao, N.; Liang, H.; Su, Z.; Zhang, D. Experimental Study on Hypersonic Double-Wedge Induced Flow Based on Plasma Active Actuation Array. *Aerospace* **2024**, *11*, 60. [CrossRef]
27. Cai, H.; Zhang, Z.; Li, Z.; Li, H. Noise Prediction and Plasma-Based Control of Cavity Flows at a High Mach Number. *Aerospace* **2023**, *10*, 922. [CrossRef]
28. Li, Z.; Wang, H.; Chen, C.; Huang, H. Effects of different constrained boundary structures on the evolution of shock waves and vortexes in muzzle jets. *J. Mech. Sci. Technol.* **2023**, *37*, 5239–5249.
29. Qin, Y.; Huang, H.X.; Tang, X.B.; Tan, H.-J.; Li, F.-B. On the double-sided shock diffractions in quiescent and supersonic crossflows. *Phys. Fluids* **2024**, *36*, 016111.
30. Li, Z.; Wang, H. Evolution of Shock Waves during Muzzle Jet Impinging Moving Bodies under Different Constrained Boundaries. *Aerospace* **2023**, *10*, 908. [CrossRef]
31. Li, Z.; Wang, H. Mechanism of Evolution of Shock Wave of Muzzle Jet under Initial Interference and Its Simplified Model. *Aerospace* **2024**, *11*, 381. [CrossRef]

aerospace MDPI

Article

Transient Flow Evolution of a Hypersonic Inlet/Isolator with Incoming Windshear

Simin Gao [1], Hexia Huang [1,*], Yupeng Meng [2], Huijun Tan [1,*], Mengying Liu [1] and Kun Guo [2]

1 Key Laboratory of Inlet and Exhaust System Technology, Ministry of Education, College of Energy and Power Engineering, Nanjing University of Aeronautics and Astronautics, Nanjing 210016, China; gaosimin@nuaa.edu.cn (S.G.); mengyingliu@nuaa.edu.cn (M.L.)
2 Beijing Power Machinery Research Institute, Beijing 100074, China; xxshchwshry@sina.com (Y.M.); wwwwguokun@163.com (K.G.)
* Correspondence: huanghexia@nuaa.edu.cn (H.H.); tanhuijun@nuaa.edu.cn (H.T.)

Abstract: In this paper, a novel flow perturbation model meant to investigate the effects of incoming wind shear on a hypersonic inlet/isolator is presented. This research focuses on the transient shock/boundary layer interaction and shock train flow evolution in a hypersonic inlet/isolator with an on-design Mach number of 6.0 under incoming wind shear at high altitudes, precisely at an altitude of 30 km with a magnitude speed of 80 m/s. Despite the low intensity of wind shear at high altitudes, the results reveal that wind shear significantly disrupts the inlet/isolator flowfield, affecting the shock wave/boundary layer interaction in the unthrottled state, which drives the separation bubble at the throat to move downstream and then upstream. Moreover, the flowfield behaves as a hysteresis phenomenon under the effect of wind shear, and the total pressure recovery coefficients at the throat and exit of the inlet/isolator increase by approximately 10% to 12%. Furthermore, this research focuses on investigating the impact of wind shear on the behavior of the shock train. Once the inlet/isolator is in a throttled state, wind shear severely impacts the motion of the shock train. When the downstream backpressure is 135 times the incoming pressure (p_0), the shock train first moves upstream and gradually couples with a cowl shock wave/boundary layer interaction, resulting in a more significant separation at the throat, and then moves downstream and decouples from the separation bubble at the throat. However, if the downstream backpressure increases to 140 p_0, the shock train enlarges the separation bubble, forcing the inlet/isolator to fall into the unstart state, and it cannot be restarted. These findings emphasize the need to consider wind shear effects in the design and operation of hypersonic inlet/isolator.

Keywords: wind shear; hypersonic inlet/isolator; shock wave/boundary layer interaction; shock train; unsteady simulation

Citation: Gao, S.; Huang, H.; Meng, Y.; Tan, H.; Liu, M.; Guo, K. Transient Flow Evolution of a Hypersonic Inlet/Isolator with Incoming Windshear. *Aerospace* **2023**, *10*, 1021. https://doi.org/10.3390/aerospace10121021

Academic Editor: Sergey Leonov

Received: 17 October 2023
Revised: 4 December 2023
Accepted: 6 December 2023
Published: 9 December 2023

1. Introduction

The hypersonic inlet/isolator, an essential component of the scramjet engine, plays a significant role in the stable and safe operation of aircraft [1–7]. During flights, because of the complex wind field environment [8–11], aircraft frequently encounter atmospheric disturbances, such as gusts, wind shear, three-dimensional wind, vortices, and continuous turbulence. While vortex and gust disturbances are common at low altitudes, hypersonic vehicles typically operate at high altitudes exceeding 20 km, where the impact of gusts is negligible. The primary wind perturbation at these altitudes is wind shear, changes in the wind vector (wind direction and wind speed) in the horizontal and/or vertical directions [12,13]. When an aircraft encounters wind shear, it experiences temporary changes in flight attitude, which may subsequently alter the internal flow characteristics of the hypersonic inlet, including historical effects and hysteresis phenomena.

Presently, most studies on the impact of wind field disturbances on inlet/isolators concentrate predominantly on low-altitude disturbances, including gusts and wind shear.

Kozakiewicz et al. analyzed the influence of changes in gust speed and direction on the development of vortices in the pitot subsonic inlet of an F-16, finding that gusts could potentially result in compressor stalls and unstable engine operations [14]. Wu et al. performed a simulation study on the flowfield and aerodynamic responses of a serpentine inlet exposed to non-stationary horizontal sinusoidal gusts and found that the gusts not only significantly change the flow structure but also play an unfavorable role in the total pressure distortion of the serpentine inlet [15]. Hussei et al. focused on the impact of side gusts (with an intensity of 56 m/s) on the supersonic inlet and discovered that they significantly affect the shock wave and airflow characteristics within the inlet, potentially leading to unstable flow phenomena such as shock wave oscillation [16]. Etkin et al. proposed a wind shear model and analyzed the flight characteristics of an aircraft in wind shear conditions [17–19]. Xie et al. developed a dynamic model of an aircraft to simulate the impact of wind fields on its flight state, considering atmospheric disturbances, discontinuous gusts, and wind shear and specifically examined parameters such as angle of attack, sideslip angle, aerodynamic force, and aerodynamic moment [20]. Zhao conducted research on the effect of high-altitude wind shear on the flight of a rocket and found that wind shear significantly increased the wind attack angle [21]. Based on results from Zhao et al. [21], Yang et al. developed a wind shear model at an altitude ranging from 20 km to 78 km and obtained changes in the angle of attack regarding aircraft affected by wind shear [22].

While most current studies focus on the effect of low-altitude wind shear on low-speed aircraft, wind shear also exists when hypersonic vehicles fly at high altitudes. To be exact, the magnitudes of wind shear in the near space (altitude > 20 km) are in the order of O(~100 m/s) with a time scale in the order of O(~100 ms) [22]. The equivalent angle of attack can be as high as 3 degrees for a hypersonic vehicle operating at Mach 6, and the time scale is so short that it induces flow response problems. Thus, it is essential to further study the effect of wind shear perturbation on the flowfield and performance of hypersonic inlet/isolators. This paper is organized as follows: Section 2 introduces the wind shear perturbation model and the hypersonic inlet/isolator model. Section 3 delineates the details of the numerical simulation setup and validates the reliability of the simulation method. Based on this, Section 4 provides a detailed discussion on the transient flow evolution of shock wave/boundary layer interaction and shock train under wind shear. Finally, Section 5 discusses the key results obtained from the current study.

2. Introduction of the Hypersonic Inlet/Isolator and the Wind Shear Model

2.1. Description of the Hypersonic Inlet/Isolator

In this study, a mixed-compression hypersonic inlet/isolator with a capture height, y_{cowl}, of 334.11 mm for a design Mach number of 6.0 is described. The external compression system contains an oblique shock with a deflection angle of 8.73 degrees and a series of compression waves to decelerate the incoming hypersonic flow with minimal total pressure loss. In essence, this is a design methodology for the hypersonic inlet/isolator that emphasizes high compression efficiency. Throughout the entire compression process, the entropy of the main flow outside the boundary layer remains constant. The throat has a minimum flow area of H_{th} = 53 mm. The isolator has an expansion angle of 0.2 degrees and the value of $L_{iso} = 15.1H_{th}$. The main geometrical parameters of the inlet/isolator model are listed in Table 1. Note that the coordinate origin is set at the leading point of the ramp. The X-axis and Y-axis are shown in Figure 1.

Table 1. Model design parameters.

Parameters	Value
δ, deg	17.13
β, deg	8.73
H_{th}, mm	53
L_{cowl}	$5.57H_{th}$
y_{th}	$5.3H_{th}$
x_{cowl}	$20.38H_{th}$
y_{cowl}	$6.3H_{th}$
L_{iso}	$15.1H_{th}$

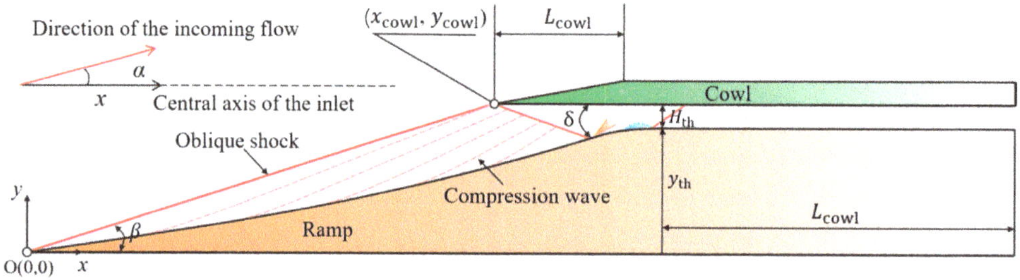

Figure 1. Inlet/isolator model.

2.2. Model for the Wind Shear

Zhao et al. [21] proposed a statistical analysis method for high-altitude wind fields. They concluded that the primary impact of wind shear on rocket aircraft is a large angle of attack induced by wind. Based on Zhao et al.'s statistical method, Yang [22] proposed a wind shear model for near space (20–78 km) based on five-year wind field data (2009 to 2013) from the Modern-Era Retrospective Analysis for Research and Applications (MERRA) [23]. The results demonstrated that the primary impact of wind shear on aircraft is the angle of attack induced by the wind. The intensity of wind shear varies from month to month, with the largest wind shear intensity in January. Taking 30 km as an example, the magnitude speed of wind shear in January is 80 m/s, and the angle of attack can change by up to 2.65 degrees when the aircraft flies at Mach 6. Based on this, the variation in the angle of attack (α) due to wind shear over time (t) in January is shown by Equation (1), drawn in the form of a normal distribution, with the change plotted in Figure 2.

$$\alpha(t) = 2.552\left\{\frac{\left[\frac{1}{0.029\sqrt{2\pi}}e^{-\frac{(t-0.05)^2}{2*0.029^2}}\right]}{10} - 0.33\right\} \quad (1)$$

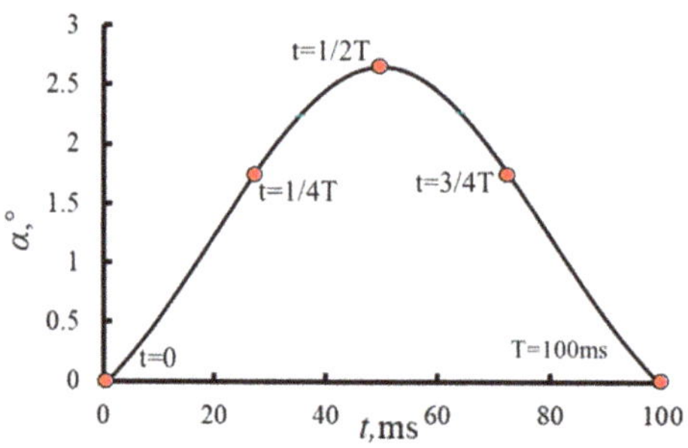

Figure 2. Distribution of wind angle of attack, α.

3. Numerical Method

This section discusses the numerical method, which includes the governing equations, the turbulence model, the boundary conditions, and the computational grid.

3.1. Computational Method

All the calculations are performed by the commercial computational fluid dynamics (CFD) software ANSYS FLUENT 18.0. It has demonstrated robust applicability in subsonic, supersonic, and hypersonic flow simulations, particularly in research related to hypersonic inlets, and been widely used by researchers globally.

Computational fluid dynamics (CFD) methods used to predict hypersonic flows include Reynolds-Averaged Navier–Stokes (RANS), Large Eddy Simulations (LES), Direct Numerical Simulation (DNS), etc. RANS demonstrates good accuracy in predicting macroscopic flows and strongly separated flows, but it has not consistently been accurate in capturing small-scale turbulence and predicting aerothermodynamic loading over the surface of hypersonic vehicles [24]. On the other hand, LES is suitable for capturing small-scale vortices, but it requires a large number of grids and has a lengthy computation cycle for unsteady simulation calculations [25]. DES is a suitable method for a genuinely time-dependent problem, but the accuracy is very dependent on the number of grids [26]. DNS excels in predicting the laminar–turbulent transition of boundary layers, which is particularly relevant in research focusing on aerodynamic heating, drag, and vehicle operation [27]. In our paper, we mainly focus on macroscopic large-scale flow, such as the shock wave/boundary layer interaction, shock train, etc., which differs from vortices in the turbulent boundary layer. Since these flow structures result from the inertial and macroscopic motion of fluid, the RANS method was chosen for simulation given the requirements of computation and accuracy.

This study utilizes the k-ω SST turbulence model developed by Menter [28]. The control equation is discretized using an upwind scheme with second-order accuracy. Molecular viscosity is calculated using Sutherland's formula. For the numerical calculation of the unsteady flowfield with dynamic changes in angle of attack, a steady flowfield was first calculated with an angle of attack of 0°, an incoming Mach number of 6, a static pressure of 1197 Pa, and a static temperature of 225.51 K. This served as the initial flowfield for unsteady numerical simulations. In these simulations, the dynamic changes in α are achieved using User-Defined Functions (UDFs), where the incoming flow conditions changed according to the wind shear model described in Equation (1). The time step for unsteady numerical simulations was set as 1 µs. The number of time intervals was determined by the period of the wind shear model, so the maximum number of iterations in each time interval was set to 250 steps.

3.2. Grid Generation

The model shown in Figure 1 was filled by hexahedral grids using the ICEM CFD software 21.0, as depicted in Figure 3. The grids are refined near the wall, ensuring a wall surface, y^+, of less than or equal to one. Given the complexity of the flows at the lip and throat, the grid in these regions was also refined to clearly capture the shock waves and separation flow.

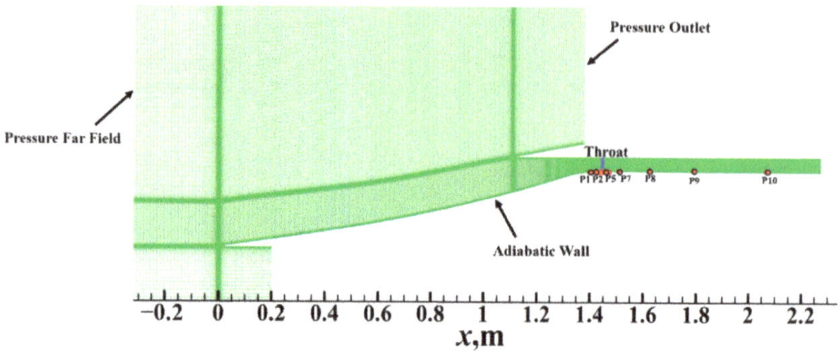

Figure 3. Computational grid and domain.

3.3. Boundary Conditions

In the simulations, the pressure far-field boundary was applied for the incoming freestream, and the wall surface condition was set to a non-slip adiabatic wall. In addition, the wind shear model mentioned above was applied to the incoming direction and loaded using User-Defined Functions (UDFs). The pressure and temperature at the pressure far field were set to correspond to the atmospheric pressure and temperature, respectively, at an altitude of 30 km.

The static pressure outside the inlet/isolator, including the static pressure at the inlet/isolator exit, was designated as the far-field static pressure to achieve an unthrottled state. Table 2 shows the freestream conditions used in the simulations. To monitor the transient flow evolution of the hypersonic inlet/isolator, ten monitoring points, labeled P1–P10, were set during the simulation, as shown in Figure 3, and the positions are listed in Table 3.

Table 2. Inlet/isolator data for the design conditions.

Parameter	Value
On-design freestream Mach number	6.0
Air model	Ideal gas
Altitude, km	30
Presser, Pa	1197.003
Temperature, K	226.509

Table 3. Position of the monitoring points.

Point	P1	P2	P3	P4	P5	P6	P7	P8	P9	P10
x, mm	1411	1425	1433	1440	1444	1446	1455	1625	1792	2065
y, mm	281	281	282	282	282	282	282	283	283	283

3.4. Inlet/Isolator Performance Parameters

The inlet/isolator performance is characterized in terms of the following parameters:

(1) The total pressure recovery coefficient (TPR) is the ratio of total pressure at the inlet/isolator exit (P_{out}*) to the freestream total pressure (P_{∞}*). The total pressure loss is the sum of shock and viscous losses. The total pressure at the exit is calculated with the mass-weighted average.

$$TPR = \frac{P^*_{out}}{P^*_{\infty}} \tag{2}$$

(2) The pressurization rate (PR) is one of the main indicators that characterize the compression characteristics of the inlet/isolator. It is defined as the ratio of inlet/isolator

exit static pressure (P_{out}) to inlet/isolator static pressure (P_∞). The static pressure at the exit is calculated with the mass-weighted average.

$$PR = \frac{P_{out}}{P_\infty} \quad (3)$$

3.5. Validation of the Numerical Method and Grid Sensitivity

This paper primarily focuses on the shock system and the shock train phenomenon in the hypersonic inlet/isolator. Consequently, this section validates the numerical method. The ability of the numerical method employed in this study to capture shock waves and separation flow is validated by comparing numerical results with experimental results. In the experiment, a hypersonic inlet with an isolator under Mach 4.92 conditions was investigated. The flow region was filled with structured quadrilateral meshes, with the mesh refined near the wall. To more clearly capture the shock wave and separation structure of the duct, the mesh of the duct was further refined. The incoming flow conditions in the simulation are identical to the experimental conditions [29]. Figure 4 compares the numerical Schlieren images obtained with different turbulence models provided by FLUENT and experimental Schlieren images of the inlet/isolator. As shown in Figure 4, only the SST κ-ω can predict such large-scale separation precisely; therefore, the turbulence model is selected to model the turbulence flow in the inlet/isolator.

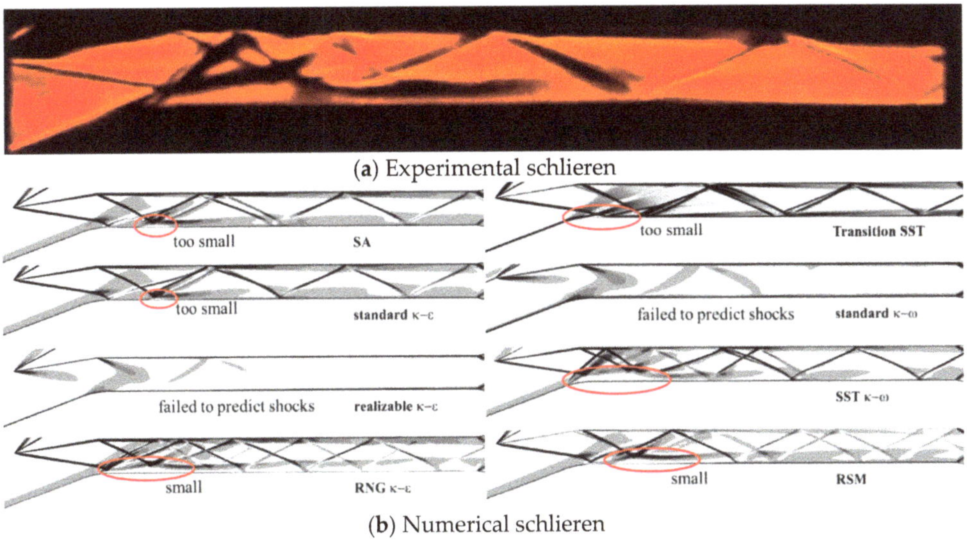

Figure 4. Comparison of flow structures in the inlet and the isolator, where red circles indicate the separation zones.

The numerical flow structures, including the two cowl-induced shocks, the cowl-shocks/boundary layer interactions on the ramp, the reflected shocks, and the reattachment shocks in the isolator, are almost identical to the experimental results. Figure 5, furthermore, compares the numerical wall static pressure distributions with the experimental results. Generally, the numerical wall pressure values and the overall distribution pattern are in good agreement with the experimental results. Therefore, the numerical method introduced in this paper is suitable for the flow simulation of a hypersonic inlet/isolator.

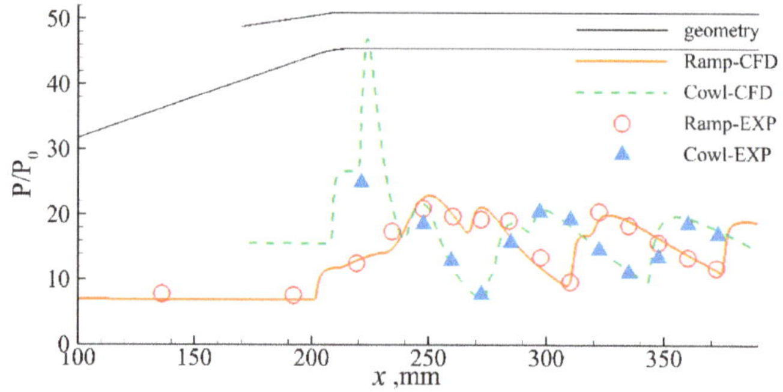

Figure 5. Comparison of pressure distribution on walls.

To ascertain grid dependencies and determine the optimal mesh in the numerical simulation, two-dimensional structured grids of the inlet/isolator were generated with six different grid numbers: 60,000, 100,000, 120,000, 150,000, 170,000, and 200,000. The y^+ distribution curves corresponding to different grids are shown in Figure 6, and it can be seen that the y^+ values meet the requirement of the turbulence model [30]. Figure 7 shows the pressure distribution curves corresponding to different grids. After reaching a grid quantity of 150,000, noticeable changes in the positions of flow structures, such as separation and shock waves, can be observed compared with the coarser grid quantities. Figure 8 illustrates that, when the grid number is below 100,000, the separation bubble does not exhibit a triangular shape. Figure 9 compares the total pressure recovery (TPR) and pressurization rate (PR) results for the inlet/isolator with the six different grids. Only when the grid number exceeds 150,000 is the predicted shock wave structure insensitive to the grid number. While finer meshes require higher computational costs, the mesh with a grid number of 150,000 was selected for subsequent wind shear simulation calculations.

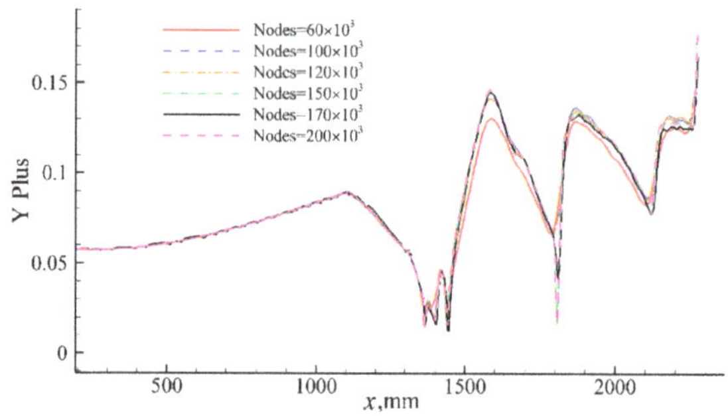

Figure 6. y^+ distribution with different grids.

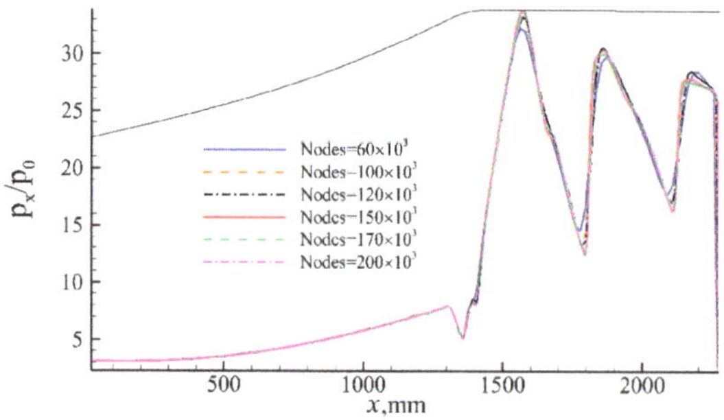

Figure 7. Pressure distribution with different grids.

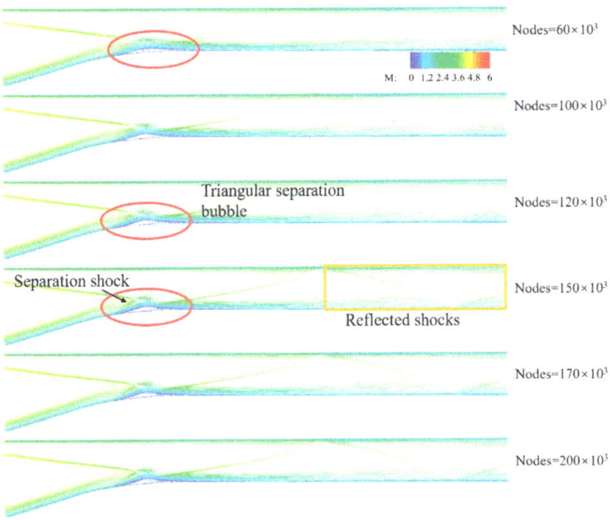

Figure 8. Mach number contours with different grids.

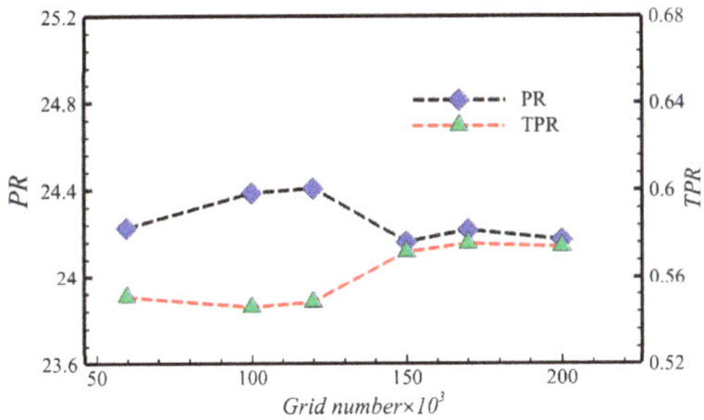

Figure 9. Performance parameters with different grids.

3.6. Numerical Dissipation Verification

It is important to note the numerical dissipation and numerical viscosity at each time step in a simulation using a computational fluid dynamics (CFD) solver. In this section, a sufficiently long computational domain is established, as depicted in Figure 10, to verify whether numerical dissipation will reduce the perturbation in the simulation process. The wind shear in January is applied as an incoming perturbation.

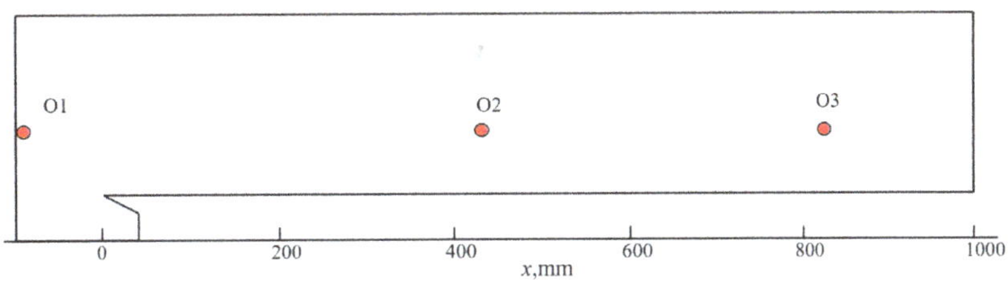

Figure 10. Numerical dissipation verification model.

The perturbation model is introduced in Section 2, and the details are shown in Figure 3. The changes in the angle of attack, α, at three points located at different positions along the flow direction are monitored. As can be seen in Figure 11, the overall change in α remains constant along the flow direction, which indicates that the perturbation of the incoming flow is not dissipated, but there is a lag due to the propagation. Therefore, this unsteady numerical method proves to be feasible for this study.

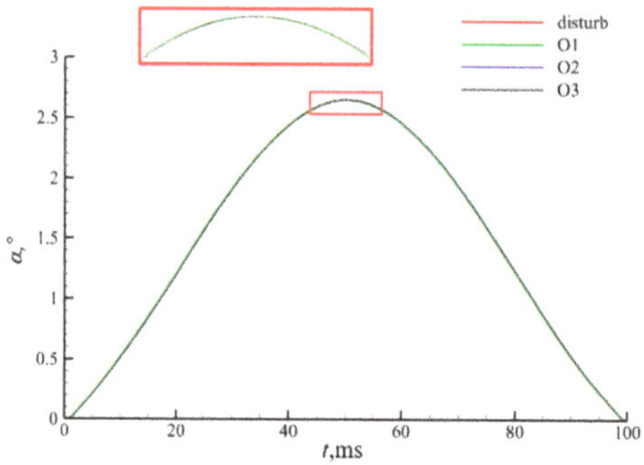

Figure 11. Perturbation dissipation of the monitoring points.

4. Results and Discussion

4.1. Effect of Wind Shear on Hypersonic Inlet/Isolator under Unthrottled Conditions

In order to investigate the effect of wind shear on the inlet/isolator shock/boundary layer interference, an unsteady calculation of the inlet/isolator in the through flow state was conducted as a basis for the subsequent study of the shock train.

The Mach number contour of the inlet/isolator at t = 0 ms is shown in Figure 12. In the flowfield, it can be seen that the external compression system contains an oblique shock and a series of compression wave systems. The cowl shock separates the flow on the ramp surface. Because of the separation bubble, a separation shock forms and interacts with the reflected oblique shock. The separated flow re-attaches downstream, and a triangular separation bubble is formed. An expansion fan and a reattachment shock are also produced. Multiple shocks are reflected in the isolator, which corresponds to the pressure on the ramp in Figure 13. The Mach number contours near the throat, with incoming wind shear, are depicted in Figure 14, and it can be seen that the separation at the throat moves downstream and then upstream. From t = 0 ms to 50 ms, the strength of the compression wave system and the cowl shock decreases, as evidenced by the pressure on the ramp decreasing during this period in Figure 13. Although the intensity of the cowl shock decreases, the separation bubble at the throat grows as the reattachment point moves downstream. This might be due to the weakening effect of the shoulder expansion fan on the separation bubble as the impingement point of the cowl shock moves downstream [31,32]. Furthermore, as shown in Figure 15, the pressure at P2 changes the most because P2 is located near the reattachment shock. Hence, as the separation bubble moves, the pressure value of P2 changes significantly. The time scale of the wind shear is 100 ms, as shown in Figure 2, and the incoming flow of the inlet/isolator returns to the initial state at t = 100 ms. However, Figures 12 and 13 reveal that the pressure distribution and separation at t = 100 ms differ from those at t = 0 ms, even though the incoming flow conditions at these two moments are identical. In particular, while the positions of the reattachment point and the peak vertex of the separation bubble change little, the separation point moves upstream at t = 100 ms. This could be because the hysteresis effect is more pronounced in low-velocity flows as the perturbation intensity recovers to zero. The inflection points of the pressure at the monitoring points vary at different moments and are delayed compared with the inflection point of the incoming flow perturbation. This observed phenomenon can be attributed to the presence of a hysteresis effect.

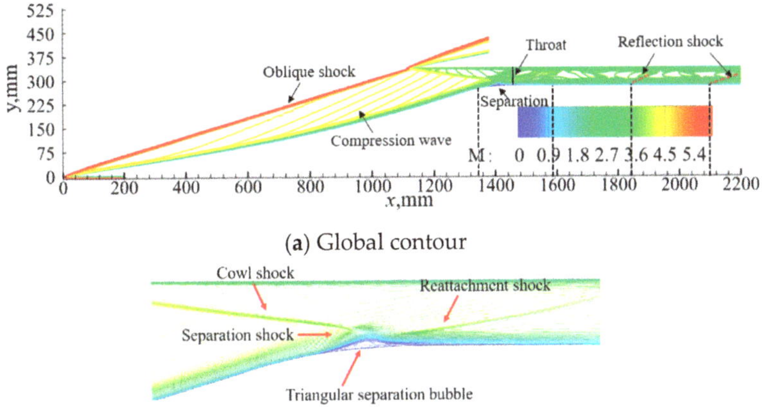

(**a**) Global contour

(**b**) Locally enlarged contour

Figure 12. Mach number contour of the inlet/isolator at t = 0 ms.

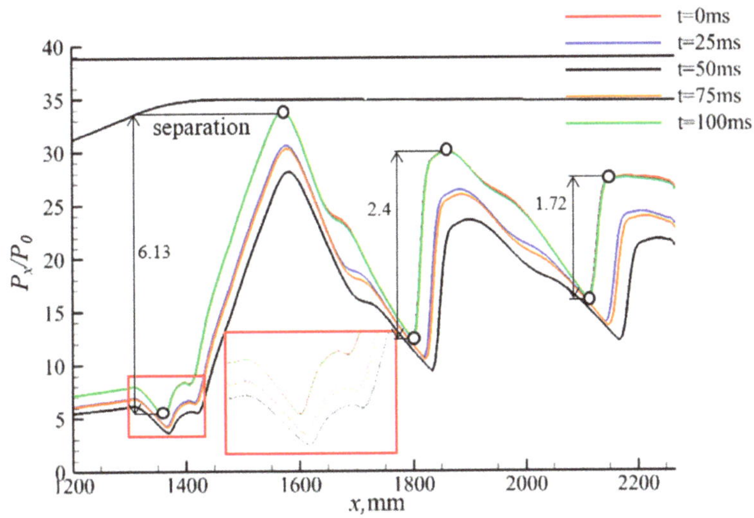

Figure 13. The pressure distribution along the ramp at different moments.

Figure 16 compares the changes in various parameters at the inlet/isolator throat with incoming wind shear, indicating a lag time of about 0.96 ms between results from the two of them. Figure 17 provides pressure variation at different heights of the throat, with the dotted lines indicating the times corresponding to the inflection points of the three pressure curves. The pressure variation of the monitored points near the upper and lower walls exceeds that of the mainstream. Moreover, the time corresponding to the inflection points is delayed by 0.96 ms, corroborating the lag time observed in Figure 16a. This is because the flow velocity in the boundary layer is slower than that of the mainstream, resulting in a slower perturbation propagation speed and a longer propagation time.

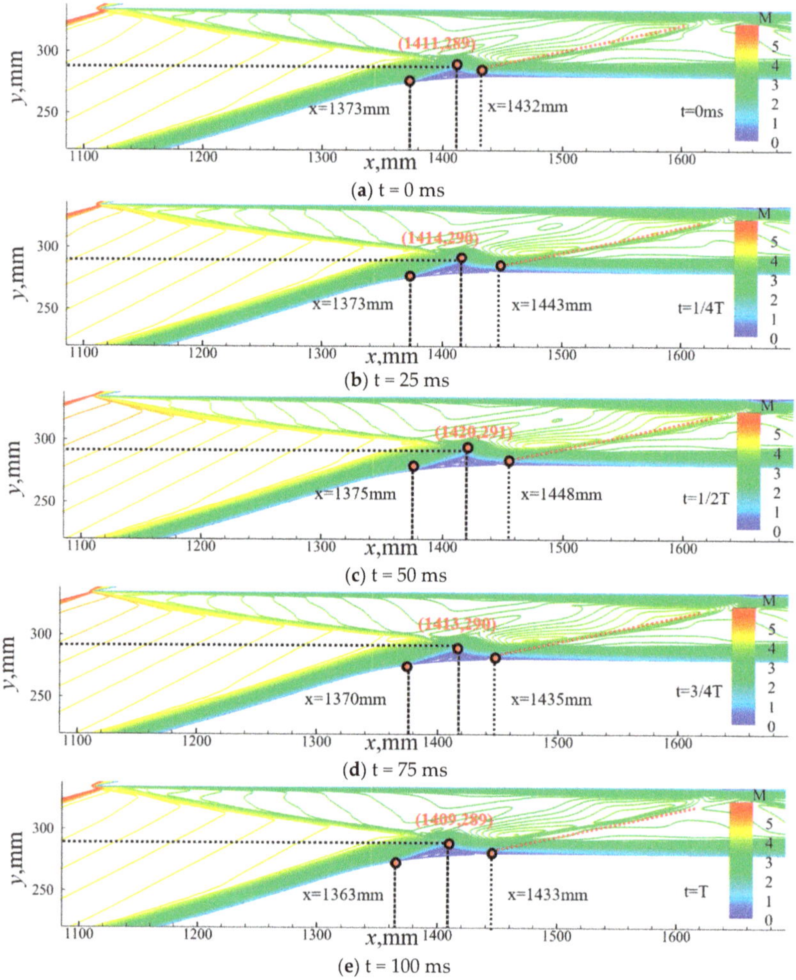

Figure 14. Enlarged Mach number contours at different moments.

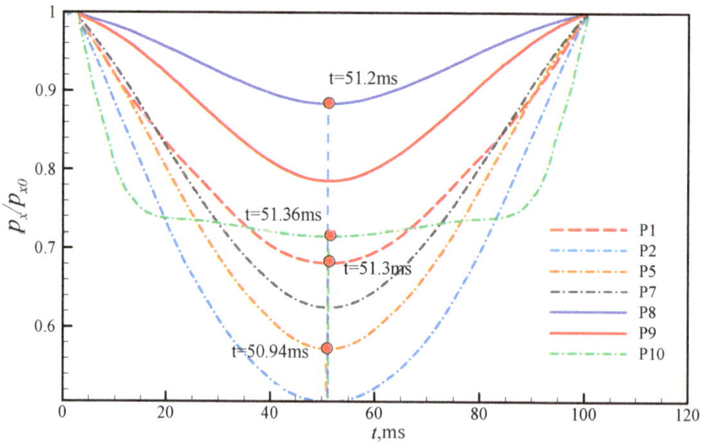

Figure 15. Pressure variations in monitor points over time.

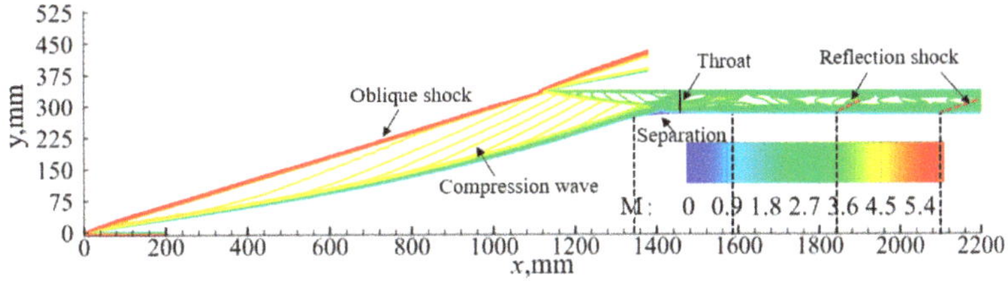

(**a**) Global contour

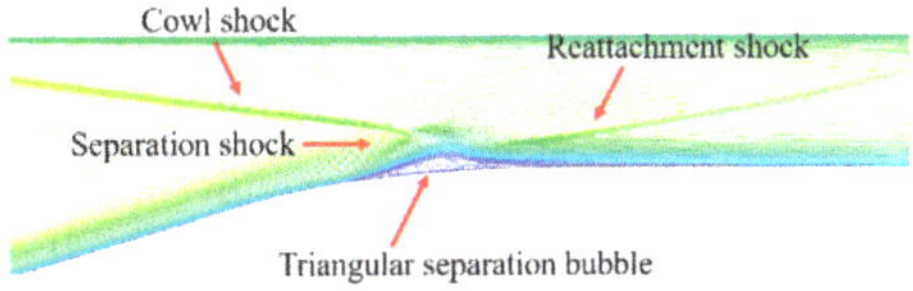

(**b**) Locally enlarged contour

Figure 12. Mach number contour of the inlet/isolator at t = 0 ms.

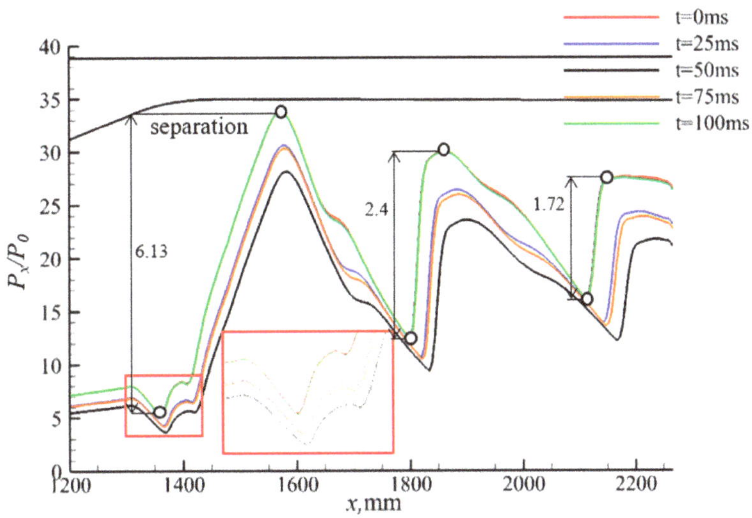

Figure 13. The pressure distribution along the ramp at different moments.

Figure 16 compares the changes in various parameters at the inlet/isolator throat with incoming wind shear, indicating a lag time of about 0.96 ms between results from the two of them. Figure 17 provides pressure variation at different heights of the throat, with the dotted lines indicating the times corresponding to the inflection points of the three pressure curves. The pressure variation of the monitored points near the upper and lower walls exceeds that of the mainstream. Moreover, the time corresponding to the inflection points is delayed by 0.96 ms, corroborating the lag time observed in Figure 16a. This is because the flow velocity in the boundary layer is slower than that of the mainstream, resulting in a slower perturbation propagation speed and a longer propagation time.

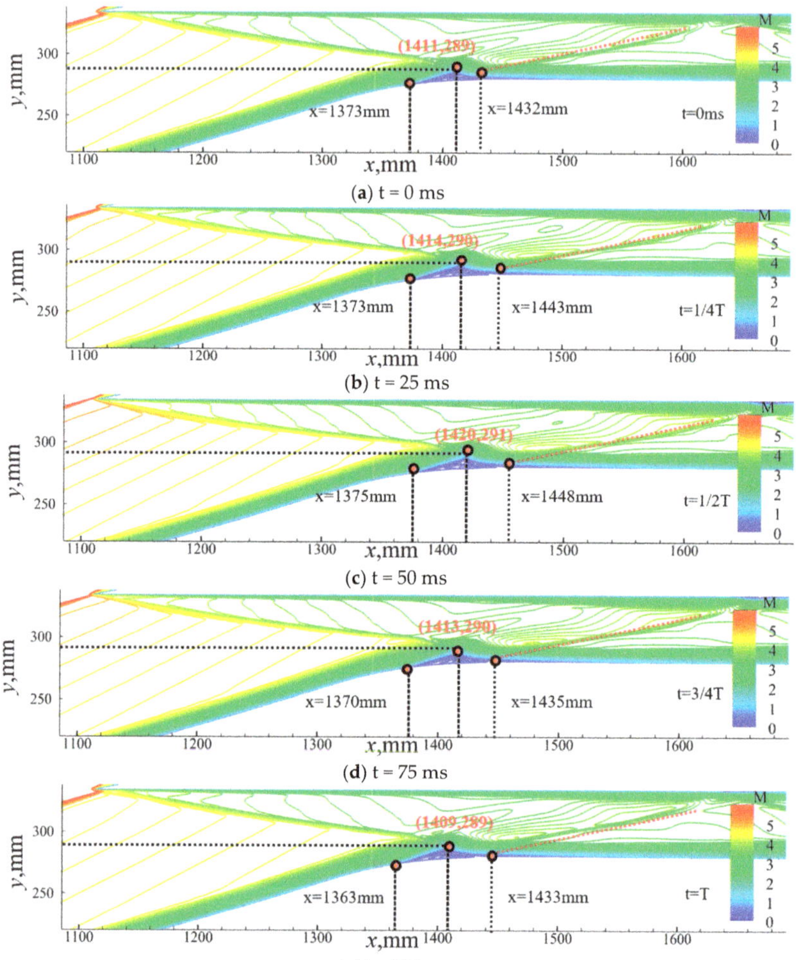

(**a**) t = 0 ms

(**b**) t = 25 ms

(**c**) t = 50 ms

(**d**) t = 75 ms

(**e**) t = 100 ms

Figure 14. Enlarged Mach number contours at different moments.

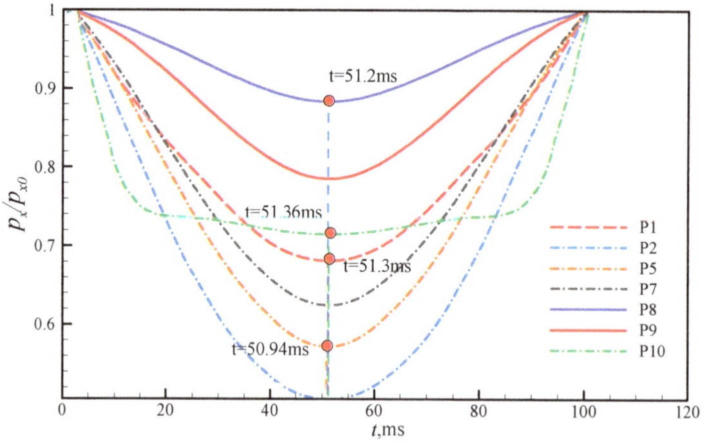

Figure 15. Pressure variations in monitor points over time.

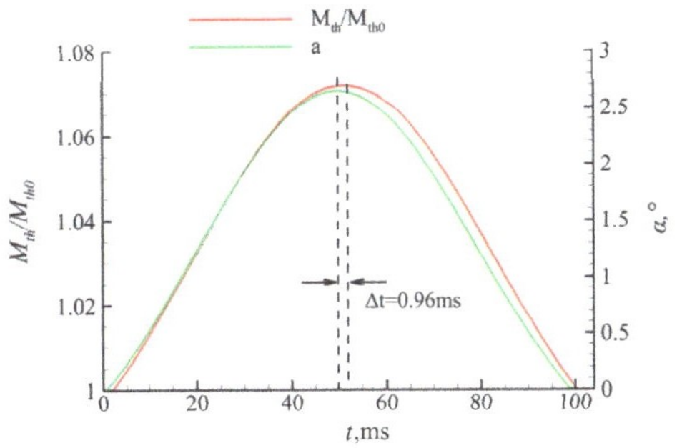

(**a**) Mach number in the throat.

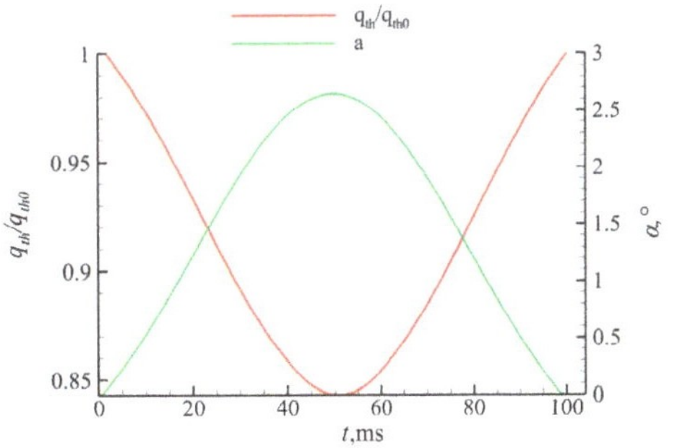

(**b**) Flow rates in the throat.

Figure 16. Variations in Mach numbers and flow rates in the throat.

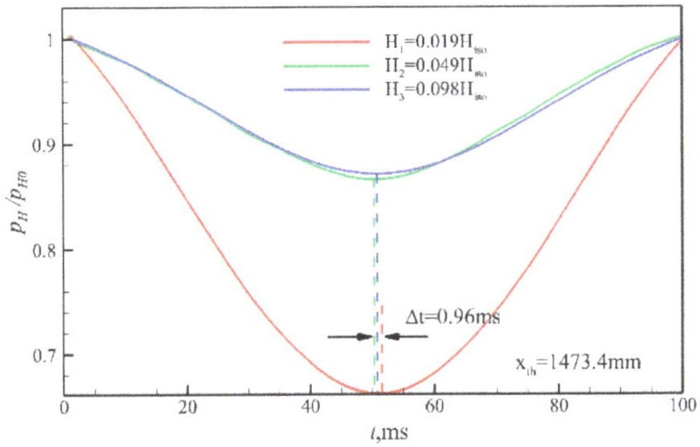

Figure 17. Pressure changes at different heights of the throat.

Figure 18 illustrates the pressure of the throat along the y direction, indicating that the pressure decreases with y. In the pressure distribution, there are three characteristic points, i.e., CP1, CP2, and CP3. Based on the pressure contours in Figure 19, the three feature points can be analyzed in detail. From the wall to y = 330 mm, the pressure is almost constant, corresponding to the thickness of the boundary layer on the upper wall surface. The expansion wave originating from the shoulder leads to a pressure decrease as y decreases to about 315 mm. The first characteristic point (CP1) at y = 315 mm is due to the shock wave generated after the interaction between the cowl shock and the separation shock, causing the pressure to increase as y decreases. The second characteristic point (CP2) appears at y = 310 mm, which is due to the expansion wave resulting from the separation of the bubble, leading the pressure to decrease as y decreases. The third characteristic point (CP3) appears at y = 295 mm, where the pressure begins to increase because of the reattachment shock generated downstream of the separation bubble. From y = 290 mm to the wall, the pressure is almost constant, corresponding to the thickness of the boundary layer on the lower wall surface. In addition, as time passes, the y-value of the third feature point decreases significantly, which can also be observed from the Mach number contour at the shoulder in Figure 14. This is because the separation bubble gradually develops downstream over time, driving the reattachment shock to move downstream.

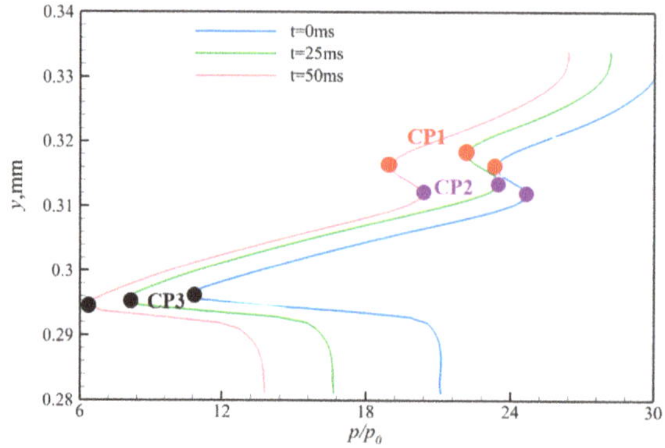

Figure 18. Pressure distribution inside the throat area at different times.

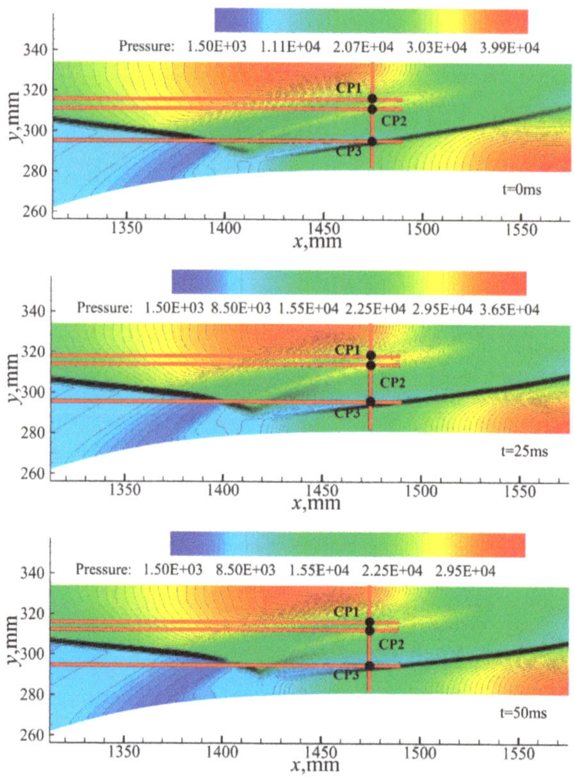

Figure 19. Pressure contours at different moments.

It can be seen in Figure 20 that, under the influence of wind shear, the total pressure recovery coefficient of the inlet/isolator increases, which is mainly due to the weakening of the cowl shock. The total pressure recovery coefficient of the exit and throat change by about 11.5% and 10.7%. This indicates that the impact of wind shear on the inlet/isolator performance needs to be considered when a hypersonic inlet/isolator operates in near space.

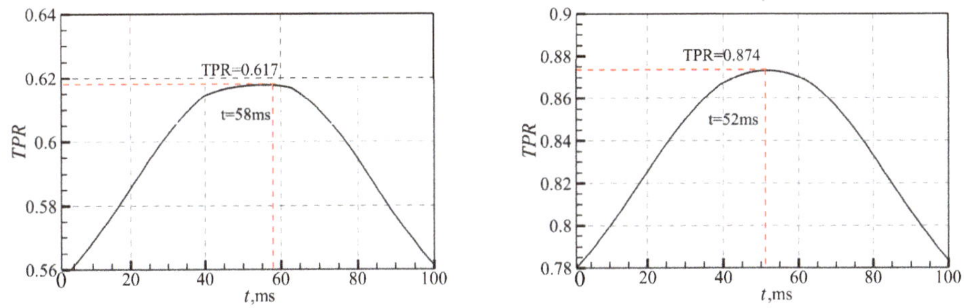

(**a**) Total pressure recovery factor at exit (**b**) Total pressure recovery factor at throat

Figure 20. Variations in the total pressure recovery factor.

4.2. Effect of Wind Shear on Shock Train

The inlet/isolator's maximum sustainable backpressure is a crucial characteristic. Moreover, the throttling process often results in complex shock train phenomena [33–35]. For the inlet/isolator in this study, the maximum sustainable backpressure ratio is 195 times the incoming static pressure (p_0) without wind shear. Once the backpressure exceeds this value, the shock train will be expelled out of the duct, and the inlet/isolator will fall into the unstart state. This section analyzes the impact of wind shear on the shock train under various throttling conditions.

When the backpressure is 135 p_0, under incoming wind shear, the transient flowfield within the inlet/isolator is numerically analyzed. Figures 21 and 22 present the Mach number and X-velocity contours of the inlet/isolator at different times. Wind shear induces the downstream shock train to continuously approach the throat, while the throat's separation moves downstream. The reason is the compression strength of the external compression system decreases with wind shear. For the duct, the pressure at the entrance becomes smaller, while the exit remains unchanged, which eventually leads to the shock train being pushed upstream. At t = 45 ms, the lower wall separation zone of the shock train couples with the throat's separation bubble. As the shock train moves upstream, the separation at the upper wall expands. This phenomenon happens because, as the shock train moves upstream, the shock causing separation changes from the right-running reflected shock to the left-running reflected shock [36].

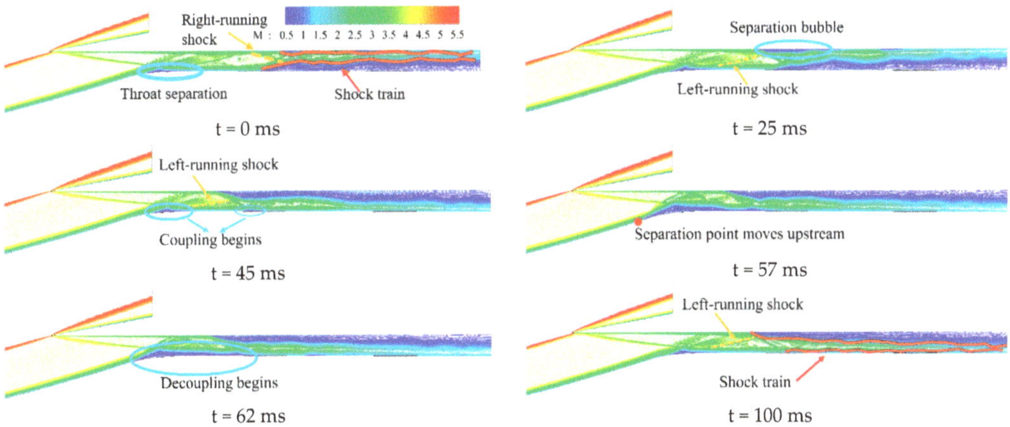

Figure 21. Changes in Mach number contours.

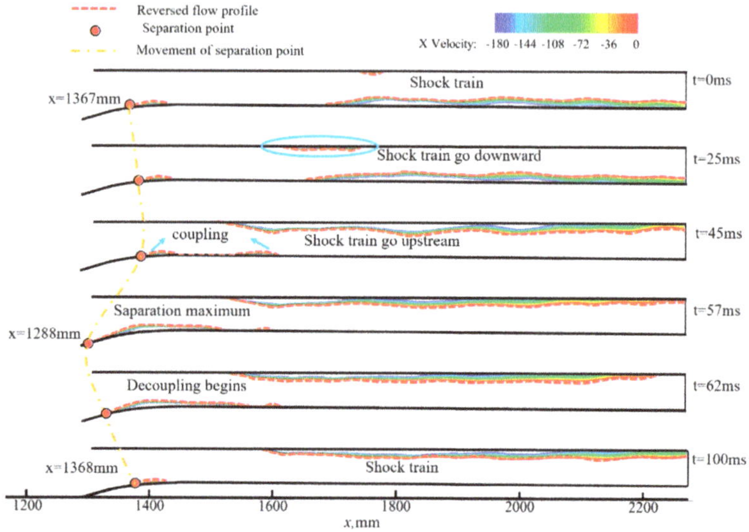

Figure 22. Reversed flow in the inlet/isolator.

The change in the angle of attack induced by wind shear starts to decrease after t = 50 ms. The change in the angle of attack progressively recovers to 0°, the coupling separation bubble at the throat begins to decouple, and the shock train moves downstream. However, because of the hysteresis effect, it becomes evident that the shock train will not return to its initial state. Figure 23 illustrates the surface pressure distribution along the lower wall at different times. Figures 24 and 25 show the distributions of the surface friction coefficient (Cf) along the upper and lower walls at different times. The pressure and Cf values both change significantly from x = 1300 mm, indicating that the cowl shock/boundary layer interaction changes. At t = 57 ms, the first rising point of the pressure is located at the furthest upstream position, as is the zero-value position of the Cf curve of the lower wall. As the separation at the throat at this moment is at maximum, which can be seen from the instantaneous Mach number contour, the separation begins further upstream than the expansion wave. In the region downstream of x = 1500 mm, the change in surface pressure is substantial, which is induced by the upstream movement of the shock train. Simultaneously, the direction of the friction drag of the upper and lower walls changes, as does the separation in the isolator on the upper wall, which is also reflected in the change in the exit Mach number distribution in Figure 26. This is because the shock train moves upstream and downward at the same time, suggesting that wind shear alters the reverse pressure gradient.

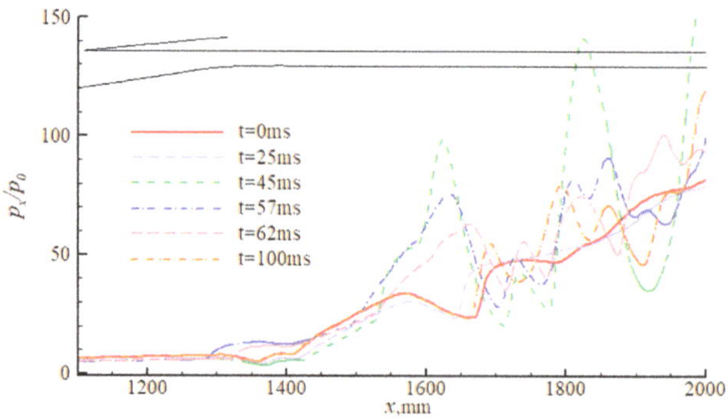

Figure 23. Pressure changes along the lower wall.

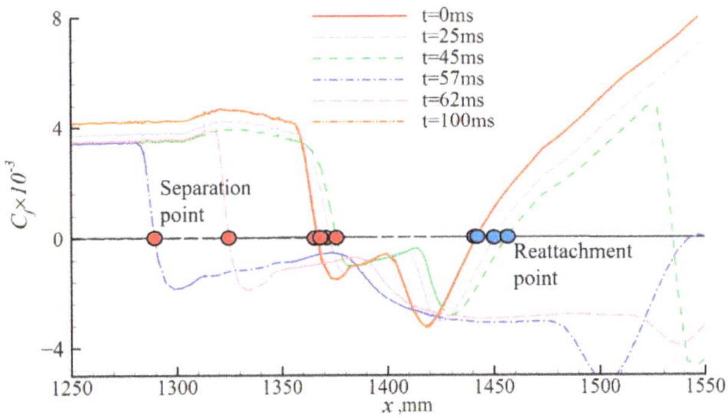

Figure 24. Distribution of frictional resistance coefficients on the lower wall.

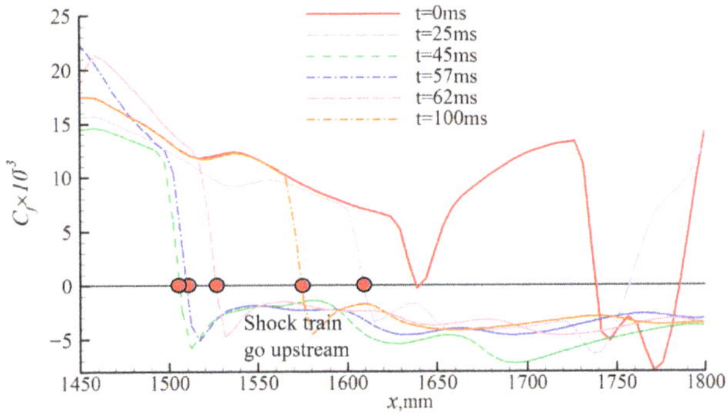

Figure 25. Distribution of frictional resistance coefficients on the upper wall.

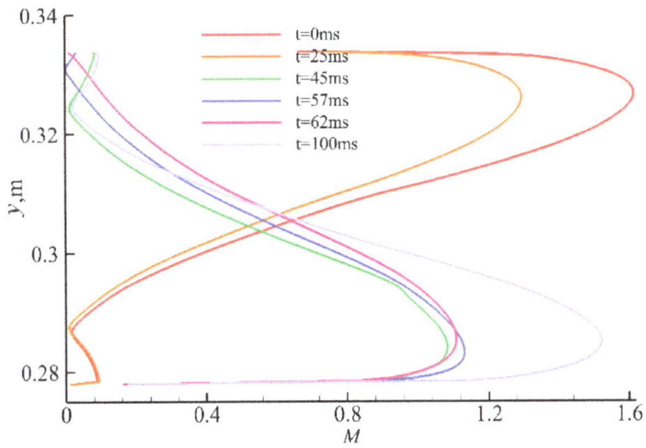

Figure 26. Distribution of Mach numbers at the exit.

Figure 27 presents the Mach number contours of the inlet/isolator at different times when the downstream backpressure is 140 p_0, clearly showing that wind shear causes the shock train to move upstream and ultimately causes the unstart state. It can be seen that the separation point of the separation bubble at the throat moves upstream, which is further demonstrated in Figure 28. Moreover, the Cf value changes for the upper and lower walls, as shown in Figures 29 and 30. This occurs because, as the shock train moves upstream, it approaches the upper wall, leading to a lower exit Mach number in Figure 31. At t = 40.6 ms, the downstream separation zone and separation bubble couple at the throat, and the separation bubble at the throat enlarges rapidly over time. The angle of attack change caused by wind shear begins to decrease after t = 50 ms, and the separation bubble at the throat starts to decouple from the downstream separation zone at t = 50.6 ms. However, unlike the condition when the backpressure is 0.675 times the limit back pressure, the Mach number contour at 52.2 ms clearly shows that the decoupling has failed, forcing the inlet/isolator to fall into the unstart state. Obviously, under this condition, the influence of wind shear on the inlet/isolator is profound, directly causing the inlet/isolator to fall into the unstart state, and it cannot be restarted.

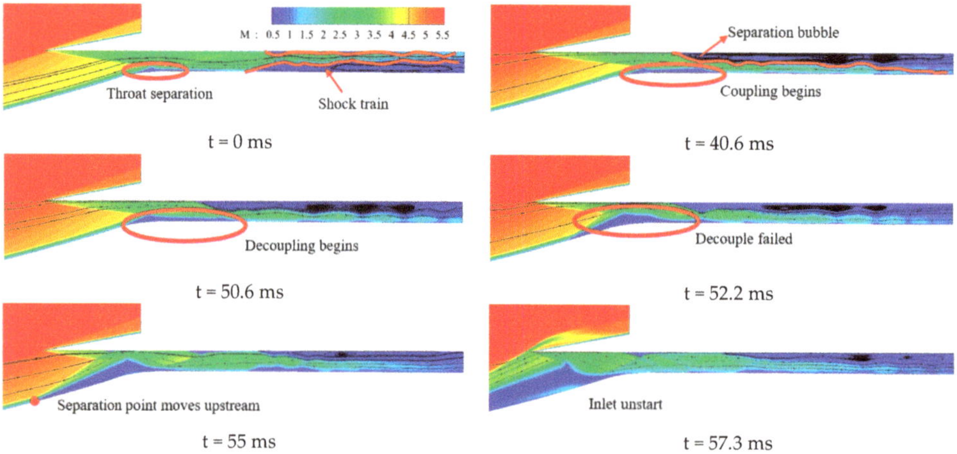

Figure 27. Changes in Mach number contours.

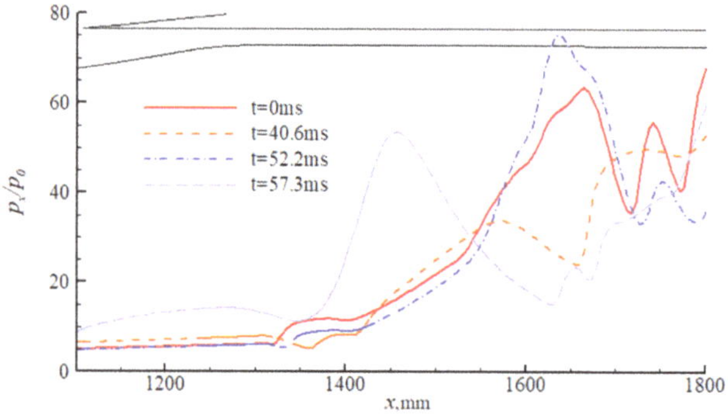

Figure 28. Pressure changes along the lower wall.

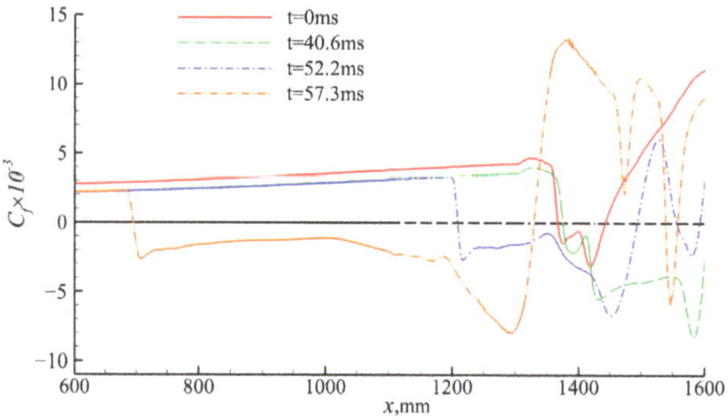

Figure 29. Distribution of frictional resistance coefficients on the lower wall.

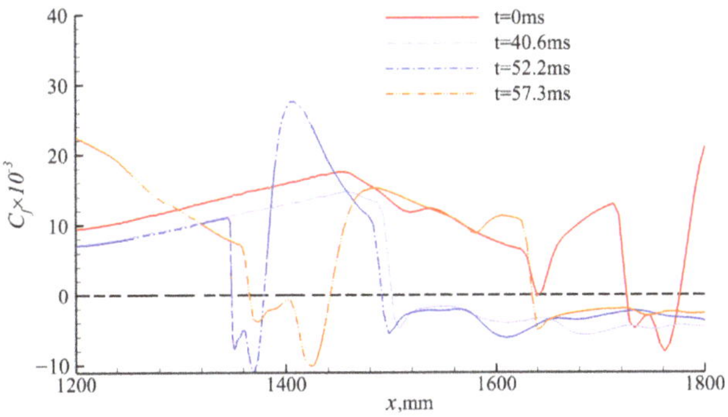

Figure 30. Distribution of frictional resistance coefficients on the upper wall.

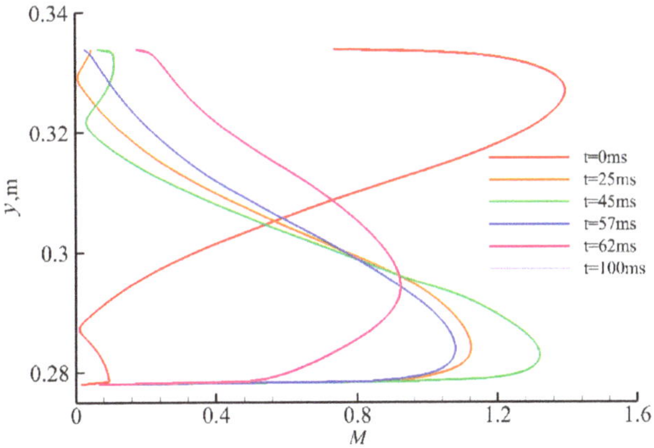

Figure 31. Distribution of Mach numbers at the exit.

Based on the above analysis, it is evident that wind shear substantially affects the inlet/isolator's ability to resist back pressure, weakening it by about 30%. Therefore, the presence of wind shear needs to be considered when designing an inlet/isolator.

5. Conclusions

Herein, the unsteady simulations of a hypersonic inlet/isolator with an on-design Mach number of 6.0 under incoming wind shear at an altitude of 30 km with a magnitude speed of 80 m/s are performed. The transient flow characteristics of the shock wave/boundary layer interaction (SWBLI) and shock train are investigated in detail.

A mixed-compression hypersonic inlet/isolator with an oblique shock and a series of compression waves is designed. The incoming wind shear perturbation is modeled as a variation in the angle of attack. Under the unthrottled state, wind shear significantly disrupts the flowfield of the inlet/isolator, affecting the interaction between the shock wave and the boundary layer. It drives the continuous movement of the separation bubble at the throat, both downstream and upstream, exhibiting hysteresis phenomena, which are also demonstrated by the pressure variation in the different positions of the inlet/isolator. Regarding the performance of the inlet/isolator, the total pressure recovery coefficients at the throat and exit increase by approximately 10% to 12%, respectively. Under throttled conditions, wind shear significantly impacts the isolator's shock train. When the downstream backpressure is 135 times the incoming pressure (p_0), the shock train first moves upstream and gradually interacts with the cowl shock wave/boundary layer, apparently enlarging the separation bubble at the throat. As the shock train moves upstream, the separation at the upper wall expands because the shock that interacts with the shock train switches from the right-running reflected shock to the left-running reflected shock. Then, the shock train moves downstream and decouples from the separation bubble at the throat. Though the perturbation amplitude reduces to zero, because of the hysteresis effect, the flowfield in the inlet/isolator cannot fully recover to the initial state. When the downstream backpressure increases to 140 p_0, the shock train expands the separation zone in both the upper and lower walls, ultimately forcing the inlet/isolator to fall into the unstart state, and it cannot be restarted. Therefore, it is necessary to pay more attention to the adverse effects of wind shear on inlet/isolator performance, which directly affects the internal flow.

Author Contributions: Conceptualization, S.G. and H.H.; methodology, S.G., H.T. and H.H; validation, S.G., K.G. and H.H.; formal analysis, S.G.; investigation, S.G.; resources, S.G. and Y.M.; data curation, S.G. and H.H; writing—original draft preparation, S.G. and M.L.; writing—review and editing, H.H., H.T. and S.G.; visualization, S.G.; supervision, H.H. and H.T. All authors have read and agreed to the published version of the manuscript.

Funding: This research was funded by the National Natural Science Foundation of China (Grant Nos. 12272177, 12025202, and U20A2070), the National Science and Technology Major Project (No. J2019-II-0014-0035), the Young Talent Lift Project (2021-JCJQ-QT-064), the 1912 Project (No. 2019-JCJQ-DA-001-164), the Advanced Jet Propulsion Innovation Center, AEAC (Project ID. HKCX202-02-005), the Key Laboratory of Inlet and Exhaust System Technology (Grant No. CEPE2020012), and the Defense Industrial Technology Development Program (JCKY2019605D001).

Data Availability Statement: The data within the article are available upon request.

Acknowledgments: The authors would like to thank the High-Performance Computing Platform of Nanjing University of Aeronautics and Astronautics for providing the computing platform. The authors are grateful to Li Xin and Qin Yuan for their assistance in facilitating valuable discussions and providing suggestions. The manuscript greatly benefited from the helpful and constructive comments of three anonymous reviewers.

Conflicts of Interest: The authors declare that they have no known competing financial interests or personal relationships that could appear to influence the work reported in this paper.

References

1. Van Wie, D.M.; D'Alessio, S.M.; White, M.E. Hypersonic Airbreathing Propulsion. *Johns Hopkins APL Tech. Dig.* **2005**, *26*, 430–437.
2. Ferri, A. Review of scramjet propulsion technology. *J. Aircraft.* **1968**, *5*, 3–10. [CrossRef]
3. McClinton, C.; Rausch, D.; Sitz, J.; Reukauf, P. Hyper-X program status. In Proceedings of the 10th AIAA/NAL-NASDA-ISAS International Space Planes and Hypersonic Systems and Technologies Conference, Kyoto, Japan, 24–27 April 2001.
4. Moses, P. X-43C Plans and Status. In Proceedings of the 12th AIAA International Space Planes and Hypersonic Systems and Technologies, Norfolk, VA, USA, 15–19 December 2003.
5. Huang, H.X.; Tan, H.J.; Li, F.B.; Tang, X.B.; Qin, Y.; Xie, L.B.; Xu, Y.Y.; Li, C.M.; Gao, S.M.; Zhang, Y.; et al. A review of the shock-dominated flow in a hypersonic inlet/isolator. *Prog. Aerosp. Sci.* **2023**, *2023*, 100952. [CrossRef]
6. Rozario, D.; Zouaoui, Z. Computational Fluid Dynamic Analysis of Scramjet Inlet. In Proceedings of the 45th AIAA Aerospace Sciences Meeting and Exhibit, Reno NV, USA, 8–11 January 2007.
7. Van Wie, D.; Kwok, F.; Walsh, R. Starting characteristics of supersonic inlets. In Proceedings of the 32nd Joint Propulsion Conference and Exhibit, Lake Buena Vista, FL, USA, 1–3 July 1996.
8. Ma, R.P.; Liao, H.Z. The characteristics of winds at height of 20-80 km in the Chinese area. *Chin. J. Space Sci.* **1999**, *19*, 334–341. [CrossRef]
9. Fleming, E.L.; Chandra, S.; Barnett, J.J.; Corney, M. Zonal mean temperature, pressure, zonal wind and geopotential height as functions of latitude. *Adv. Space Res.* **1990**, *10*, 11–59. [CrossRef]
10. Fleming, E.L.; Chandra, S.; Burrage, M.D.; Skinner, W.R.; Hays, P.B.; Solheim, B.H.; Shepherd, G.G. Climatological mean wind observations from the UARS high-resolution Doppler imager and wind imaging interferometer: Comparison with current reference models. *J. Geophys. Res. Atmos.* **1996**, *101*, 10455–10473. [CrossRef]
11. Johnson, D.; Roberts, B.; Vaughan, W.; Justus, C. Atmospheric Models For Engineering Applications. In Proceedings of the 41st Aerospace Sciences Meeting and Exhibit, Reno, NV, USA, 6–9 January 2003.
12. Luers, J. *A Model of Wind Shear and Turbulence in the Surface Boundary Layer*; NASA: Washington, DC, USA, 1973.
13. Smith, O.E. *Vector Wind and Vector Wind Shear Models 0 to 27 km Altitude for Cape Kennedy, Florida, and Vandenberg AFB, California*; NASA: Washington, DC, USA, 1976.
14. Kozakiewicz, A.; Frant, M. Analysis of the gust impact on inlet vortex formation of the fuselage-shielded inlet of an jet engine powered aircraft. *J. Theor. Appl. Mech.* **2013**, *51*, 993–1002.

15. Sun, S.; Wu, Z.; Huang, H.; Bangga, G.; Tan, H. Aerodynamic Response of a Serpentine Inlet to Horizontal Periodic Gusts. *Aerospace* **2022**, *9*, 824. [CrossRef]
16. Halwas, H.K.; Aggarwal, S. Effect of Side Gust on Performance of External Compression Supersonic Inlet. *J. Aircr.* **2019**, *56*, 569–582. [CrossRef]
17. Etkin, B. Effect of Wind Gradient on Glide and Climb. *J. Aeronaut. Sci.* **1947**, *14*, 365–367. [CrossRef]
18. Etkin, B. Turbulent Wind and Its Effect on Flight. *J. Aircr.* **1981**, *18*, 327–345. [CrossRef]
19. Etkin, B.; Hughes, P.C.; Zhu, S. Equivalent deterministic inputs for random processes. *J. Guid. Control Dyn.* **1984**, *7*, 477–482. [CrossRef]
20. Xie, R.; Wang, X.M.; Gong, J.Y. Dynamic Modeling and Simulation for a Supermaneuverable Aircraft in Disturbance of Wind Field. In Proceedings of the 2012 Fifth International Conference on Intelligent Computation Technology and Automation, Zhangjiajie, China, 12–14 January 2012; pp. 147–150.
21. Zhao, R.L.; Chen, Z.G.; Fu, W.X. Wind shear and rocket design. *J. Astronaut.* **1998**, *19*, 106–109.
22. Yang, J.F.; Xiao, C.Y.; Hu, X.; Cheng, X. Wind shear characteristics in near space and their impacts on air vehicle. *J. Beijing Univ. Aeronaut. Astronaut.* **2019**, *45*, 57–65. [CrossRef]
23. Woollen, J.; Sienkiewicz, M.; Ruddick, A.G.; Robertson, F.R.; Reichle, R.; Redder, C.R.; Pegion, P.; Pawson, S.; Owens, T.; Molod, A.; et al. MERRA: NASA's Modern-Era Retrospective Analysis for Research and Applications. *J. Clim.* **2011**, *24*, 3624–3648. [CrossRef]
24. Soltania, M.R.; Daliria, A.; Younsib, J.S. Effects of shock wave/boundary-layer interaction on performance and stability of a mixed-compression inlet. *Sci. Iran.* **2016**, *23*, 1811–1825. [CrossRef]
25. Kianvashrad, N.; Knight, D. Large Eddy Simulation of Hypersonic Turbulent Boundary Layers. *Fluids* **2021**, *6*, 449. [CrossRef]
26. Trapier, S.; Deck, S.; Duveau, P. Delayed Detached-Eddy Simulation of Supersonic Inlet Buzz. *AIAA J.* **2012**, *46*, 118–131. [CrossRef]
27. Zhong, X.L.; Wang, X.W. Direct Numerical Simulation on the Receptivity, Instability, and Transition of Hypersonic Boundary Layers. *Annu. Rev.* **2012**, *44*, 527–561. [CrossRef]
28. Menter, F.R. Two-equation eddy-viscosity turbulence models for engineering applications. *AIAA J.* **1994**, *32*, 1598–1605. [CrossRef]
29. Simeonides, G.; Haase, W.; Manna, M. Experimental, analytical, and computational methods applied to hypersonic compression ramp flows. *AIAA J.* **1994**, *32*, 301–310. [CrossRef]
30. Mansour, A.; Laurien, E. Numerical error analysis for three-dimensional CFD simulations in the tworoom model containment THAI+: Grid convergence index, wall treatment error and scalability tests. *Nucl. Eng. Des.* **2018**, *326*, 220–233. [CrossRef]
31. Zhang, Y.; Tan, H.J.; Zhuang, Y.; Wang, D.P. Influence of Expansion Waves on Cowl Shock/Boundary Layer Inte8action in Hypersonic Inlets. *J. Propuls. Power* **2014**, *30*, 1183–1191. [CrossRef]
32. Huang, H.X.; Sun, S.; Tan, H.j.; Ning, L.; Wang, J. Characterization of Two Typical Unthrottled Flows in Hypersonic Inlet/Isolator Models. *J. Aircr.* **2015**, *52*, 1715–1721. [CrossRef]
33. Huang, H.X.; Tan, H.J.; Zhuang, Y.; Sheng, F.J.; Sun, S. Progress in Internal Flow Characteristics of Hypersonic Inlet/Isolator. *J. Propuls. Technol.* **2018**, *39*, 2252–2273. [CrossRef]
34. Huang, H.X.; Tan, H.J.; Sun, S.; Wang, Z.Y. Behavior of Shock Train in Curved Isolators with Complex Background Waves. *AIAA J.* **2018**, *56*, 329–341. [CrossRef]
35. Tan, H.J.; Sun, S.; Huang, H.X. Behavior of shock trains in a hypersonic inlet/isolator model with complex background waves. *Exp. Fluids* **2012**, *53*, 1647–1661. [CrossRef]
36. Huang, H.X.; Tan, H.J.; Sun, S.; Sheng, F.J. Unthrottled Flows with Complex Background Waves in Curved Isolators. *AIAA J.* **2017**, *55*, 2942–2955. [CrossRef]

aerospace

MDPI

Article

The Influence of External Flow Field on the Flow Separation of Overexpanded Single-Expansion Ramp Nozzle

Yang Yu [1,*], Yuepeng Mao [2], Tao Yu [1], Yalin Yang [1], Shulin Xu [3] and Sijia Liang [3]

1 School of Aeronautics, Chongqing Jiaotong University, Chongqing 400074, China; yutao@mails.cqjtu.edu.cn (T.Y.); 622210991013@mails.cqjtu.edu.cn (Y.Y.)
2 Computational Aerodynamics Institute, China Aerodynamics Research and Development Center, Mianyang 621010, China; yuepengmao@126.com
3 The Green Aerotechnics Research Institute of Chongqing Jiaotong University, Chongqing Key Laboratory of Green Aviation Energy and Power, Chognqing 401120, China; xutreeview@126.com (S.X.); liangsijia17@163.com (S.L.)
* Correspondence: yuyang227@vip.sina.com

Abstract: Flow separation and transitions of separation patterns are common phenomena of nozzles working with a wide Mach range. The maximum thrust method is applied to design the single-expansion ramp nozzle (SERN) for specific operating conditions. The nozzle is used to numerically simulate the transition processes of separation patterns under the linear change in the external flow Mach number and the actual trajectory take-off condition of a rocket-based combined cycle (RBCC), to investigate the mechanism through which the external flow field influences the separation pattern transition during acceleration. The computational fluid dynamics (CFD) method is briefly introduced, followed by experimental validation. Then, the design procedure of SERN is described in detail. The simulation results indicate that as the external Mach number increases, the flow field in the nozzle undergoes transitions from RSS (ramp) to FSS, and finally exhibits a no-flow separation pattern. The rate at which the external Mach number varies has little effect on the transition principle of the nozzle flow separation patterns, but it has a significant effect on the critical Mach number of the transition points. The external flow field of the nozzle has an airflow accumulation effect during acceleration, which can delay the transition of the flow separation pattern.

Keywords: overexpansion state; flow separation; single-expansion ramp nozzle (SERN); external flow; numerical simulation

Citation: Yu, Y.; Mao, Y.; Yu, T.; Yang, Y.; Xu, S.; Liang, S. The Influence of External Flow Field on the Flow Separation of Overexpanded Single-Expansion Ramp Nozzle. *Aerospace* **2023**, *10*, 958. https://doi.org/10.3390/aerospace10110958

Academic Editor: Konstantinos Kontis

Received: 24 September 2023
Revised: 6 November 2023
Accepted: 10 November 2023
Published: 13 November 2023

1. Introduction

Speed is an important index to measure the performance of aircraft, so hypersonic flight technology has attracted much attention [1,2]. The scramjet engine is one of the most ideal power sources for hypersonic vehicles because of its simple configuration and excellent efficiency at high Mach numbers. However, the scramjet operates within a narrow range of Mach numbers and needs to reach a high flight Mach number before startup. As a result, hypersonic vehicles powered by scramjet engines are forced to use other types of booster devices, drastically limiting flexibility and increasing operating costs. Therefore, researchers proposed the concept of combined cycle engines [3], which successfully combine several propulsion units and exploit the characteristics of each unit to maintain high thrust and specific impulse throughout the flight process.

The RBCC engine combines the advantages of rocket engine with high thrust-to-weight ratio and ramjet with high specific impulse to operate over a wide Mach range. Considering the requirement for the integrated installation of the aircraft and the nozzle, the afterbody of the aircraft is typically designed as the expansion ramp of SERN. This not only helps reduce the weight of the aircraft but also provides a certain self-stabilization for the engine during off-design operations [4]. With the support of numerous experiments, the GTX scheme [5] and the ISTAR scheme [6] for RBCC were determined to achieve the thermal throat via combustion and to reach the required thrust and specific impulse using the SERN with a fixed geometric structure. The SERN is the main thrust component of the hypersonic propulsion system, so the design of the nozzle is very important [7]. The design point of the nozzle is slightly smaller than the cruise Mach number [8], which makes the SERN work in a serious overexpansion state at low flight Mach number. Due to the reverse pressure gradient, flow separation is bound to occur. The shock wave/boundary layer interaction (SWBLI) will cause unsteady flow in the nozzle [9,10]. The separation caused by the shock wave will also produce unsteady force, reducing the service life of the nozzle [11].

Concerning overexpansion flow separation, early research focused mostly on the axisymmetric rocket nozzle. Studies have suggested that there are two distinct patterns of separation, namely free shock separation (FSS) and restricted shock separation (RSS) [12,13], with FSS being the more common pattern. In addition to the traditional axisymmetric rocket nozzle, symmetric and asymmetric flow separation structures were also observed in two-dimensional convergence–diffusion channel experiments [14]. Similarly, FSS and RSS also exist in SERN, with RSS (ramp) being the most common. FSS requires an extremely narrow range of nozzle pressure ratios (NPR), so some nozzles do not appear. Restricted shock separation with the separation bubble forming on the flap (RSS (flap)) will be present in SERN with a lengthy flap. In a previous study [15], the authors observed the transition from RSS (ramp) to RSS (flap) during the nozzle shutdown process, as illustrated in Figure 1. The transition was instantaneous, which resulted in a mutation in nozzle performance and hindered stable flight.

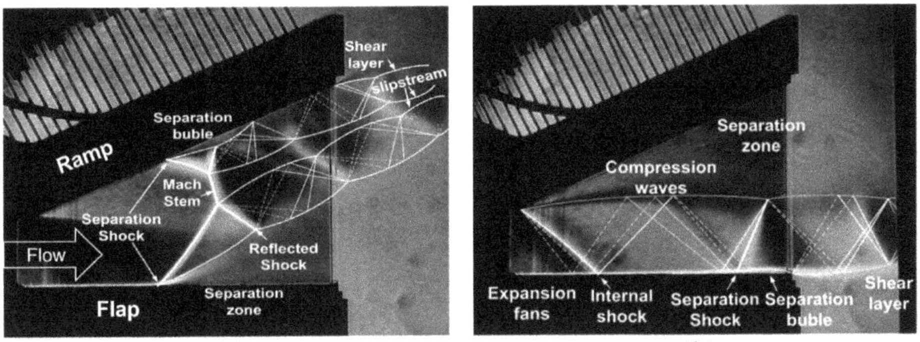

Figure 1. Flow separation patterns during SERN shutdown: (**a**) shock wave structure in RSS (ramp) pattern; (**b**) shock wave structure in RSS (flap) pattern.

In view of the problems caused by the nozzle working in the overexpansion state, several early groups [16–19] conducted research on the flow separation prediction, the transition of separation patterns, and the side load of the axisymmetric rocket nozzle, concluding that the primary source of the side load was separation transition. Watanabe et al. [17] found significant side loads in LE-7A experiments, as well as the vulnerability of some experimental components to damage during nozzle startup and shutdown. Since then, an increasing number of papers have been published studying the transition of separation patterns and its effect on nozzle performance [20–22]. Martelli et al. [21] numerically investigated a subscale parabolic overexpanded rocket nozzle and attempted to explain the hysteresis cycle between the separation patterns by analyzing the numerical solutions. He et al. [22] observed an abnormal transition process from no-flow separation at severe overexpansion to RSS and finally to FSS, even at the design condition. In addition, reducing the gas density or mass flow in severely overexpanded conditions leads to a decrease in the adverse pressure gradient across the separation shock or a weaker shock system, exhibiting a strong coupling relationship with flow separation behavior.

The majority of relevant studies were conducted in a static environment; however, the overexpanded separation zone inside the nozzle would interact with the external airflow [23,24]. Mousavi et al. [25] compared the shock wave position at different Mach numbers of the external flow, finding that the shock wave was positioned closer to the nozzle exit at high Mach numbers. Lee et al. [26] numerically studied the flow separation mechanism of a thrust-optimized parabolic nozzle in high-altitude experiments, and the observed flow characteristics showed that the separation pattern and the transition process were very different from those in sea-level experiments. In previous projects conducted by the authors [27,28], the impact of NPR and external Mach numbers on the flow separation patterns of an overexpanded SERN was extensively examined. The results showed that FSS became the most common separation pattern considering the external flow field, which was in contrast to the results under the static state.

The operation of the RBCC engine is governed by the combined effect of the NPR and flight Mach number. Specifically, the acceleration process of aircraft startup entails specific rules regarding the variation process and rate of the flight Mach number. To the authors' knowledge, no research has been carried out on the flow field under the aforementioned conditions and the effect of external flow acceleration on the nozzle performance. Therefore, the work presented in this paper is instructive for investigating the interaction between the internal and external flow of the nozzle.

2. Numerical Simulation Methods and Validation

The test model adopts the asymmetric nozzle shown in Figure 2. The design of the expansion contour is based on the maximum thrust theory. The sketch of the nozzle is shown in Figure 3. The angle at the ramp exit is 0°, the area of the throat (A_t) is 419.9 mm^2, and the aspect ratio of the throat is 2.5. The area of the nozzle exit (A_{exit}) is 1290.58 mm^2, the expansion length is 92 mm, and the contraction length is 30 mm. This model was used to conduct a cold airflow wind tunnel experiment, and the details of the experimental scheme are provided in a previous study [29].

Figure 2. Tested SERN model in the experiment.

The simulation model was a subscale two-dimensional SERN, and the geometry was generated using ICEM v12.0 software with a structured mesh, as shown in Figure 4. The node distributions in the x and y directions of regions 1, 2, 3, 4, 5, and 7 were 80 × 80, 150 × 80, 150 × 80, 120 × 80, 40 × 50, and 150 × 50, respectively. The node distribution of regions 6 and 8 were 120 × 50. The grid described above is defined as the medium one, the grid with double the nodes in the x and y directions is defined as the fine grid, and the mesh with half the nodes is defined as the coarse grid.

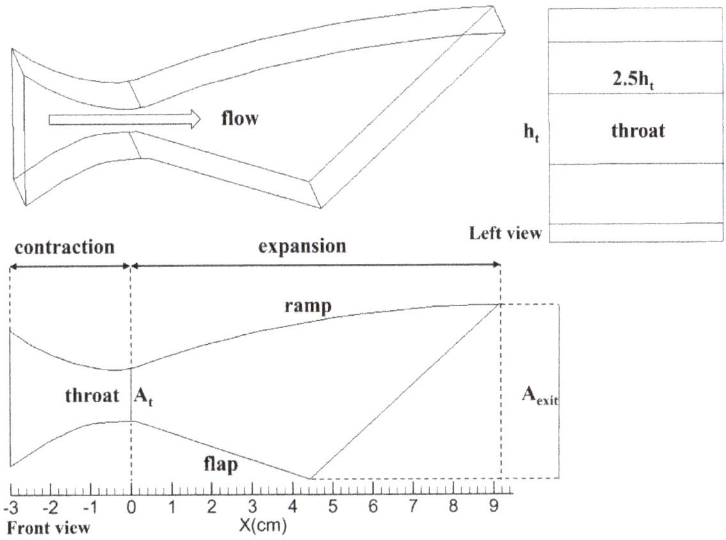

Figure 3. Sketch of the SERN model.

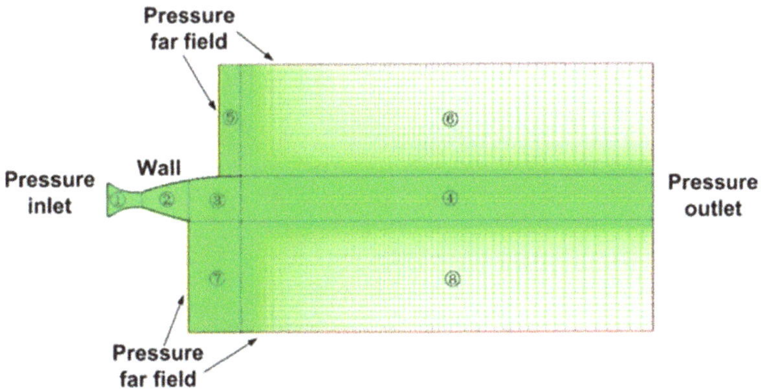

Figure 4. Mesh of the SERN for numerical simulation.

Numerical simulations were carried out using the commercial software FLUENT v12.0, and the two-dimensional Navier–Stokes equations were solved, including the conservation equations for mass, momentum, and energy. In the computational process, the implicit method was used for time advancement to accelerate the convergence. The default Roe-FDS format was used to compute the flux, and the second-order upwind format was used to discretize the control equation. In order to ensure stability and correctness, the Courant number was maintained below 5 in the two-dimensional steady simulation. In the iterative process, the convergence of the computation was assessed by monitoring the mass flow at the inlet and outlet of the nozzle. On the other hand, the residual values of parameters such as the continuity equation, the velocity components, and the energy equation had to be reduced to less than 10^{-4}. The viscous model is the renormalization group (RNG) k–ε model, which is a well-validated turbulence model for predicting the separation flow field of the nozzle [27–30]. The simulation model has the same conditions as the experiment, and the settings are listed in Table 1. Due to the complex flow structure formed by separation and reattachment in SERN, y^+ cannot be controlled in a small range. The y^+ value varied between 10 and 80 in the simulation, thus meeting the requirements of the turbulence model.

Table 1. Settings for numerical simulation.

Property	Setting
Materials	Ideal gas, compressible
Dimensionality	2D
Discretization method	Second-order upwind
Solution method	Density-based solver
Solution formulation	Implicit
Time dependence	Steady
Turbulent model	k-epsilon RNG
Near-wall treatment	Standard wall function
Pressure—inlet	Total pressure = 124,008.5 Pa, temperature = 296.5 K
Pressure—far-field	*Ma* = 0, static pressure = 35,422.69 Pa, temperature = 296.5 K
Pressure—outlet	Total pressure = 35,422.69 Pa, temperature = 296.5 K
Wall	Adiabatic

Figure 5 shows the pressure distribution of the expansion ramp. The horizontal axis was normalized using throat height, while the vertical axis was normalized using the ambient static pressure. The simulations accurately predicted the separation point and reattachment process; however, there was a slight discrepancy in the pressure rates before and after the shock wave. It can be observed from Figure 5 that the fine grid results are in good agreement with those of the medium grid. However, a larger deviation is observed between the results obtained from the coarse grid and the medium grid. At approximately X/ht = 6.4, the maximum relative error between the two grids reached 3%. Thus, the medium grid is suitable for obtaining accurate solutions. Consequently, the medium grid was chosen for subsequent research. Figure 6 presents a comparison between the experimental Schlieren and the numerical results; the CFD method allows for the accurate simulation of the flow-field structure. Simulation data also facilitate the presentation of flow-field details more conveniently.

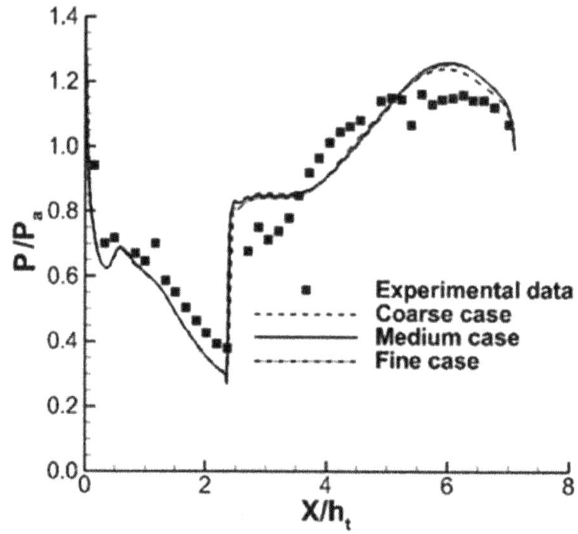

Figure 5. Wall pressure distribution of the expansion ramp.

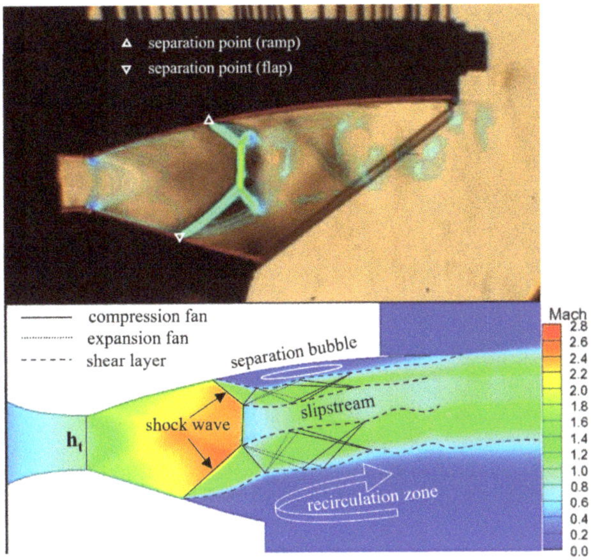

Figure 6. Comparison of the Schlieren image and the CFD results, *NPR* = 3.5.

3. Results and Discussion

3.1. Nozzle Design

The objective of the RBCC is flight within an extremely wide range of Mach numbers, enabling the aircraft to take off from the ground and accelerate to hypersonic cruising states. Throughout this process, there is a significant variation in the operating pressure ratio of the nozzle, which starts at around 2 and increases to over 600. The SERN was designed according to the experimental data of the Glenn Research Center [31] shown in Table 2.

Table 2. Experimental data of RBCC engine.

Ma_∞	Altitude (km)	Ambient Static Pressure, P_a (Pa)	Combustor Total Pressure, P_c* (Pa)	*NPR*
0	0	101,325		
2	8.3	34,061.1	166,899.4	4.9
2.5	11.3	21,781.0	228,700.5	10.5
3	13.6	15,084.1	303,190.4	20.1
4	17.3	8473.5	558,403.6	65.9
5	20.1	5414.4	904,746.2	167.1
6	24.4	2811.0	705,561.0	251
9	28.2	1580.5	1,134,799.0	718
10	31.1	1017.6	1,996,229	1961.7
12	33.5	710.4	3,368,361.6	4741.5

For the Strutjet engine, a previous study [32] demonstrated that the cruise Mach number of the aircraft reaches 8. In order to take into account the low-Mach number performance at ramjet takeover and Mach 8 cruise state, the nozzle optimization was selected at Mach 6. When operating in injection mode or subsonic-combustion ramjet mode, the RBCC engine with fixed geometry construction lacks a contraction section in the nozzle that facilitates the acceleration of the airflow to sonic speed. Therefore, the throat serves as the nozzle inlet. The thermal throat is formed through organized combustion, and the number of the airflow reaches 1 via heat injection. However, the combustion chamber channel usually has an expansion angle, which makes the inlet Mach number of the nozzle slightly larger than 1. When the RBCC operates in supersonic-combustion ramjet (scramjet) mode, the airflow Mach number before the nozzle inlet may reach 2 or higher [33].

Assuming no chemical reaction occurs within the nozzle channel, the wall of the nozzle is adiabatic, and the total pressure loss is ignored; thus, $P_{in}^* = P_{exit}^*$, $T_{in}^* = T_{exit}^*$.

According to the flow conservation relationship, the following equations can be obtained:

$$K_{in}\frac{P_{in}^*}{\sqrt{T_{in}^*}}q(Ma_{in})A_{in} = K_{exit}\frac{P_{exit}^*}{\sqrt{T_{exit}^*}}q(Ma_{exit})A_{exit} \quad (1)$$

$$K = \left(\frac{\gamma}{R}\right)^{\frac{1}{2}}\left(\frac{2}{\gamma+1}\right)^{\frac{\gamma+1}{2(\gamma-1)}} \quad (2)$$

$$q(Ma) = Ma\left[\frac{2}{\gamma+1}\left(1+\frac{\gamma-1}{2}Ma^2\right)\right]^{-\frac{\gamma+1}{2(\gamma-1)}} \quad (3)$$

where * denotes the stagnation flow parameters; subscripts *in* and *exit* represent the nozzle inlet and outlet, respectively; A is the local area; R is the gas constant; and γ is the specific heat ratio of the gas.

The design *NPR* of the nozzle was determined as follows:

$$NPR_{design} = \frac{P_{in}^*}{P_{exit}} = \frac{P_{exit}^*}{P_{exit}} = \left(1-\frac{\gamma_{exit}-1}{2}Ma_{exit}^2\right)^{\frac{\gamma_{exit}}{\gamma_{exit}-1}} \quad (4)$$

Using the data in Table 1, and ignoring combustion in the nozzle and changes in the specific heat ratios of the inlet and outlet, a relationship between the area ratio of the nozzle outlet to the inlet and the inlet Mach number was established for each design NPR, as illustrated in Figure 7.

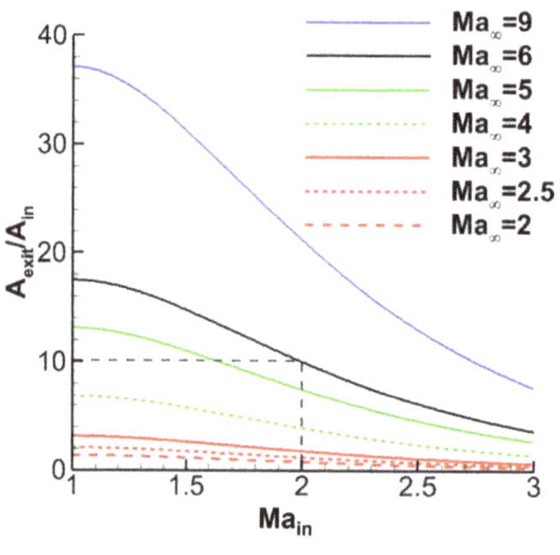

Figure 7. The link between the design area ratio of SERN and the inlet Mach number under various flying situations.

For the Strutjet engine, the working Mach number of 2.5–6 is the subsonic-combustion ramjet mode, and the working Mach number of over 6 is the scramjet mode. When the engine is in the subsonic-combustion ramjet mode, it has a thermal throat, and the Mach number of the nozzle inlet is approximately 1. During the design process, it was considered that the scramjet mode occurred at Mach 6. When the Mach number of the nozzle inlet was 2, the design area ratio was 10.01. Again, for the subsonic-combustion ramjet mode, the nozzle with the area ratio of 10.01 was under-expanded at Mach 5 operation, whereas it was slightly overexpanded at Mach 4 operation and significantly overexpanded below Mach 3. To achieve the desired performance at low speed, the nozzle's design Mach number was 6, and the design area ratio was 10.01. Additionally, the performance of the cruising state with Mach 8 should be considered, and the NPR under this circumstance was determined using the interpolation method. Finally, the nozzle was designed using the maximum thrust theory, with a design NPR of 562.3, corresponding to the operating state of Mach 8. After obtaining the profile, it was truncated according to the condition that the area ratio of outlet to inlet was 10.01, and the resulting nozzle profile is displayed in Figure 8. Table 3 details the specific design parameters for the maximum thrust nozzle.

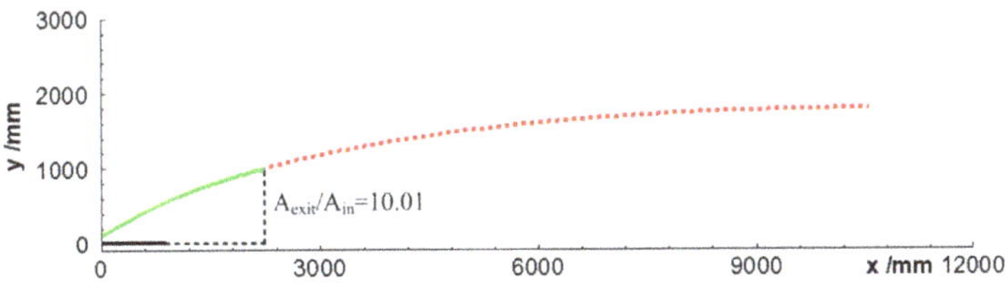

Figure 8. The designed profile of the minimum length of the nozzle for Mach 8.

Table 3. Specific design parameters of maximum thrust nozzle profile at Mach 8.

Design Parameters	Value
Total pressure of the inlet P_D* (Pa)	888,732.02
Total temperature of the inlet T_D* (K)	2000
Static pressure of the inlet P_D (Pa)	115,250.77
Ambient pressure P_a (Pa)	1580.53
Height of inlet h_t (mm)	100
The ratio of specific heat γ	1.33

The thrust coefficient is defined with a surplus impulse of the nozzle outlet as follows:

$$\text{Thrust confficient} = \frac{\text{actual surplus impulse of nozzle outlet}}{\text{ideal surplus impulse of nozzle outlet}} \tag{5}$$

The surplus impulse function is expressed as follows:

$$I = \dot{m}v + A(p - p_a) \tag{6}$$

where $\dot{m}$ denotes the mass flow rate.

As seen in Table 4, the nozzle maintained a high level of thrust performance under the Mach numbers of 4–9.

Table 4. Thrust coefficient of the nozzle under different Mach number flight conditions.

Flight Mach Numbers	Thrust Coefficient
3	0.8011
4	0.9636
5	0.9784
6	0.9770
8	0.9747
9	0.9725

When the Mach number was <3, the thrust performance of the nozzle was poor. Particularly, when the Mach number <2, due to the actual area ratio of the nozzle outlet to inlet being too large, the gas in the nozzle experienced severe overexpansion, and the positive impulse could not be obtained at the nozzle outlet, preventing the nozzle from producing effective thrust. When the Mach numbers were in the range of 0–3, however, the RBCC engine operated in the rocket injection mode, and partial thrust could be provided via the rocket to compensate for the loss caused by the overexpansion state.

3.2. Simplified Acceleration Process

There were also different separation flow fields for the designed SERN in the severe overexpansion state, and the transition of separation patterns occurred under changing conditions. Figure 9 shows the Mach contour of the flow field under various working NPRs at sea level. As can be seen, the airflow in the nozzle was along the expansion ramp, resulting in RSS (ramp); there was no FSS or RSS (flap). Thus, the sea level condition with only a change in NPR did not cause a transition of the flow separation pattern in the nozzle.

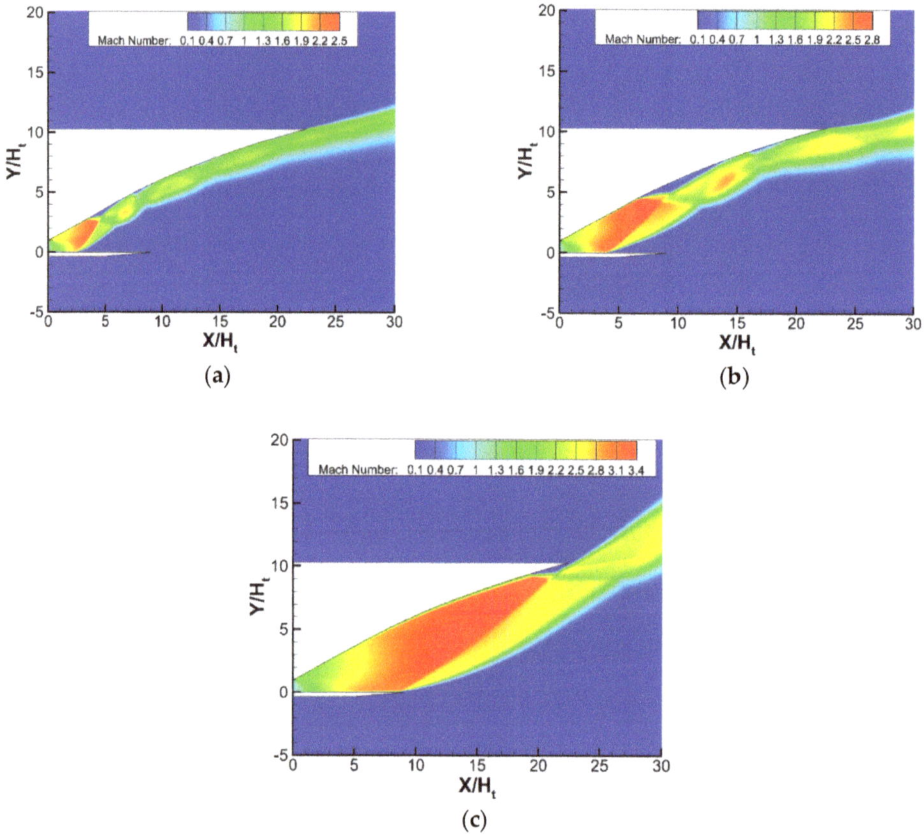

Figure 9. Mach contours of the flow field under different *NPR*, *Ma* = 0: (**a**) *NPR* = 4.9; (**b**) *NPR* = 10.5; (**c**) *NPR* = 20.1.

Figure 10 provides the Mach contours of the flow field under real flight conditions. As can be seen, the nozzle exhibits FSS pattern under conditions (a) and (b), and there is no separation on the flap but a tail shock. Under condition (c), the shock wave on the flap does not shoot to the ramp and hence does not induce the formation of separation shock wave and separation bubble there. If the aircraft takes off from the ground and accelerates, RSS (ramp) and FSS separation patterns appear sequentially in the nozzle, accompanied by the corresponding transition of the separation patterns.

The simplified acceleration process was simulated under the condition that the variation rate of Mach number was 1/s, and the results are shown in Figure 11. With the increase in external Mach number, the separation shock wave on the flap steadily moved backward, compressing the flap recirculation zone. The separation shock on the flap rapidly moved to the trailing edge until the jet shear layer interacted with the boundary of outflow, as illustrated in Figure 11b. Subsequently, the jet deflected downward, the flap recirculation zone disappeared, and the separation bubble on the ramp expanded, forming a large recirculation zone. The ambient gas gradually entered the recirculation zone, and the separation shock on the ramp moved forward to the nozzle inlet. As illustrated in Figure 11c, the lower shear layer of the jet fully interacted with the outflow to form a slipstream. Finally, as shown in Figure 11d, the flow field remained steady, and the transition process was completed.

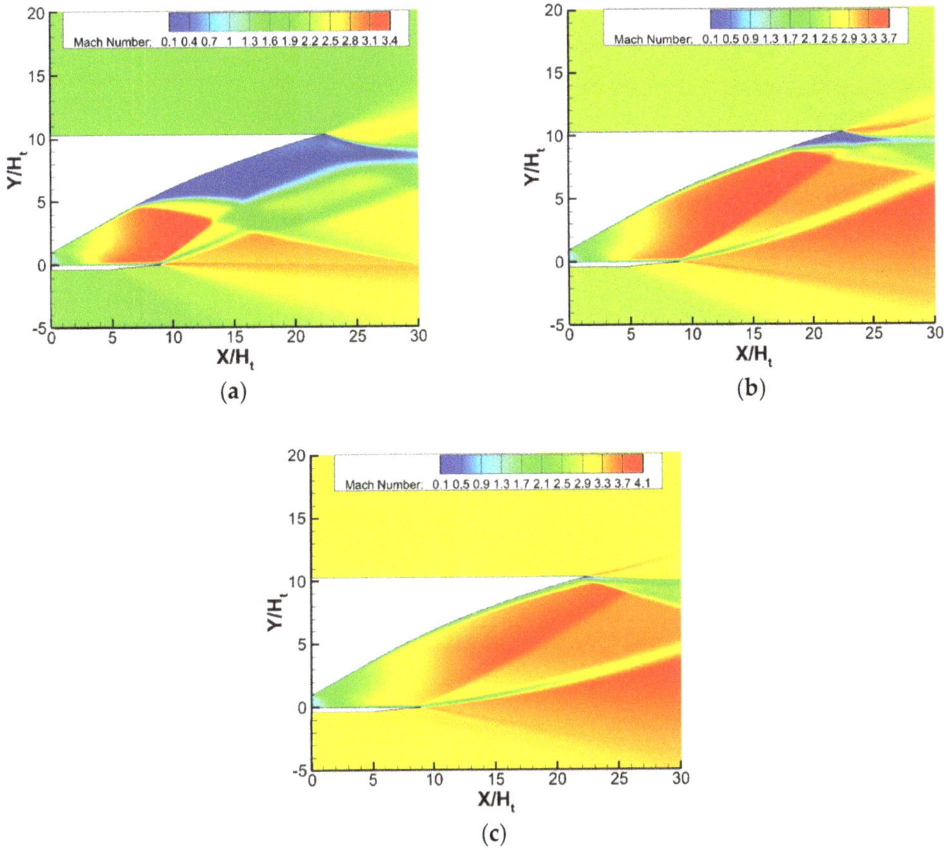

Figure 10. Mach contour of the flow field under different flight conditions: (**a**) *NPR* = 4.9, *Ma* = 2.0; (**b**) *NPR* = 10.5, *Ma* = 2.5; (**c**) *NPR* = 20.1, *Ma* = 3.0.

Figure 12 illustrates the fluctuation in nozzle performance during the transition process. The thrust, lift, and moments are nondimensionalized using the initial values. Significant alterations occurred in these parameters during the transition. At the peak of the performance degradation, the decreases in thrust, lift, and moments were 11.8%, 31.27%, and 37.30%, respectively. Upon completion of the transition, the decreases in thrust, lift, and moments were 5.56%, 15.93%, and 26.65%, respectively.

The above results verify that the transition of separation patterns will occur in the SERN during startup and acceleration, but the real flight conditions of the RBCC engine are complex. The interactions between the external flow and the jet, the recirculation zone, and the boundary layer are critical during the transition process induced by external Mach number. Different accelerations can have an effect on the Mach number of the airflow at the nozzle exit, hence altering the transition process.

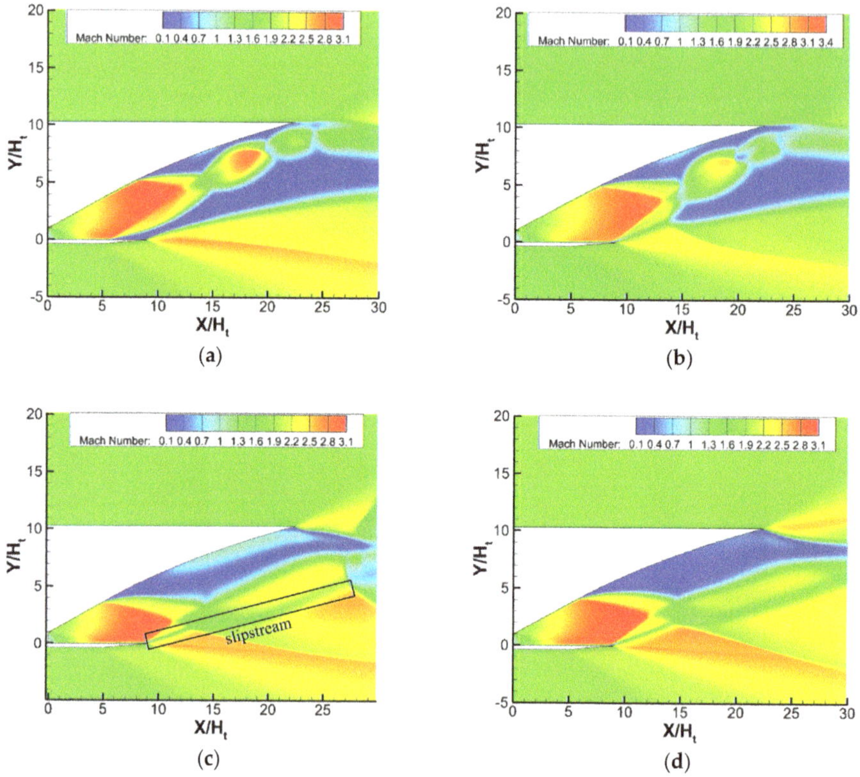

Figure 11. The transition from RSS (ramp) to FSS during the acceleration process: (**a**) t_0, Ma = 1.86; (**b**) t_0 + 0.004 s, Ma = 1.864; (**c**) t_0 + 0.009 s, Ma = 1.869; (**d**) t_0 + 0.02 s, Ma = 1.88.

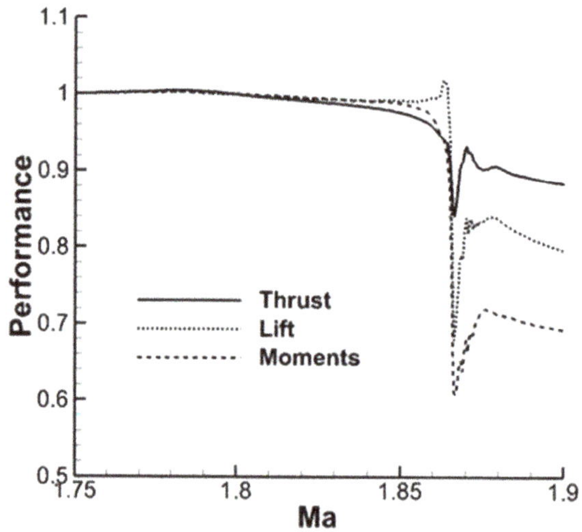

Figure 12. Variation in nozzle performance during the transition from RSS (ramp) to FSS.

3.3. Effect of Acceleration

The numerical simulation results with external Mach number increase rates of 0.2/s, 0.5/s, 1/s, and 2/s are chosen for discussion. When the rate is 0.2/s, it is close to the real

flight situation of the RBCC engine. The transition from RSS (ramp) to FSS was simulated with an initial external Mach number of 1.75 and NPR of 4.9.

Figure 13 demonstrates the relationship between nozzle thrust and Mach number for different increase rates of Mach number. Meanwhile, to explore the influence of the Mach number change rate on the transition duration, the minimum thrust points under Mach number increase rates of 0.5/s, 1/s, and 2/s are adjusted to coincide with the point of 0.2/s case, and the time at these points is marked as t_0, as shown in Figure 14. According to the analysis of the results, RSS (ramp)–FSS transition occurred under the four cases, indicating that the external Mach number variation rate did not affect the principle of the separation flow field and had a negligible effect on the duration of the transition process. With an increase in the Mach number, the critical Mach number corresponding to the transition points also gradually increased, and the variation in peak thrust during the transition process similarly increased.

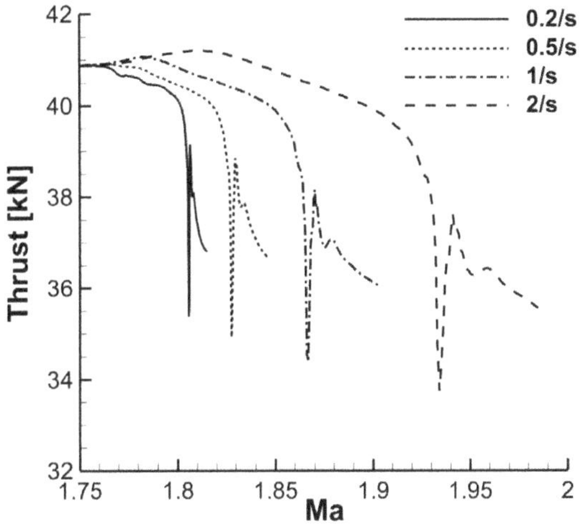

Figure 13. Variation in the nozzle thrust with Mach number for different increase rates of Mach number.

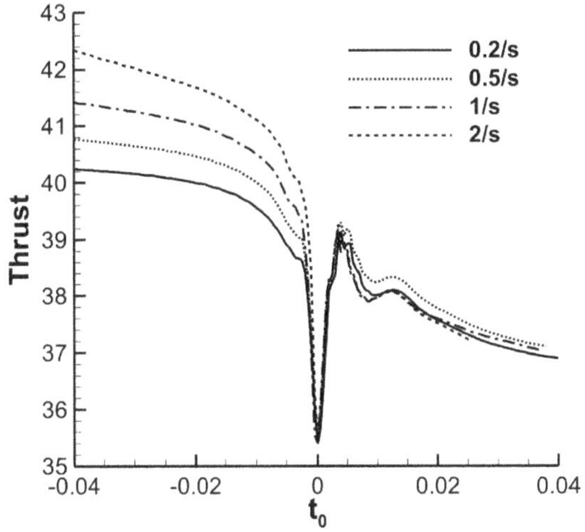

Figure 14. Variation in nozzle thrust with time for different increase rates of Mach number, after an adjustment.

To further discuss the influence of the external Mach number change rate, the constant Mach number contour and constant pressure contour for the flow field at an external Mach number of 1.8 are presented in Figures 15 and 16. Among the four simulated cases, apart from the varying rate of change in the external Mach number, all other settings remained identical. Due to the acceleration of the flight, the external flow field underwent an airflow accumulation process, that is, when the fluid with a lower Mach number had not yet completely flown through the outer wall of the aircraft, the gas with a higher Mach number began to enter the computational domain. Consequently, a higher rate of change in the external Mach number led to a more pronounced accumulation effect in the external flow field, indicating that the aircraft accelerated to a higher Mach number condition while the Mach number at the nozzle exit remained comparatively low. As a result, the faster the change in the Mach number, the more noticeable the effect of external flow-field accumulation.

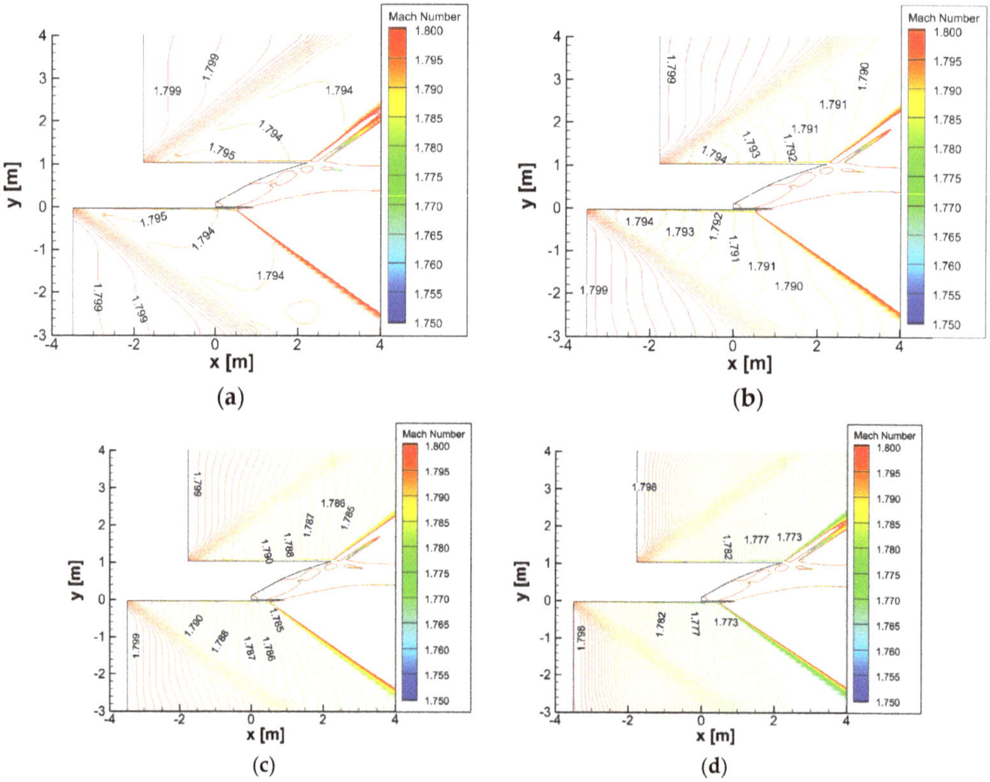

Figure 15. Constant Mach number contour at Mach 1.8 with different increase rates of Mach number: (**a**) 0.2/s; (**b**) 0.5/s; (**c**) 1/s; (**d**) 2/s.

These cases simulated the process of accelerated flight at a specific altitude, where the boundary static pressure remained constant. However, due to the accumulation of airflow, the gas with a higher Mach number upstream exerted a certain compressive effect on the gas with a lower Mach number downstream. Combined with observations from Figure 16, it can be inferred that a higher rate of change in the external Mach number led to a more pronounced compressive effect, resulting in higher static pressure at the nozzle exit. Consequently, the actual nozzle pressure ratio decreased, which contributed to delaying the transition of separation patterns.

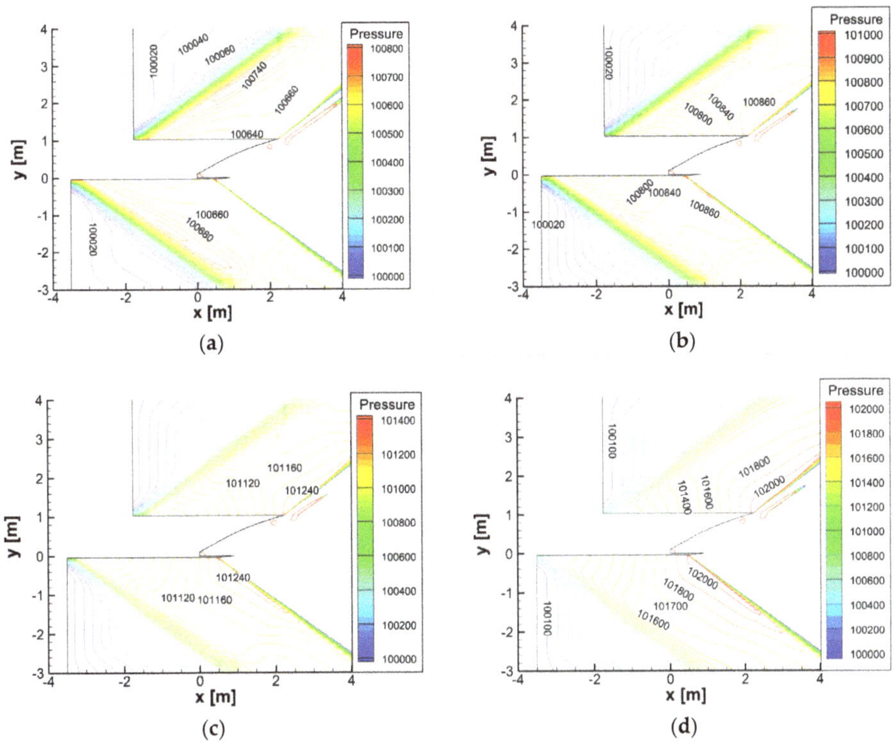

Figure 16. Constant pressure contour at Mach 1.8 with different increase rates of Mach number: (**a**) 0.2/s; (**b**) 0.5/s; (**c**) 1/s; (**d**) 2/s.

3.4. Real Take-off Acceleration Process

RBCC-powered booster aircraft typically follow a two-stage trajectory: inconstant dynamic pressure climbing and constant dynamic pressure climbing. The separation pattern transition phenomena of the nozzle during take-off acceleration were explored in this work using the booster trajectory design of Xue et al. [34]. During the take-off phase, the aircraft flew along a direct ascending path, reaching the flight condition Mach 3.5 and achieving ideal dynamic pressure in the ramjet mode of the RBCC engine. The airplane then began climbing in the direction of the constant dynamic pressure path. Understanding the link between flight altitude and Mach number is required to determine the environmental parameters and obtain the boundary conditions for the nozzle flow separation pattern transition simulations. Figure 17 illustrates the relationship between the Mach number and the flight altitude with flight time in a direct ascending path.

The method described in ISO 2533—Standard Atmosphere (1975) was used to calculate the ambient gas parameters. Static pressure in the environment is expressed as follows:

$$p = \begin{cases} 100 \times (3.731444 - 8.41728H)^{5.25588}, -1450 < H \leq 11000\text{m} \\ 100 \times 226.32 \times \exp(1.7345737 - 1.5768852 \times 10^{-4}H), 11000 < H \leq 20000\text{m} \\ 100 \times (1.2386515/(1 + 5.085177 \times 10^{-6}H))^{34.16321878}, 20000 < H \leq 32000\text{m} \\ 100 \times (1.9630052/(1 + 2.013364 \times 10^{-5}H))^{12.20114957}, 32000 < H \leq 47000\text{m} \end{cases} \tag{7}$$

Ambient temperature is expressed as follows:

$$T = \begin{cases} 301.15 - 6.5 \times (H - 2), -2 < H \leq 0\text{km} \\ 288.15 - 6.5H, 0 < H \leq 11\text{km} \\ 216.65, 11 < H \leq 20\text{km} \\ 216.65 + (H - 20), 20 < H \leq 32\text{km} \\ 228.65 + 2.8 \times (H - 32), 32 < H \leq 47\text{km} \end{cases} \tag{8}$$

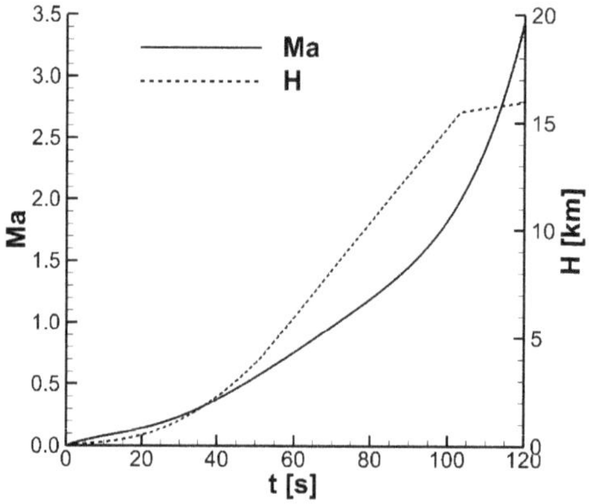

Figure 17. Relationship between the Mach number and the flight altitude with flight time in a direct ascending path.

The boundary conditions in the numerical simulation were set in accordance with the relationship between the Mach number and NPRs shown in Table 2. When the Mach number ≤ 2, the given NPR = 4.9; when 2 < Mach number < 2.5, the NPR increased linearly from 4.9 to 10.5; when 2.5 < Mach number ≤ 3, the NPR increased linearly from 10.5 to 20.1; and when 3 < Mach number ≤ 4, the NPR increased linearly from 20.1 to 65.9. When the flight Mach number ≤ 3.5, the operational modes of the RBCC were the injection mode or subsonic-combustion ramjet mode, with the nozzle inlet Mach numbers around 1. During the numerical simulation, the influence of nonuniform parameters at the nozzle inlet was neglected. Instead, a uniform inlet condition with Mach number = 1 was imposed using far-field boundary conditions, while other boundary conditions were varied using user-defined functions (UDFs). The computational process began with an external Mach number of 0.1 and continued until there was no flow separation inside the nozzle. As illustrated in Figure 17, the Mach number change rate was slow throughout the actual ascending phase, ranging from 0.02/s to 0.03/s. In the simulation process, the Mach number change rate increased by 50 times due to the consideration of the computational amount, resulting in a Mach number change rate of approximately 1/s.

As illustrated in Figure 18, the nozzle performance curve exhibits a declining tendency during the direct ascending path and then begins to climb at approximately Mach 2.5. The Mach contours of the nozzle flow field with external Mach numbers of 0.5, 1.0, 1.5, and 2.0 are shown in Figure 19. According to the presented flow fields, when Mach number < 2, the nozzle entered RSS (ramp) mode, and as the external Mach number increased, the separation points on the ramp and flap steadily moved backward, away from the nozzle inlet. When the external Mach number was small, the design NPR was high, and its actual working NPR was relatively low, the separation points of the upper and lower walls were close to the throat and distant from the trailing edge of the nozzle. As a result, there is considerable room for airflow development after the separation point, and the RSS (ramp) pattern continues to be observed in the flow field of the nozzle over a broad Mach number range.

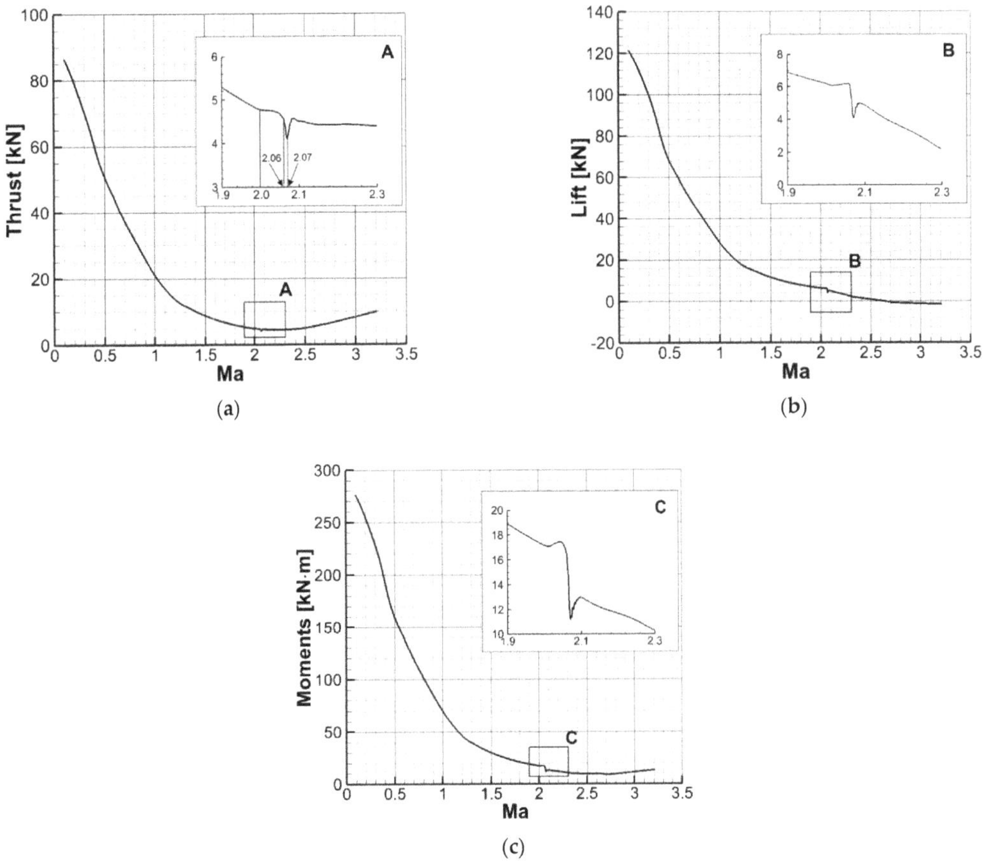

Figure 18. Variation in the nozzle performance with Mach number in the direct ascending path: (**a**) thrust; (**b**) lift; (**c**) moments.

In addition, a dramatic change can be clearly detected on the performance curve, corresponding to the separation pattern transition from RSS (ramp) to FSS depicted in Figure 20. Although the flow-field variation in the separation pattern transition process is comparable to that in the linearly increasing Mach number seen in Figure 11, the critical Mach numbers of the two transition processes are considerably different.

As demonstrated in Figures 19 and 20, during the transition process from RSS (ramp) to FSS, the flap separation shock rapidly moved toward the trailing edge, followed by a subsequent downward deflection of the jet, resulting in drastic changes in the flow field. After the transition from RSS (ramp) to FSS, a larger recirculation zone remained on the expansion ramp, as shown in Figure 21a. As the external Mach number increased, the separation shock on the expansion ramp continued to move downstream, suggesting the possibility of a similar phenomenon, where the ramp separation shock rapidly moved toward the trailing edge, causing significant changes in the flow-field structure. However, as illustrated in Figure 18, there is a continuous and smooth change in the nozzle following the RSS (ramp)–FSS transition process, with no mutation. Figure 21 depicts the process through which the flow field changed following the transition from RSS (ramp) to FSS. As the external Mach number increased, the separation shock wave on the ramp gradually moved downstream, compressing the recirculation zone until it disappeared. Throughout the process, no flow separation occurred.

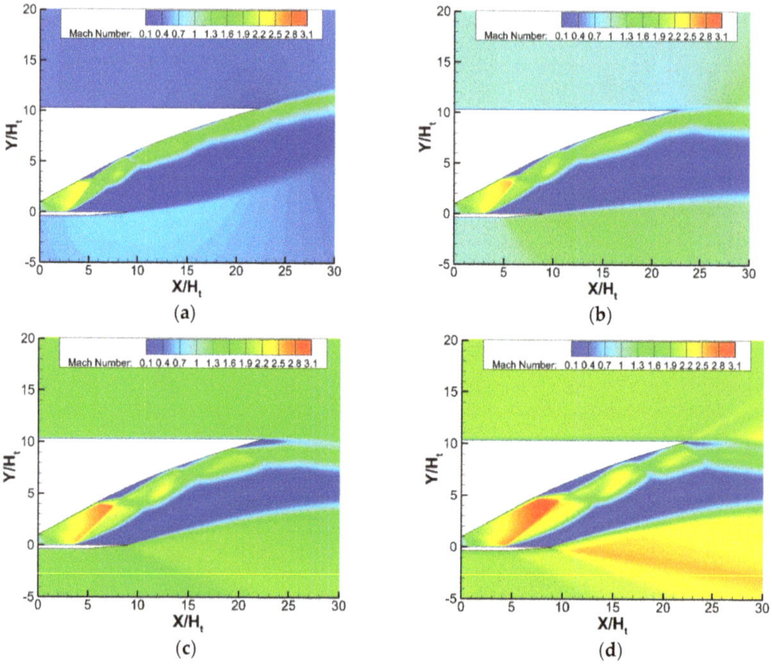

Figure 19. Mach contour of the flow field in RSS (ramp) at different Mach numbers: (**a**) $Ma = 0.5$; (**b**) $Ma = 1.0$, (**c**) $Ma = 1.5$; (**d**) $Ma = 2.0$.

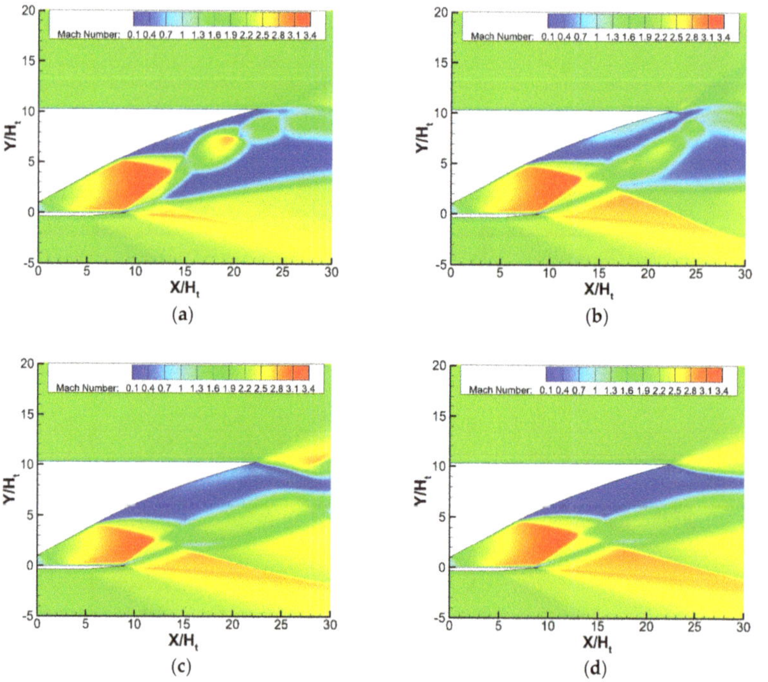

Figure 20. The transition from RSS (ramp) to FSS in the direct rising path: (**a**) t_0, $Ma = 2.0606$; (**b**) $t_0 + 0.003$ s, $Ma = 2.0693$; (**c**) $t_0 + 0.008$ s, $Ma = 2.0840$; (**d**) $t_0 + 0.015$ s, $Ma = 2.1048$.

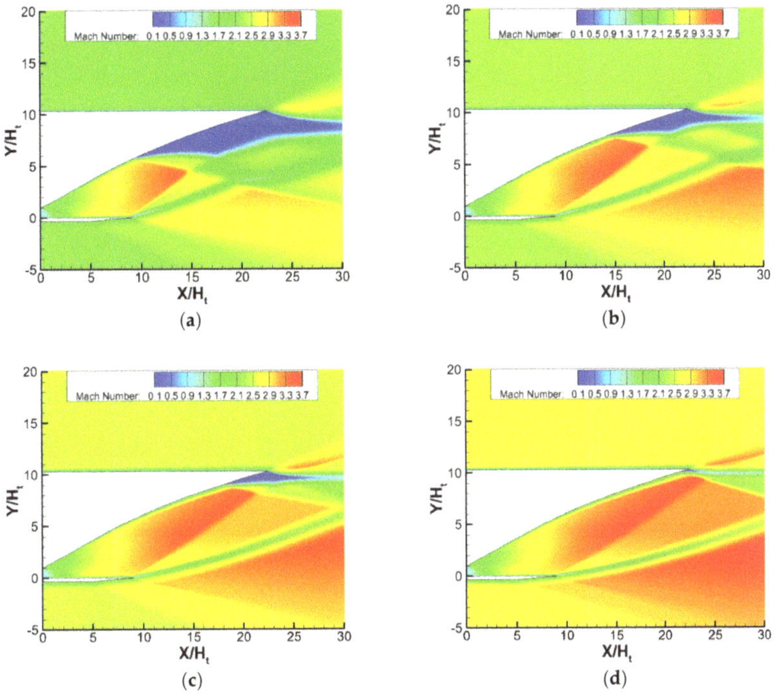

Figure 21. Variation in the flow field during the transition from FSS to no-flow separation pattern: (**a**) $Ma = 2.2$; (**b**) $Ma = 2.4$; (**c**) $Ma = 2.6$; (**d**) $Ma = 2.8$.

In comparison to the transition from RSS (ramp) to FSS, the transition from FSS to a no-flow separation pattern did not result in a performance mutation. First, the recirculation zone on the expanded ramp in FSS was larger than that on the flap in RSS (ramp). Environmental flow entered the ramp recirculation zone more smoothly, allowing the zone to develop more steadily as well. Second, tail shock on the flap severely restricted the development of the separation shock at the expansion ramp when the external Mach number was high, as did the expansion of the recirculation zone, and the effect became more pronounced as the external Mach number increased. Due to the acceleration effect of outflow passing through the flap cowl, with the increase in flight Mach, the tail shock here gradually deflected downward and eventually stopped, which caused flow separation on the ramp, thus ensuring a stable transition process.

4. Conclusions

In this paper, considering the low-Mach number performance at ramjet takeover and the Mach 8 cruise state, the nozzle optimization design was considered to be at Mach 6. Through the numerical method, the flow separation pattern transition processes under the circumstances of linear change in the external Mach number and the real trajectory were simulated. Detailed analyses were carried out for the processes to further understand the influence of changes in the external flow field on the flow field inside the nozzle. The results are as follows:

(1) The external flow Mach number had a significant effect on the overexpansion flow field of the RBCC nozzle. With an increase in the external Mach number, sequential transitions of RSS (ramp) to FSS and FSS to no-flow separation pattern occurred.
(2) The transition principle of the flow separation patterns in the real ascending path was similar to the case with external flow varying linearly, but the Mach number corre-

sponding to the transition points was considerably different. The variation rate of the external Mach number affected the nozzle performance during the transition process.

(3) The higher the variation rate of the external flow Mach number, the more obvious the airflow accumulation effect of the external flow field, which caused an increase in the static pressure at the outlet and a decrease in the real nozzle pressure ratio, delaying the transition of flow separation patterns.

Author Contributions: Conceptualization, Y.Y. (Yang Yu); methodology, Y.Y. (Yang Yu); software, Y.Y. (Yang Yu) and Y.M.; investigation, Y.Y. (Yang Yu), Y.M. and T.Y.; resources, Y.Y. (Yang Yu); writing—original draft preparation, Y.Y. (Yang Yu); writing—review and editing, Y.M., T.Y., Y.Y. (Yalin Yang), S.X. and S.L. All authors have read and agreed to the published version of the manuscript.

Funding: This research was funded by the National Natural Science Foundation of China under Grant No. 11802124, the Jiangsu Province Key Laboratory of Aerospace Power System Foundations (CEPE2020003), and the National Natural Science Foundation of Chongqing (cstc2021jcyj-msxmX1120).

Data Availability Statement: Some or all data, models, or codes generated or used during this study are available from the corresponding author upon reasonable request.

Conflicts of Interest: The authors declare no conflict of interest.

Nomenclature

SERN	Single-expansion ramp nozzle
RBCC	Rocket-based combined cycle
CFD	Computational fluid dynamics
RSS	Restricted shock separation
RSS (ramp)	Restricted shock separation with separation bubble forming on the ramp
RSS (flap)	Restricted shock separation with separation bubble forming on the flap
FSS	Free shock separation
SWBLI	Shock wave/boundary layer interaction
NPR	Nozzle pressure ratio
UDF	User-defined functions
A_t	Area of the nozzle throat
A_{in}	Area of the nozzle inlet
A_{exit}	Area of the nozzle exit
P_a	Ambient pressure
P^*	Total pressure
h_t	Height of the nozzle throat
Ma_∞	Mach number of freestream
Ma_{in}	Mach number of nozzle inlet

References

1. Shi, L.; Yang, Y.; Yang, X.; Zhao, G.; Chen, F.; He, G. Start Limits and Positive Control of RBCC Inlet. *J. Aerosp. Eng.* **2022**, *35*, 04021128. [CrossRef]
2. Wang, Y.; Zhang, Q.; Tang, Z.; Tian, Z.; Zheng, Y. Integrated Thermal Protection System Design for Hypersonic Vehicle Based on New Thermal–Mechanical Method. *J. Aerosp. Eng.* **2022**, *35*, 04021121. [CrossRef]
3. Hiraiwa, T.; Ito, K.; Sato, S.; Ueda, S.; Tani, K.; Tomioka, S.; Kanda, T. Recent Progress in Scramjet/Combined Cycle Engines at JAXA, Kakuda Space Center. *Acta Astronaut.* **2008**, *63*, 565–574. [CrossRef]
4. Gronland, T.; Berens, T. Nozzle/Afterbody Integration of Hypersonic Vehicles by Means of Secondary Air Injection. In Proceedings of the International Aerospace Planes and Hypersonics Technologies, Chattanooga, TN, USA, 3–7 April 1995; International Space Planes and Hypersonic Systems and Technologies Conferences. American Institute of Aeronautics and Astronautics: Reston, VA, USA, 1995.
5. Traci, R.M.; Farr, J.L., Jr.; Laganelli, T.; Walker, J. *A Thermal Management Systems Model for the NASA GTX RBCC Concept*; NASA: Washington, DC, USA, 2002.
6. Lee, J.; Krivanek, T. Design and Fabrication of the ISTAR Direct-Connect Combuster Experiment at the Hypersonic Tunnel Facility. In Proceedings of the 43rd AIAA Aerospace Sciences Meeting and Exhibit, Reno, NV, USA, 10–13 January 2005; American Institute of Aeronautics and Astronautics: Reston, VA, USA, 2005.

7. Ju, S.; Yan, C.; Wang, X.; Qin, Y.; Ye, Z. Optimization Design of Energy Deposition on Single Expansion Ramp Nozzle. *Acta Astronaut.* **2017**, *140*, 351–361. [CrossRef]
8. Yu, K.; Xu, J.; Lv, Z.; Song, G. Inverse Design Methodology on a Single Expansion Ramp Nozzle for Scramjets. *Aerosp. Sci. Technol.* **2019**, *92*, 9–19. [CrossRef]
9. Hadjadj, A.; Onofri, M. Nozzle Flow Separation. *Shock Waves* **2009**, *19*, 163–169. [CrossRef]
10. Zebiri, B.; Piquet, A.; Hadjadj, A.; Verma, S.B. Shock-Induced Flow Separation in an Overexpanded Supersonic Planar Nozzle. *AIAA J.* **2020**, *58*, 2122–2131. [CrossRef]
11. Tomita, T.; Takahashi, M.; Sasaki, M.; Sakamoto, H.; Takahashi, M.; Tamura, H. Experimental Evaluation of Side-Loads in LE-7A Prototype Engine Nozzle. *Shock Waves* **2009**, *19*, 213–228. [CrossRef]
12. Nave, L.; Coffey, G. Sea Level Side Loads in High-Area-Ratio Rocket Engines. In Proceedings of the 9th Propulsion Conference; American Institute of Aeronautics and Astronautics, Las Vegas, NV, USA, 5–7 November 1973.
13. Frey, M.; Hagemann, G. Flow Separation and Side-Loads in Rocket Nozzles. In Proceedings of the 35th Joint Propulsion Conference and Exhibit, Los Angeles, CA, USA, 20–24 June 1999; American Institute of Aeronautics and Astronautics: Reston, VA, USA, 1999.
14. Verma, S.; Chidambaranathan, M.; Hadjadj, A. Analysis of Shock Unsteadiness in a Supersonic Over-Expanded Planar Nozzle. *Eur. J. Mech. B Fluids* **2018**, *68*, 55–65. [CrossRef]
15. Yu, Y.; Xu, J.; Yu, K.; Mo, J. Unsteady Transitions of Separation Patterns in Single Expansion Ramp Nozzle. *Shock Waves* **2015**, *25*, 623–633. [CrossRef]
16. Frey, M.; Hagemann, G. Restricted Shock Separation in Rocket Nozzles. *J. Propuls. Power* **2000**, *16*, 478–484. [CrossRef]
17. Watanabe, Y.; Sakazume, N.; Tsuboi, M. LE-7A Engine Nozzle Problems during Transient Operations. In Proceedings of the 38th AIAA/ASME/SAE/ASEE Joint Propulsion Conference & Exhibit, Indianapolis, IN, USA, 7–10 July 2002; American Institute of Aeronautics and Astronautics: Reston, VA, USA, 2002.
18. Östlund, J.; Damgaard, T.; Frey, M. Side-Load Phenomena in Highly Overexpanded Rocket Nozzles. *J. Propuls. Power* **2004**, *20*, 695–704. [CrossRef]
19. Baars, W.J.; Tinney, C.E.; Ruf, J.H.; Brown, A.M.; McDaniels, D.M. Wall Pressure Unsteadiness and Side Loads in Overexpanded Rocket Nozzles. *AIAA J.* **2012**, *50*, 61–73. [CrossRef]
20. Verma, S.B.; Haidn, O. Study of Restricted Shock Separation Phenomena in a Thrust Optimized Parabolic Nozzle. *J. Propuls. Power* **2009**, *25*, 1046–1057. [CrossRef]
21. Martelli, E.; Nasuti, F.; Onofri, M. Numerical Calculation of FSS/RSS Transition in Highly Overexpanded Rocket Nozzle Flows. *Shock Waves* **2010**, *20*, 139–146. [CrossRef]
22. He, M.; Qin, L.; Liu, Y. Numerical Investigation of Flow Separation Behavior in an Over-Expanded Annular Conical Aerospike Nozzle. *Chin. J. Aeronaut.* **2015**, *28*, 983–1002. [CrossRef]
23. Chutkey, K.; Viji, M.; Verma, S.B. Interaction of External Flow with Linear Cluster Plug Nozzle Jet. *Shock Waves* **2018**, *28*, 1207–1221. [CrossRef]
24. Li, J.; Chen, S.; Cai, F.; Yan, C. Numerical Investigation of Vented Plume into a Supersonic Flow in the Early Stage of Rocket Hot Separation. *Aerosp. Sci. Technol.* **2020**, *107*, 106249. [CrossRef]
25. Mousavi, S.M.; Pourabidi, R.; Goshtasbi-Rad, E. Numerical Investigation of over Expanded Flow Behavior in a Single Expansion Ramp Nozzle. *Acta Astronaut.* **2018**, *146*, 273–281. [CrossRef]
26. Lee, C.; Choi, K.; Kim, C.; Han, S. Computational Investigation of Flow Separation in a Thrust-Optimized Parabolic Nozzle during High-Altitude Testing. *Comput. Fluids* **2020**, *197*, 104363. [CrossRef]
27. Yu, Y. Over-Expanded Separation Transitions of Single Expansion Ramp Nozzle in the Accelerating and Decelerating Processes. *Aerosp. Sci. Technol.* **2020**, *98*, 105674. [CrossRef]
28. Yu, Y.; Mao, Y.; Yu, T.; Xu, S. Numerical Investigation on Mechanism of External Flow Field–Induced Separation Pattern Transition. *J. Aerosp. Eng.* **2023**, *36*, 04022123. [CrossRef]
29. Yu, Y.; Xu, J.; Mo, J.; Wang, M. Principal Parameters in Flow Separation Patterns of Over-Expanded Single Expansion RAMP Nozzle. *Eng. Appl. Comput. Fluid Mech.* **2014**, *8*, 274–288. [CrossRef]
30. Galván, S.; Reggio, M.; Guibault, F. Assessment Study of K-ε Turbulence Models and Near-Wall Modeling for Steady State Swirling Flow Analysis in Draft Tube Using Fluent. *Eng. Appl. Comput. Fluid Mech.* **2011**, *5*, 459–478. [CrossRef]
31. Smith, T.D.; Ill, F.C.; Rice, T.; Blaha, B. *Development of an Integrated Nozzle for a Symmetric, RBCC Launch Vehicle Configuration i35*; Glenn Research Center: Cleveland, OH, USA, 2000.
32. Campbell, B.T.; Siebenhaar, A.; Nguyen, T. Strutjet Engine Performance. *J. Propuls. Power* **2001**, *17*, 1227–1232. [CrossRef]
33. Tang, X.; He, G.; Qin, F. Investigation on Combustion Performance of Axisymmetric RBCC Combustor in Scramjet Mode. *J. Northwestern Polytech. Univ.* **2014**, *32*, 29–34.
34. Xue, R.; Hu, C.; Lv, X.; Qing, F. RBCC Constant Dynamic Pressure Booster Trajectory Design and Propellant Mass Flowrate Analysis for TSTO Transportation System. *J. Solid Rocket Technol.* **2013**, *36*, 155–160.

aerospace

Article

Experimental Investigation of the Shock-Related Unsteadiness around a Spiked-Blunt Body Based on a Novel DMD Energy Sorting Criterion

Yifan Wang [1], Jinglei Xu [1,*], Qihao Qin [1,2], Ruiqing Guan [1] and Le Cai [1]

1 State Key Laboratory of Mechanics and Control of Aeronautics and Astronautics Structures, Nanjing University of Aeronautics and Astronautics, Nanjing 210016, China; yifan@nuaa.edu.cn (Y.W.); qinqh@nuaa.edu.cn (Q.Q.); reaching_guan@nuaa.edu.cn (R.G.); nuaacl@nuaa.edu.cn (L.C.)
2 AECC Aero-Engine Control System Institute, Wuxi 214063, China
* Correspondence: xujl@nuaa.edu.cn

Abstract: In this study, we propose a novel dynamic mode decomposition (DMD) energy sorting criterion that works in conjunction with the conventional DMD amplitude-frequency sorting criterion on the high-dimensional schlieren dataset of the unsteady flow of a spiked-blunt body at Ma = 2.2. The study commences by conducting a comparative analysis of the eigenvalues, temporal coefficients, and spatial structures derived from the three sorting criteria. Then, the proper orthogonal decomposition (POD) and dynamic pressure signals are utilised as supplementary resources to explore their effectiveness in capturing spectral characteristics and spatial structures. The study concludes by summarising the characteristics and potential applications of DMD associated with each sorting criterion, as well as revealing the predominant flow features of the unsteady flow field around the spiked-blunt body at supersonic speeds. Results indicate that DMD using the energy sorting criterion outperforms the amplitude and frequency sorting criteria in identifying the primary structures of unsteady pulsations in the flow field, which proves its superiority in handling an experimental dataset of unsteady flow fields. Moreover, the unsteady pulsations in the flow field around the spiked-blunt body under supersonic inflow conditions are observed to exhibit multi-frequency coupling, with the primary frequency of 3.3 kHz originating from the periodic motion of the aftershock.

Keywords: dynamic mode decomposition; spiked-blunt body; flow unsteadiness; high-speed schlieren; spectral analysis

Citation: Wang, Y.; Xu, J.; Qin, Q.; Guan, R.; Cai, L. Experimental Investigation of the Shock-Related Unsteadiness around a Spiked-Blunt Body Based on a Novel DMD Energy Sorting Criterion. *Aerospace* **2024**, *11*, 188. https://doi.org/10.3390/aerospace11030188

Academic Editor: Zhijin Wang

Received: 8 January 2024
Revised: 8 February 2024
Accepted: 14 February 2024
Published: 27 February 2024

1. Introduction

The phenomenon of flow unsteadiness is typically observed ahead of a wide range of axisymmetric forebodies, particularly when operating at supersonic and hypersonic velocities [1,2]. Various configurations, including mixed compression inlets [3,4], double cones [5,6], forward-facing cavities [7], axially positioned cavities [8], wall protrusions [9], and spiked/aerodisk forebodies [10–12], have been identified as significant sources of flow unsteadiness. This unsteadiness spans a broad spectrum of flow Reynolds numbers, which affects laminar and turbulent flow states [13–15]. In general, the most pronounced form of shock-related unsteadiness is referred to as 'buzzing' [16,17]. This phenomenon is primarily induced by inviscid, unsteady shock processes [18] and is known for its potential to inflict severe damage to the structure of the vehicle. By contrast, another form of flow unsteadiness, which is predominantly influenced by the interactions between the viscous shock wave and the turbulent boundary layer, is observed to be less severe when compared with the buzzing phenomenon [19,20].

During the supersonic flight, the forebody of the aircraft encounters substantial post-shock total pressure and undergoes localised aerothermal ablation phenomena [21]. Currently, the most widely adopted measures for reducing drag and providing thermal protection to supersonic aircraft forebodies involve the incorporation of slender cylindrical rods at the stagnation point of the incoming flow to form a 'spiked-blunt body' configuration [1]. In terms of flow stability, spiked forebodies generally exhibit a steady flow pattern. A necessary requirement to maintain flow stability dictates that at least a single streamline within the shear layer, known as the dividing streamline [22], must stagnate on the forebody [23]. However, the flow may exhibit unsteadiness under certain circumstances, depending on factors such as spike length, freestream Mach number, and forebody configuration. Two distinct modes of flow instability are associated with the use of a mechanical spike device: violent pulsation and mild oscillation, both characterised by self-sustained periodic variations in the flow field structures. The violent mode is commonly referred to as the 'pulsation mode', while the mild mode is termed the 'oscillation mode' [24]. The two modes are typically observed with flat or highly blunt conical forebodies. On the contrary, hemispherical and spherically blunted forebodies equipped with mechanical spikes generally exhibit steady flow. However, manifestations of the mild oscillation mode have been recorded in these scenarios [25–27]. The introduction of an aerodisk at the spike tip has been demonstrated to stabilise the flow for certain spike lengths [28]. Typical flow field structures during steady and unsteady states are depicted in Figure 1. The investigation of the flow unsteadiness of the spiked-blunt body is critical for preserving the structural integrity of the vehicle and flight control during actual flight.

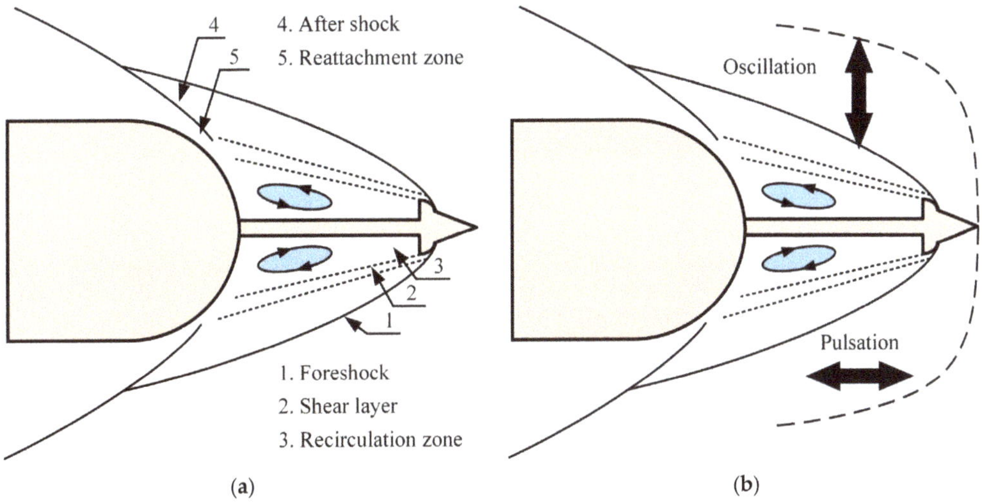

Figure 1. Typical flow field structures over a spiked-blunt body in (**a**) the steady state and (**b**) the unsteady state. Although spikes and blunt bodies have different configurations, the fundamental structures of the flow field are similar.

Dynamic mode decomposition (DMD) is a data-driven algorithm that is designed to extract dynamic information from experimental measurements or numerical simulations of flow fields. DMD was introduced by Schmid [29,30], and it has been widely applied in fluid mechanics to analyse the essential features of complex unsteady flows and construct lower-order dynamic models of flow fields. A fundamental aspect of the DMD method is the conceptualisation of flow evolution as a linear, dynamic process. By conducting an eigenanalysis of snapshots depicting the flow evolution process, the method provides lower-order modes that encapsulate flow field information along with their corresponding eigenvalues. A distinctive feature of the DMD method is that each mode corresponds to a singular frequency and a specific growth rate. This feature provides DMD with a significant advantage when analysing dynamic linear and cyclical flows. Furthermore, DMD allows for the direct representation of flow evolution processes through the eigenvalue of each mode, which eliminates the need for additional control equations. This simultaneous acquisition of mode characteristics and dynamic information grants DMD a unique edge over current flow field reduction methods based on system identification (using time series and input–output samples) and feature extraction (using spatial samples). Specifically, DMD facilitates the integrated modelling of space and time.

Since the introduction of DMD, it has captured the attention of numerous researchers due to its comparative simplicity [31–33]. Subsequent efforts have aimed to enhance the capabilities of conventional DMD, particularly in terms of pre-processing and post-processing. These advanced strategies encompass various methodologies, such as ensemble-averaging [34], sparsity-promoting techniques [35,36], and refined least-square methods [37–40]. Scholars have also deepened their understanding of the DMD algorithm by investigating the effects of noise [41] and contemplating considerations such as data updates [42], sub-Nyquist-rate data [43], and arbitrarily sampled systems [44]. Adaptations of DMD, including DMD with control [45] and input–output DMD [46], have been established to enable the design of control laws.

Notably, the conventional DMD has proven effective in capturing dominant flow modes in periodic flows or fully linear systems, with typical applications including investigations of boundary layer flows [47], transitional jets [48], airfoil transitions [49], and backward-facing step flow [50]. This effectiveness largely arises from the characteristics of most periodic or linear flows, where the magnitude of each mode can vary significantly, which allows for the straightforward identification of dominant modes from either the initial conditions or the norm of each mode. However, capturing the primary flow features becomes even more challenging when dealing with the experimental dataset of unsteady flow fields, such as the flow field around a spiked-blunt body under supersonic conditions. This challenge arises from the potential presence of multiple fundamental frequencies and the need for a greater number of numerically transient modes to fully approximate the samples. Accurate ranking of the importance of each mode requires consideration of not only the initial condition but also the evolution of the mode throughout the entire dataset.

In this study, we propose a novel energy sorting criterion that fully takes into account the initial conditions and the time evolution of each mode. The DMD with the energy sorting criterion and the conventional amplitude and frequency sorting criteria are applied to the experimental dataset of the unsteady flow of the spiked-blunt body at Ma = 2.2. A comparative analysis is conducted on the results obtained using the three sorting criteria from the perspectives of eigenvalues, temporal coefficients, and flow field structures. In addition, the experimental data are subjected to POD, and the results are discussed in conjunction with dynamic pressure signals. This study concludes by summarising the characteristics and application scenarios of DMD under the three sorting criteria and by revealing the main flow characteristics of the unsteady flow field of the spiked-blunt body under supersonic conditions.

2. Experimental Facility and Method

2.1. Direct-Connect Wind Tunnel

The experiment was conducted in the SCVT-1 supersonic direct-connect wind tunnel of the State Key Laboratory of Mechanics and Control of Aeronautics and Astronautics Structures at Nanjing University of Aeronautics and Astronautics. The main structure of the wind tunnel is shown in Figure 2. Given the relatively low Mach number of the experiment, a combination of an atmospheric inlet and vacuum suction was adopted. The inflow initially accelerates to supersonic through the facility nozzle, then enters the test section, and lastly exhausts into a vacuum tank with a volume of 400 m^3 through the vacuum pipeline.

The size of the test section is $400 \times 120 \times 160$ mm^3, and it is surrounded by plexiglass side walls on all sides. This design enables the execution of high-speed schlieren experiments. Calibration results indicate that the Mach number of the wind tunnel is Ma = 2.2. During the experiment, the total pressure of the wind tunnel inlet was 101 kPa, and the total temperature was maintained at 305 K. The detailed experimental conditions are meticulously documented in Table 1.

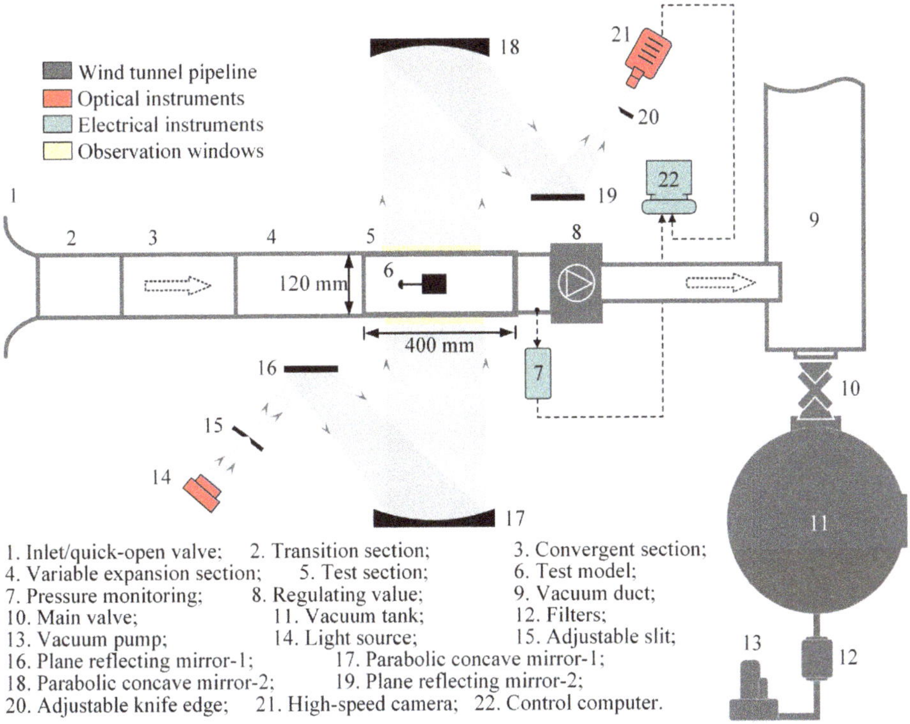

Figure 2. Schematic of the SCVT-1 supersonic wind tunnel and the Z-type schlieren system.

Table 1. Experimental conditions.

Ma_e, [-]	$T_{t,e}$, [K]	$p_{t,e}$, [Pa]	p_b, [Pa]	Re_D, [-]
2.2	305	1.01×10^3	8.5×10^3	2.6×10^5

2.2. Spiked-Blunt Body Model

In this study, we utilised an experimental model comprising an aerodome and a cylindrical blunt body, as illustrated in Figure 3a. The diameter of blunt body D was determined by applying one-dimensional isentropic flow theory and one-dimensional normal shock wave theory while considering the unstarting problem inherent to a direct-connected wind tunnel. In addition, the safe blockage ratio A_m/A_e of the wind tunnel was calculated based on the wind tunnel's operating Mach number Ma_e and the coefficient of total pressure loss σ. The detailed derivation processes for these considerations are provided in previous research [51] conducted on this wind tunnel, with the final result expressed in the form of Equation (1):

$$\frac{A_m}{A_e} = 1 - \frac{Ma_e}{\sigma}\left[\frac{2}{k+1}\left(1+\frac{k-1}{2}Ma_e^2\right)\right]^{-\frac{k+1}{2(k-1)}} \times \left\{\left(\frac{2k}{k+1}Ma_e^2 - \frac{k-1}{k+1}\right)\left[\frac{k-1}{k+1} + \frac{2}{(k+1)Ma_e^2}\right]^k\right\}^{\frac{1}{k-1}}. \quad (1)$$

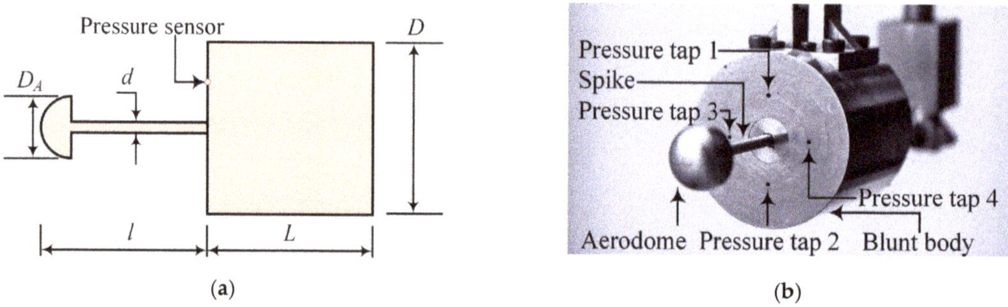

Figure 3. Schematics of the spiked-blunt body model. (**a**) Schematic of the geometry of the model. (**b**) Physical assembly drawing of the model with fixation.

To avoid the unstart phenomenon, the diameter of blunt body D was set to 40 mm, which resulted in a blockage ratio of A_m/A_e of 0.128. Based on the computational findings and the requirements of the actual experiment, the length of the cylindrical blunt body L and the spike length l were both set to D, while the spike diameter d was 0.065 D and the aerodome diameter D_A was 0.36 D. These detailed parameters are documented in Table 2.

Table 2. Dimensions of the spiked-blunt body model.

Parameter	Symbol	Value [mm]
Diameter of blunt body	D	40
Length of blunt body	L	40
Length of spike	l	40
Diameter of spike	d	2.6
Diameter of aerodome	D_A	14.4

The spike was connected to the blunt body through threading and positioned using the step surface at the base of the spike. The cylindrical, blunt body is supported from the top and affixed to the upper wall of the wind tunnel using two threaded rods. Figure 3b presents the physical assembly drawing of the model at an attack angle of 0°.

2.3. *High-Speed Schlieren System*

The current experiment employs a Z-type schlieren optical path, as illustrated in Figure 2. Parallel light passes perpendicularly through the straight test section, and a high-speed CMOS camera, the Phantom VEO 710L, is used in conjunction with a continuously adjustable LED light source (0–100 W) to capture the unsteady flow field. The exposure time of the camera is set to 1.74 µs. The sampling frequency f_s of the camera is set to 22 kHz to capture as much information as possible about the unsteady flow field. According to the Nyquist sampling theorem, this setup allows for the capture of flow field characteristic frequencies up to 11 kHz. Considering the need for resolution and the sensitivity to small density gradients and orientations, a slit configuration was chosen for the schlieren system. Meanwhile, the knife-edge direction and the slit length direction are set vertically in this experiment. As a result, the obtained schlieren images can reflect the density gradient variations in the flow field in the horizontal direction.

The experiment simultaneously monitors unsteady pressure fluctuations on the windward side of the cylinder using pressure taps. As shown in Figure 3, four measurement points are situated at half the cylindrical diameter and evenly distributed in the circumferential direction. Fluctuating temporal pressure signals are recorded using a dynamic pressure sensor (Kulite XTL190SM) and a pressure acquisition card (NI USB-9162). The sensor has a rated measurement pressure of 120 kPa with an accuracy of ±0.05% of the full range. The acquisition rate of the pressure sensors is set to 10 kHz. According to the

Nyquist sampling theorem, this configuration enables the capture of pulsating pressure signals up to 5 kHz.

2.4. Dynamic Mode Decomposition

The experimentally derived data are processed into a snapshot sequence from the 1st to Nth moment $\{x_1, x_2, x_3, \ldots, x_N\}$, where the column vector x_i represents the flow field snapshot at the ith moment. At this stage, a linear transformation relationship exists between the flow field x_{i+1} and the adjacent moment's flow field x_i, as defined by Equation (2):

$$x_{i+1} = Ax_i. \tag{2}$$

where A represents the system matrix of the high-dimensional flow field. This procedure is a linear estimating process, even if the dynamic system is nonlinear. Because a linear relationship is assumed, the dynamical characteristics are contained in the eigenvalues of matrix A. By utilising the flow field snapshots from the 1st to Nth moment, two snapshot matrices can be constructed as $X = \{x_1, x_2, x_3, \ldots, x_{N-1}\}$ and $Y = \{x_2, x_3, x_4, \ldots, x_N\}$, as per Equation (3):

$$\begin{aligned} Y &= \{x_2, x_3, x_4, \ldots, x_N\} \\ &= \{Ax_1, Ax_2, Ax_3, \ldots, Ax_{N-1}\} \\ &= AX. \end{aligned} \tag{3}$$

This study employs singular value decomposition (SVD) to perform similarity transformations on high-order operators to obtain a low-order representation of the system and extract the dominant eigenvalues and primary DMD modes.

The high-dimensional matrix A can be replaced by a low-dimensional similar matrix $\widetilde{A}$. To find the orthogonal space of the similarity transformation, SVD is performed on the matrix X, as defined by Equations (4) and (5):

$$X = U\Sigma V^H, \tag{4}$$

$$A = U\widetilde{A}U^H. \tag{5}$$

The matrix Σ is a diagonal matrix with r singular values on its diagonal, and the unitary matrices U and V satisfy $U^H U = I$ and $V^H V = I$. It should be noted that the SVD was truncated in this study for noise reduction of the experimental data. When truncating the SVD in DMD, the premise is to retain only the dominant modes or components that capture the most significant dynamics of the system while discarding the less significant modes. This truncation is often carried out based on the decay of the singular values. The calculation of the matrix $\widetilde{A}$ can be viewed as a minimisation problem of the Frobenius norm, which is used to solve for the approximate matrix, as per Equation (6):

$$A \approx \widetilde{A} = U^H Y V \Sigma^{-1}. \tag{6}$$

Given that the matrix $\widetilde{A}$ is a similarity transformation of the matrix A, it contains the primary eigenvalues of A. The eigenvalues of the matrix $\widetilde{A}$ are determined by Equation (7):

$$\widetilde{A}\Lambda_j = \lambda_j \Lambda_j, \tag{7}$$

where λ_j is the jth eigenvalue of $\widetilde{A}$ and Λ_j is the eigenvector corresponding to the eigenvalue λ_j. This procedure allows the calculation of the jth DMD mode, as defined by Equation (8):

$$\Phi_j = U\Lambda_j. \tag{8}$$

The growth rate g_j and frequency ω_j corresponding to the jth mode are given by Equations (9) and (10):

$$g_j = \mathrm{Re}\{\lg(\lambda_j)\}/\Delta t, \tag{9}$$

$$w_j = \mathrm{Im}\{\lg(\lambda_j)\}/\Delta t, \tag{10}$$

The magnitude of the amplitude of the jth mode represents the contribution of this mode to the initial snapshot $\boldsymbol{x}_1$. The modal amplitude corresponding to the jth mode is given by Equation (11):

$$\alpha_j = \boldsymbol{\Lambda}_j^{-1}\boldsymbol{U}^H x_1. \tag{11}$$

The method used to determine the dominant DMD mode is usually not unique. The current general method is to sort all the modes based on the characteristic parameters and extract a subset of the front modes to represent the primary features of the flow field. Each sorting method has its own unique advantages and disadvantages. The conventional frequency sorting criterion and amplitude sorting criterion take the modal frequency and amplitude as the characteristic parameters, as illustrated in Equations (10) and (11), respectively. They differ in their focus and application, with the former focusing more on the intrinsic dynamic properties of the flow field and the latter focusing more on the actual behaviour and response of the flow field during vibration.

For each mode, DMD calculates a corresponding temporal coefficient, which describes the evolution of that mode over time. To further analyse the flow unsteadiness, the temporal coefficient of the jth mode at time instant i is defined by Equation (12):

$$C_{ij} = (\lambda_j)^{i-1}\alpha_j. \tag{12}$$

This study proposes an energy sorting criterion specifically designed for experimental data on unsteady flow fields. This criterion considers initial conditions and the temporal evolution of each mode. In addition to assessing the temporal coefficient, it integrates the spatial elements of each mode. It defines the Frobenius norm of the space–time matrix of each mode as the total energy of that mode throughout the entire period, as shown in Equation (13):

$$E_j = \|\boldsymbol{\Phi}_j C_j\|_F^2. \tag{13}$$

To highlight the advantages of the energy sorting criterion in handling an experimental dataset of unsteady flow fields, this criterion is applied to high-speed schlieren snapshots of the unsteady flow field around a spiked-blunt body. The results are compared with those obtained using the conventional amplitude and frequency sorting criteria. Moreover, the validity of the results is verified by combining the POD results with the dynamic pressure signals.

3. Results and Discussion

3.1. Conventional Amplitude Sorting Criterion

The amplitude sorting criterion is a commonly used method in conventional DMD. This section presents the results obtained by applying DMD to 5000 consecutively acquired schlieren snapshots from a single experiment and sorting the modes according to their amplitude magnitudes. Figure 4a illustrates the distribution of eigenvalues of the modes after applying DMD. Each unit circle in the diagram denotes the eigenvalue of a mode, with the coordinates of each point representing the real and imaginary parts of the eigenvalues. Circles located on the unit circle indicate stable modes; those within the unit circle represent decaying modes; and those outside the unit circle are unstable modes. The eigenvalues appear in conjugate pairs, with most mode eigenvalues located on the unit circle. A small fraction of modes with high amplitudes are within the unit circle, which suggests that most flow fields are stable following DMD, with only a few modes starting with high initial amplitudes that subsequently decay. The eigenvalues of the first six DMD modes have been extracted and marked with red circles. These eigenvalues are all located within the unit circle. This condition indicates that the initial amplitudes of the first six modes are high, but they are expected to decay over time and are classified as decaying modes. In Figure 4b, the relationship between the growth rate and amplitude of each mode is illustrated, which reveals a general trend where stable modes have smaller amplitudes and lower growth rates, while decaying modes have larger amplitudes and higher growth rates. Among the first six modes, the first four exhibit relatively lower growth rates, which implies slower decay. The fifth and sixth modes have higher growth rates, which suggests a faster decay rate. This aspect is one of the shortcomings of the amplitude sorting criterion, which will be further discussed in the subsequent sections.

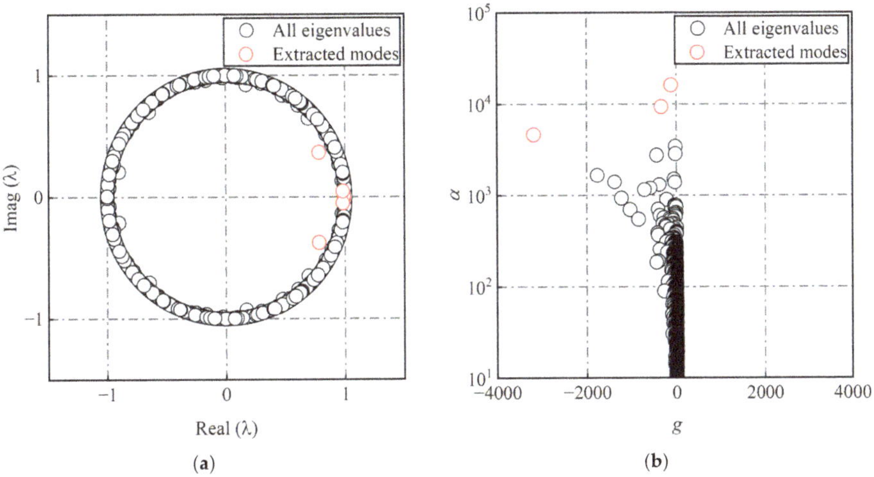

Figure 4. Modal eigenvalue analysis of the DMD using the amplitude sorting criterion. (**a**) Distribution of modal eigenvalues. (**b**) Relationship between modal growth rate and amplitude. Herein, black circles represent all eigenvalues, while red circles denote the extracted eigenvalues of the first six modes.

Each mode appears in conjugate pairs, with the same real part for the even and odd modes of a pair and a 180° phase difference in the imaginary part. Therefore, only the odd modes are used to represent a pair of conjugate modes in this study, such as mode 1–2, which represents the common real part of modes 1 and 2, as well as the imaginary part of mode 1. Figure 5 depicts the evolution of the temporal coefficients for the first six modes of the DMD, sorted by amplitude. Figures 5a and 5b represent the real part and the imaginary part, respectively. A comparison between Figure 5a,b reveals that the amplitudes and frequencies of the real and imaginary parts of the temporal coefficients for a given mode pair are nearly identical. The temporal coefficients of the first six modes all initially display high amplitudes that decay over time. Mode 1–2 exhibits the highest initial amplitudes and also the fastest rate of decay. The amplitude and decay rate progressively decrease with the increase in mode order. These observations align with the patterns of the eigenvalues of the first six modes shown in Figure 4.

Figure 6 presents the power spectral density (PSD) of the temporal coefficients for the first six modes of the DMD using the amplitude sorting criterion. Figures 6a and 6b represent the real part and the imaginary part, respectively. A comparison between Figure 6a,b reveals that the PSD of the temporal coefficients for each mode exhibits a single frequency peak. This observation shows that the temporal coefficient variation of each mode closely approximates a simple harmonic motion at a single frequency. In other words, the unsteady pulsations in the flow field have been extracted into a combination of single-frequency modes, with each mode having a unique frequency component but a relatively dispersed energy distribution. Although the real and imaginary parts of the temporal coefficients for each mode exhibit different developmental trajectories in their PSD, they share the same peak frequencies. The first three pairs of peak frequencies are 23, 171, and 1546 Hz, respectively. The mode sorting was based on the modal amplitude magnitude, with no consideration for the decay rate. Consequently, the first mode exhibits a high initial amplitude and a high decay rate. Given its low-frequency pulsation, the first mode represents low-frequency, high-amplitude noise in the flow field rather than the evolution of the main flow field.

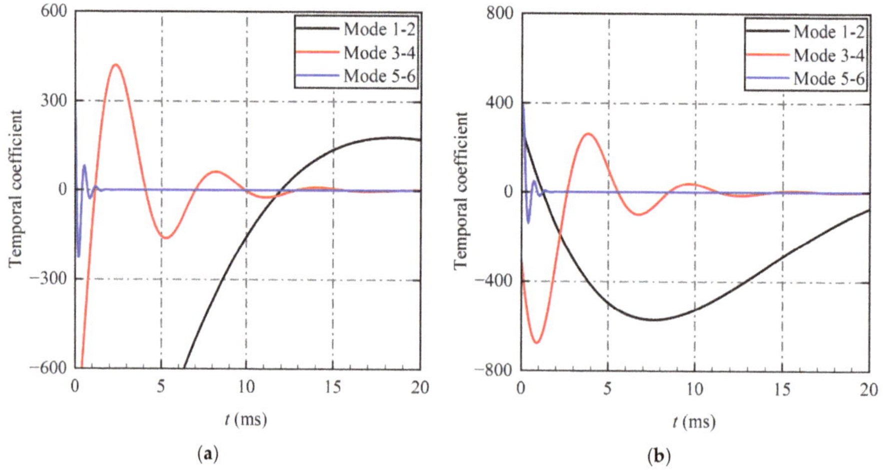

Figure 5. Evolution of the temporal coefficients amongst the first six modes of the DMD using the amplitude sorting criterion. (**a**) Real part. (**b**) Imaginary part.

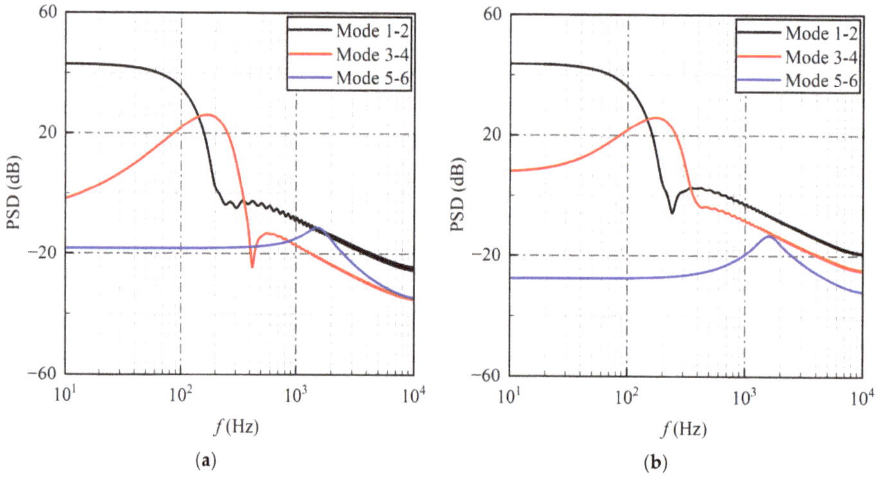

Figure 6. Power spectral density of the temporal coefficients amongst the first six modes of the DMD using the amplitude sorting criterion. (**a**) Real part. (**b**) Imaginary part.

The DMD of experimental data offers a modal representation of the primary spatial structures in the flow at various characteristic frequencies. Adjacent modes are complex conjugates of one another. In other words, every two adjacent DMD modes share identical real parts and opposite imaginary parts. Figure 7 illustrates the flow field structures of the initial six DMD modes for the spiked-blunt body by using the amplitude sorting criterion. Figures 7a–c and 7d–f represent the real and imaginary parts, respectively. For ease of description, the coordinates of the schlieren images are nondimensionalised by the cylinder diameter D, with the coordinate origin located at the spike tips. The real and imaginary parts of the same mode exhibit identical flow structures. In mode 1–2, a shallow aftershock is observed ahead of the cylindrical blunt body, which is accompanied by speckles and scratches in the field of view. They are caused by water droplets due to humidity and impurities on the glass on the day of the experiment. Considering the previous results, specifically the high initial amplitude and rapid decay rate of mode 1–2, it represents low-frequency noise in the experiment rather than the primary flow structures of interest. The flow field of mode 3–4 does not contain the low-frequency noise observed in mode 1–2 and only exhibits an aftershock ahead of the cylindrical blunt body. This aftershock is axially symmetrical along the spike and is replaced by fine flow features in the centre region of the cylindrical blunt body surface (approximately $-0.36 < y/D < 0.36$). The aftershock is attached to the shoulder of the cylindrical blunt body, which results in a three-dimensional ring-like structure. In mode 5–6, the flow field similarly displays only a bow aftershock ahead of the cylindrical blunt body. However, this aftershock appears fragmented, with a higher intensity in the central region (approximately $-0.47 < y/D < 0.47$) due to its unsteady pulsation along the flow direction, which continuously impacts the surface of the cylindrical blunt body. Notably, a very weak aftershock is present in the flow field above the aerodome in modes 3–4 and 5–6. This bow shock is a consequence of the high-speed incoming flow being obstructed by the hemispherical aerodome. Given that the aftershocks in this state are relatively stable, they exhibit very weak intensity in the modes sorted by amplitude. As indicated by the modal eigenvalues in Figure 4 and the evolution of the temporal coefficients in Figure 5, the corresponding spatial structures of the flow field for each mode will decay over time at different rates.

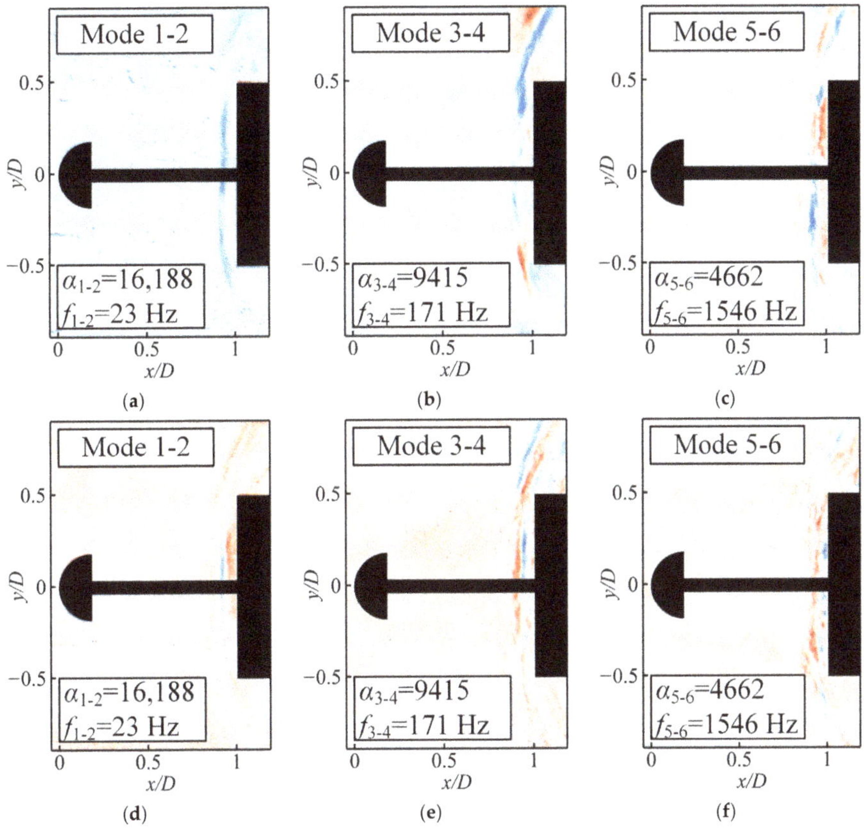

Figure 7. Flow field structures of the first six modes of the DMD using the amplitude sorting criterion for the spiked-blunt body. Specifically, (**a–c**) represent the real part, while (**d–f**) represent the imaginary part.

3.2. Conventional Frequency Sorting Criterion

The frequency sorting criterion is also a commonly used method in conventional DMD. This section presents the results of applying DMD to the same dataset as in Section 3.1 and sorting the modes based on their frequency magnitudes. In Figure 8a, the distribution of DMD mode eigenvalues reveals that these eigenvalues appear as pairs of conjugate complex numbers. Most of the mode eigenvalues are located on the unit circle, while a smaller fraction of higher amplitude modes is found within the unit circle. Therefore, the flow field for most modes is relatively stable, with only a few modes with high initial amplitudes showing decay. The eigenvalues of the first six DMD modes are extracted and marked with red circles, which signifies that these modes have low initial amplitudes but are relatively stable. Thus, they are classified as stable modes. Figure 8b presents the relationship between the growth rates and amplitudes of each mode. In general, DMD modes with smaller amplitudes and subsequently lower growth rates are stable, while those with larger amplitudes and higher growth rates are decaying modes. Among the first six modes selected in this section, all exhibit low amplitudes. Therefore, these modes are relatively stable. However, considering that the eigenvalues of the first six modes are nearly identical, clearly distinguishing their individual features is difficult. This limitation constitutes one of the drawbacks of the frequency sorting criterion. This aspect will be further discussed later in this subsection.

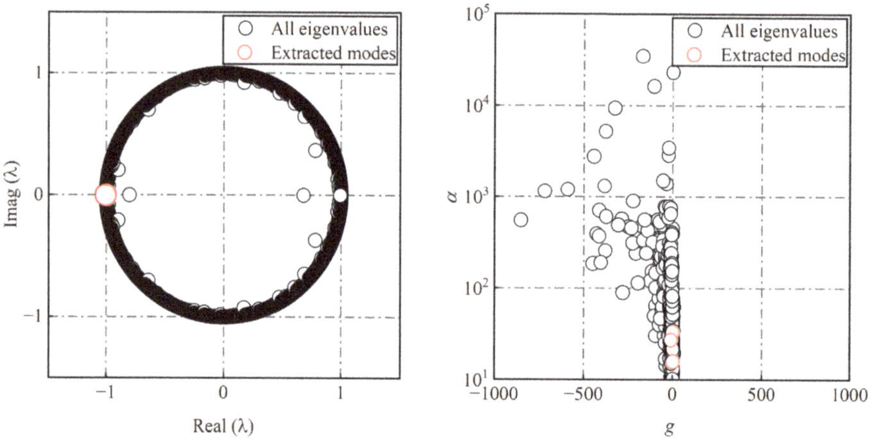

(**a**) (**b**)

Figure 8. Modal eigenvalue analysis of the DMD using the frequency sorting criterion. (**a**) Distribution of modal eigenvalues. (**b**) Relationship between modal growth rate and amplitude. Herein, black circles represent all eigenvalues, while red circles denote the extracted eigenvalues of the first six modes.

Figure 9 illustrates the evolution of the temporal coefficients for the first six modes of the DMD using the frequency sorting criterion. Figures 9a and 9b represent the real and imaginary parts, respectively. The evolution of the temporal coefficients for the first four modes remains relatively stable and consistently exhibits periodic motion with a high frequency and a small amplitude. Only the amplitude of mode 5–6 increases at a relatively low growth rate. Given that the modes are organised according to the magnitude of their frequencies, the frequencies of the temporal coefficient evolution decrease sequentially with the increase in mode sequence. However, they remain very close to each other.

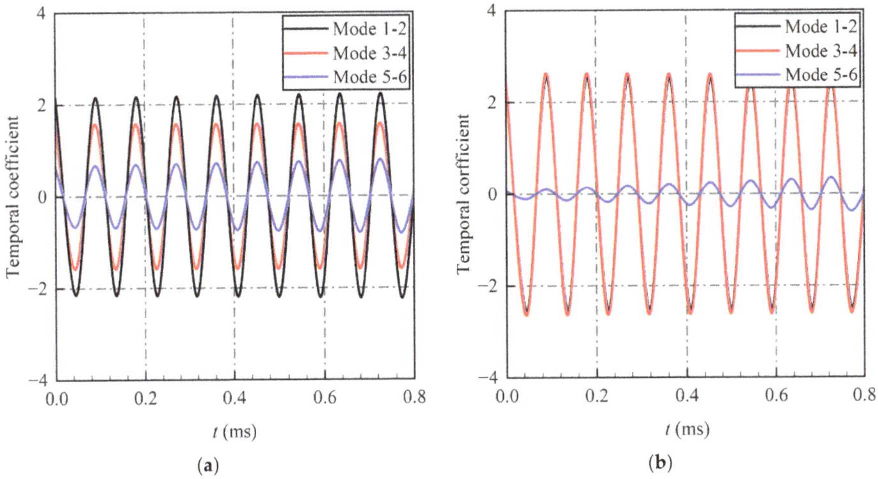

Figure 9. Evolution of the temporal coefficients amongst the first six modes of the DMD using the frequency sorting criterion. (**a**) Real part. (**b**) Imaginary part.

Figure 10 presents the PSD of the temporal coefficients among the first six modes for the DMD using the frequency sorting criterion. Figures 10a and 10b represent the real and imaginary parts, respectively. Although the trajectories of the PSD of the real and imaginary sections of the temporal coefficients differ for the same mode, they share the same peak frequency. The peak frequencies of the first six modes are extremely close to one another because the modes are sorted by their frequency magnitudes. The characteristic frequency of mode 1–2 corresponds to the highest frequency amongst the collected signals, specifically 11×10^3 Hz. This frequency is a result of setting the sampling frequency of the Phantom VEO 710L high-speed camera to 22×10^3 Hz. According to the Nyquist sampling theorem, frequencies up to 11×10^3 Hz can be collected for the flow field characteristic frequency. The characteristic frequencies of the two other pairs of modes are 10,993 and 10,988 Hz, respectively.

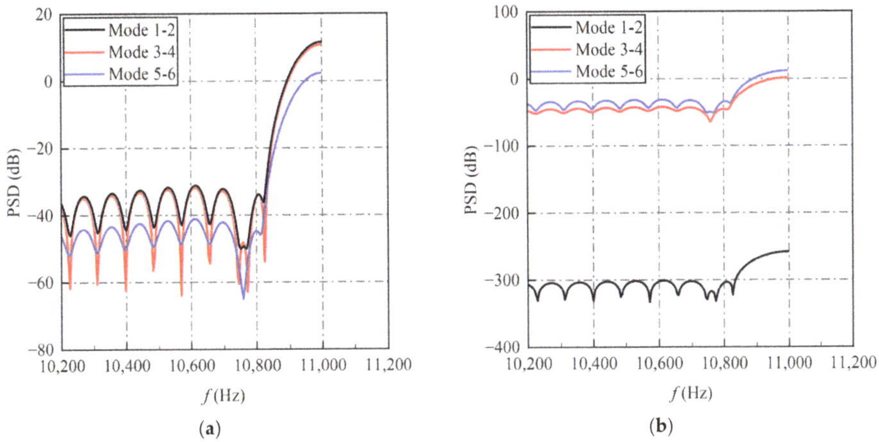

Figure 10. Power spectral density of the temporal coefficients amongst the first six modes of the DMD using the frequency sorting criterion. (**a**) Real part. (**b**) Imaginary part.

Figure 11 depicts the flow field structures of the first six modes of the DMD using the frequency sorting criterion for the spiked-blunt body. Figure 11a–c represent the real parts, and Figure 11d–f represent the imaginary parts. Given the application of the frequency sorting criterion, the frequencies of the first six modes are extremely close, which displays identical flow structures without distinctly differentiating the flow structures of each mode. The real and imaginary parts of the first six modes both exhibit the same flow structures. Specifically, the flow field is devoid of the low-frequency noise mentioned in the previous section (i.e., spots and scratches in the field of view), and only an aftershock that entirely covers the windward surface of the cylindrical blunt body is observable. This shock wave is distributed axially symmetrically along the spike axis, weaker in the central region of the cylindrical blunt body (approximately $-0.36 < y/D < 0.36$) and stronger at the shoulder positions where reattachment occurs. Notably, the aftershock appears highly fragmented, which represents a minor high-frequency pulsation structure within the typical pulsation mode. As can be inferred from the modal eigenvalue features in Figure 8 and the evolution of the time coefficients in Figure 9, the spatial structures corresponding to each mode will exhibit periodic pulsations at a high frequency (approximately 11×10^3 Hz).

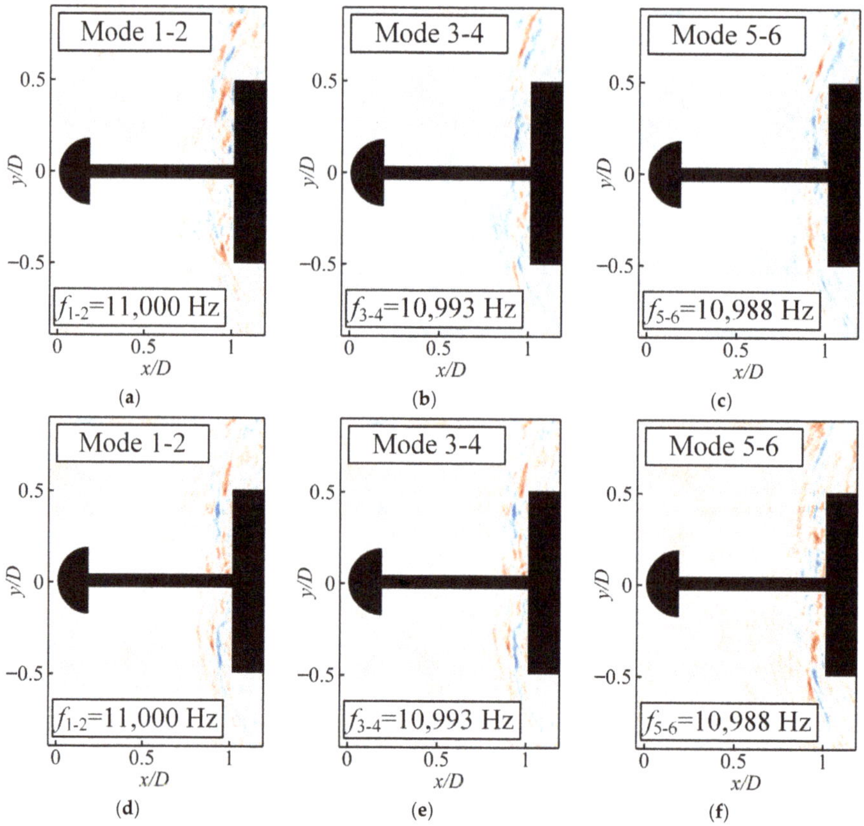

Figure 11. Flow field structures of the first six modes of the DMD using the frequency sorting criterion for the spiked-blunt body. Specifically, (**a–c**) represent the real part, while (**d–f**) represent the imaginary part.

3.3. *Novel Energy Sorting Criterion*

From the research results presented in previous sections, we observe that the amplitude sorting criterion frequently results in the selection of modes with larger initial amplitudes, but they also exhibit higher decay rates over time. Consequently, these modes evolve into flow field structures with less energy, which potentially overlooks the structures that actually constitute a significant portion of the energy. In the meantime, the frequency sorting criterion often leads to the selection of modes with nearly identical characteristic frequencies, which causes difficulty in effectively distinguishing between different modes. This similarity amongst modes results in flow structures that closely resemble each other, which prevents the reduction of high-dimensional flow fields into distinguishable multiple low-dimensional structures. Moreover, focusing solely on high or low frequencies can lead to the misinterpretation of noise at these frequencies as flow field structures. Considering these shortcomings, this study proposes an energy sorting criterion that is better suited for unsteady flow field experimental data.

To precisely quantify the contribution of each mode to the unsteady flow field, this subsection presents the results of performing DMD on the same dataset as in the previous subsections and sorting the modes based on their energy levels. Figure 12 shows the energy proportion of each mode and the cumulative energy proportion of modes for the first 1000 modes in the DMD using the energy sorting criterion. The blue circles represent the energy proportion of each mode among all modes. The first mode contributes 95.0% of the energy, which represents the mean flow field that persists throughout the time domain. The second and third modes contribute 2.7% and 0.6% of the total energy, respectively, with the energy proportions of subsequent modes significantly dropping to very low levels. The red circles represent the cumulative energy proportions of the modes. The first three modes together contribute 98.3% of the total energy, and the cumulative energy proportion reaches over 99% at the 1000th mode. The rapid convergence of the energy spectrum suggests that most of the important flow features can be identified in the initial modes. Consequently, this study primarily discusses the dominant flow modes within the first five modes.

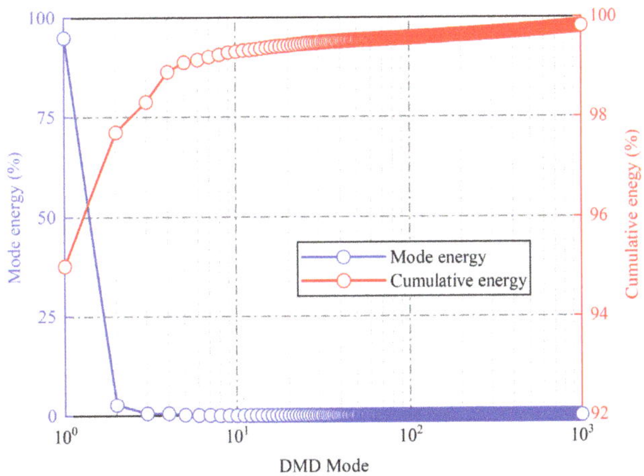

Figure 12. Energy proportion of each mode (left axis) and cumulative energy proportion of modes (right axis) for the first 1000 modes in the DMD using the energy sorting criterion.

Figure 13a displays the distribution of the modal eigenvalues of the DMD using the energy sorting criterion. The eigenvalues are observed as conjugate complex pairs, with most of them located on the unit circle. Only a few modes with high amplitudes have eigenvalues within the unit circle, which indicates that most flow fields are relatively stable following DMD, and only a few modes with high initial amplitudes exhibit decay. As the first mode represents the mean flow field, the eigenvalues of the first five modes are extracted and marked in red. These markers are positioned on the unit circle, which suggests that the first five modes are stable. Figure 13b portrays the relationship between the growth rates and amplitudes of each mode. The overall trend shows that modes with smaller amplitudes and, thus, lower growth rates are stable. In the meantime, those with larger amplitudes and higher growth rates are decaying modes. For the first five modes selected in this section, all of them are classified as stable modes. Figure 13c illustrates the relationship between the frequencies and amplitudes of each mode. The overall trend reveals that modes with larger amplitudes are low-frequency modes, while high-frequency modes have small amplitudes. Therefore, if either of the two factors is independently chosen as a sorting criterion, then the other factor is likely to be somewhat neglected. However, the first five modes selected according to this energy sorting criterion consider amplitude and frequency, which effectively address the shortcomings of the first two sorting criteria.

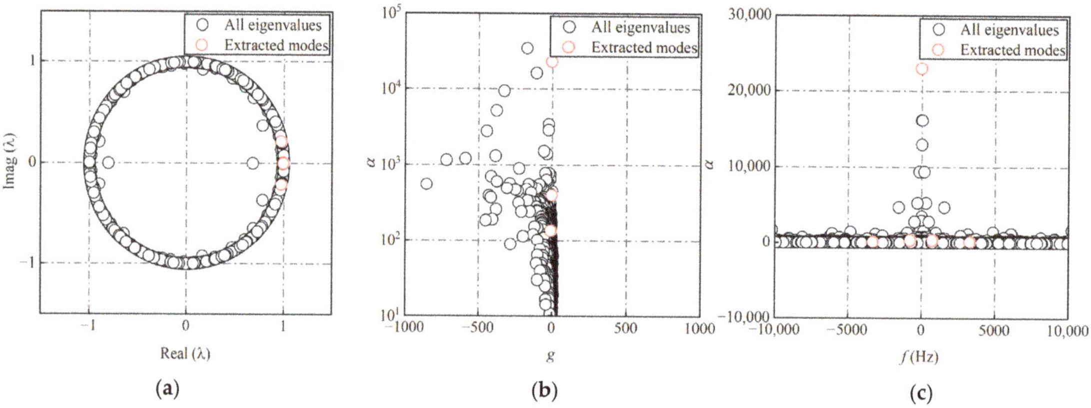

Figure 13. Modal eigenvalue analysis of the DMD using the energy sorting criterion. (**a**) Distribution of modal eigenvalues. (**b**) Relationship between modal growth rate and amplitude. (**c**) Relationship between modal frequency and amplitude. Herein, black circles represent all eigenvalues, while red circles denote the extracted eigenvalues of the first five modes.

Given that the first mode represents the mean flow field, the corresponding characteristic frequency of the temporal coefficient is zero. For this reason, the temporal coefficient results of the first mode will not be discussed here. Figure 14 presents the evolution of the temporal coefficients for modes 2–3 and 4–5 of the DMD using the energy sorting criterion. Figures 14a and 14b represent the real and imaginary parts, respectively. A comparison between Figure 14a,b reveals that the real and imaginary parts of the temporal coefficients for the same mode pair have identical amplitudes and frequencies. The evolution of the two components of the temporal coefficients of the same mode pair is relatively stable, which follows periodic motion at a single characteristic frequency. The amplitude of mode 2–3 is smaller than that of mode 4–5, but the frequency of mode 2–3 exceeds that of mode 4–5.

Figure 15 presents the PSD of the temporal coefficients for modes 2–3 and 4–5 of the DMD using the energy sorting criterion, where Figures 15a and 15b represent the real and imaginary parts, respectively. A comparison between Figure 15a,b shows that the PSD of the temporal coefficients for each mode exhibits only one frequency peak. This observation indicates that the variations in the temporal coefficients for each mode closely resemble simple harmonic motion with a single frequency, which implies that the unsteady pulsations in the flow field have been captured as a combination of single-frequency modes. Although the PSD of the real and imaginary parts of the temporal coefficients for each mode exhibit different trajectories, they share the same peak frequencies. The peak frequency for mode 2–3 is 3307 Hz, and that for mode 4–5 is 750 Hz. The DMD modes selected according to this energy sorting criterion avoid the high decay rate situation encountered when using the amplitude sorting criterion and the issues related to high- and low-frequency noise found when using the frequency sorting criterion.

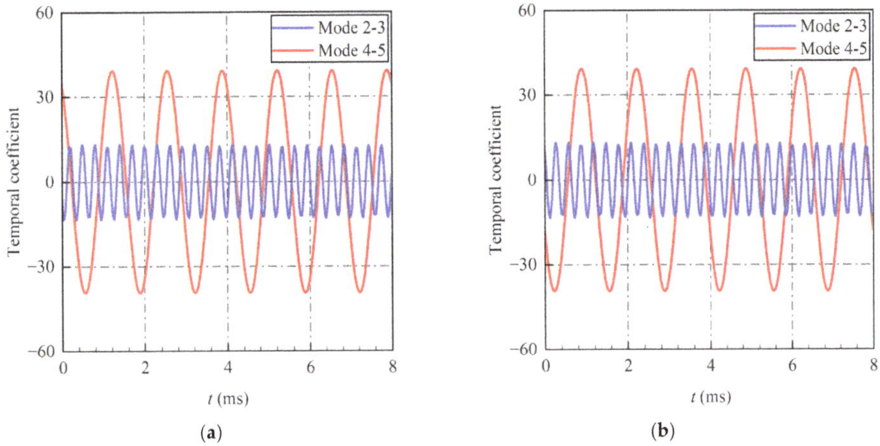

Figure 14. Evolution of the temporal coefficients amongst modes 2–3 and 4–5 of the DMD using the energy sorting criterion. (**a**) Real part. (**b**) Imaginary part.

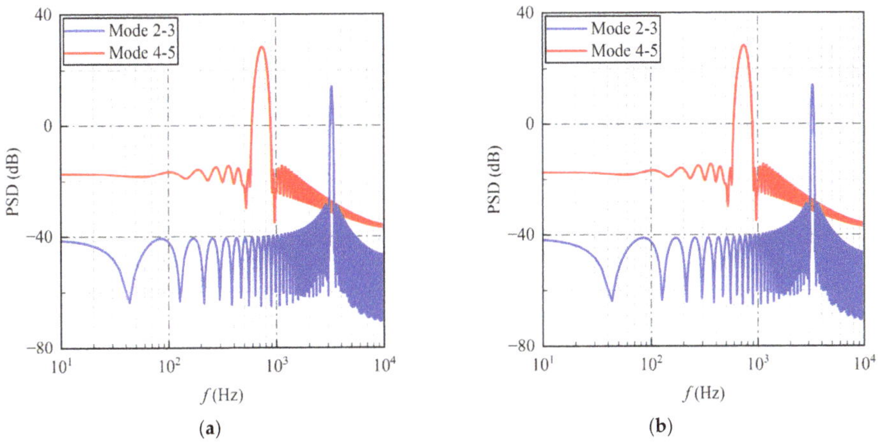

Figure 15. Power spectral density of the temporal coefficients amongst modes 2–3 and 4–5 of the DMD using the energy sorting criterion. (**a**) Real part. (**b**) Imaginary part.

Figure 16 illustrates the flow field structures of the first five modes of the DMD using the energy sorting criterion for the spiked-blunt body. Figure 16a–c represents the real parts, while Figure 16d,e represents the imaginary parts. The first mode represents the mean flow field and serves as the fundamental flow field structure for the typical/suppressed pulsation mode. This mode contributes 95.0% to the total energy of the flow evolution, with its frequency and growth rate equating to zero. Therefore, its flow structure pervades the entire unsteady pulsating process of the flow field. In the first mode, a complete aftershock appears in front of the cylindrical, blunt body. The shock wave exhibits an axisymmetric distribution and attaches to the shoulders of the cylindrical, blunt body. Behind the aerodome, a clear shear layer encapsulates the recirculation zone. The airflow within the recirculation zone undergoes reciprocating pulsation due to pressure fluctuations. Furthermore, the trace of ejected flow mass near the cylinder shoulder is observable. As the flow direction progresses, the clear profile of the shear layer gradually becomes blurred due to the unsteady motion of the downstream shear layer. This condition implies that the fluctuation of the shear layer along the flow direction gradually intensifies. This process closely resembles the evolution of the Kelvin–Helmholtz instability. The interaction between the aftershock and the shear layer is noticeable around the coordinates ($x/D = 0.95$, $y/D = 0.45$). The fluctuations in the shear layer interacting with the aftershock further amplify the unsteady motion of the shock wave. Notably, some specks and streaks are present in the first mode, which can be attributed to noise (water droplets and glass impurities) that was present during the experiment and influenced the mean flow field.

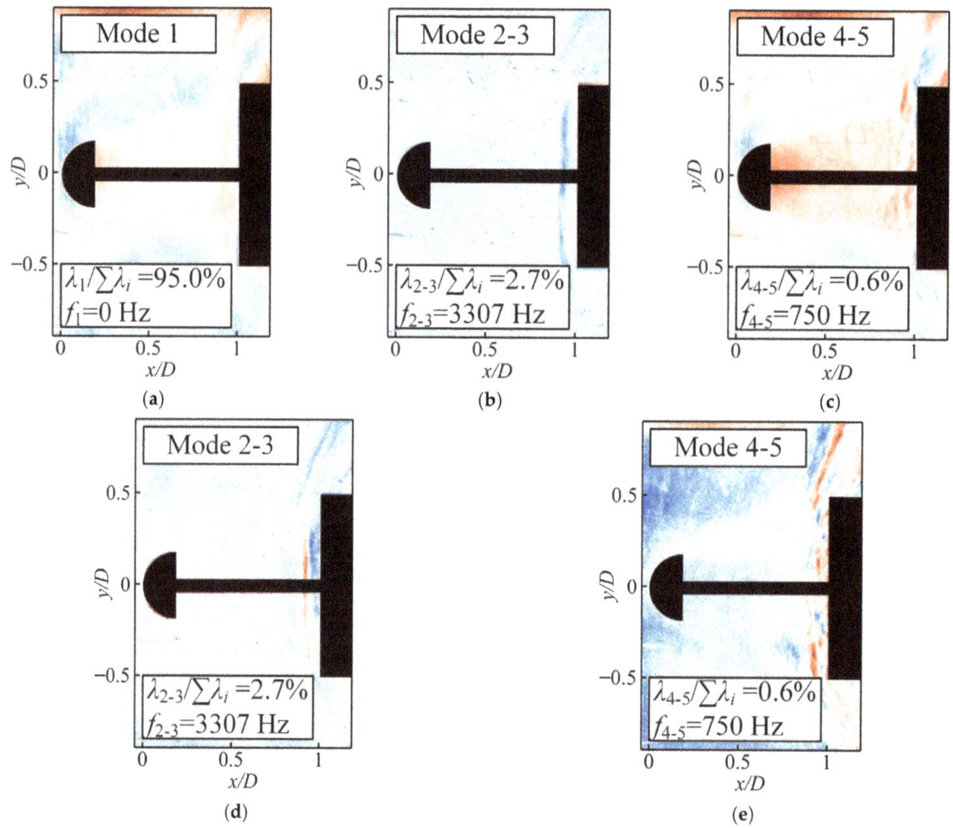

Figure 16. Flow field structures of the first five modes of the DMD using the energy sorting criterion for the spiked-blunt body. Specifically, (**a**–**c**) represent the real part, while (**d**,**e**) represent the imaginary part.

Mode 2–3 contributes 2.7% of the total energy with a characteristic frequency of 3307 Hz. The flow field structure is characterised by a single aftershock in front of the cylindrical, blunt body. This aftershock also exhibits an axisymmetric distribution, but it is stronger in the central area of the cylindrical blunt body and weaker at the shoulders. This observation is similar to the spatial structure observed in the suppressed pulsation mode as observed in a previous study [52], where the aftershock entirely covers the windward surface of the cylindrical blunt body. The primary frequency of 3.3 kHz arises from the low-amplitude pulsations of the aftershock.

Mode 4–5 accounts for 0.6% of the total energy with a characteristic frequency of 750 Hz. The primary feature of the flow field structure is the recirculation zone that dominates the windward surface of the blunt body. The aftershock only covers the shoulder of the cylindrical blunt body, and the shock wave within the range of $-0.36 < y/D < 0.36$ has been replaced by fine flow features. The aftershock appears to have been 'broken' by the recirculation zone. The aftershock attaches to the shoulder of the cylindrical blunt body, which results in a three-dimensional ring-like structure. Simultaneously, the shear layer interacts with the aftershock at approximately $x/D = 0.95$. This flow field structure, where the recirculation zone occupies the surface of the cylindrical blunt body, can remain stable for a short time. The primary frequency of 750 Hz results from the low-frequency, high-amplitude pulsations of the aftershock. These pulsations are associated with the back-and-forth movement of the airflow in the recirculation zone, which is driven by the pressure differences along the flow direction. In addition, a bow foreshock emerges in front of the aerodome, which is formed as a consequence of the aerodome blocking the high-speed inflow.

3.4. Proper Orthogonal Decomposition

This subsection utilises the results of POD for comparative study and analysis to further compare the advantages and disadvantages of the three DMD sorting criteria applied to the unsteady experimental flow field. The POD modes are orthogonal to each other in the spatial dimension, and each mode can be regarded as a perturbation to the basic flow field (i.e., the mean flow field). Figure 17 presents the energy proportion of each mode and the cumulative energy proportion of modes in the POD for assessing the influence of POD modes on the mean flow field.

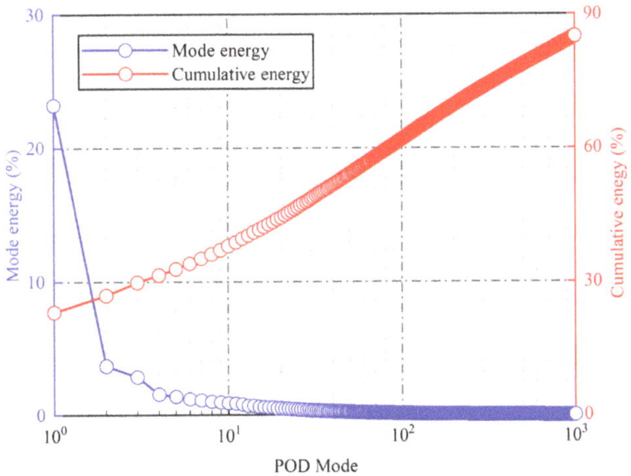

Figure 17. Energy proportion of each mode (left axis) and cumulative energy proportion of modes (right axis) in the POD for the spiked-blunt body.

The initial modes possess higher energy, with the first mode having the highest energy proportion of 23.2%, which represents the mean flow field. The energy proportions of the second and third modes are 3.7% and 2.9%, respectively, which are significantly higher than those of the subsequent modes. However, as the mode number increases, the energy proportion of each subsequent mode sharply decreases and maintains a lower magnitude. The energy proportion is nearly zero when the mode number exceeds 100. For the cumulative energy proportion of the modes, it gradually increases with the growing number of modes, and the growth rate gradually slows down. At 1000 modes, the cumulative energy proportion reaches 88.6%. The convergence of the energy spectrum indicates that the first few modes can be considered the dominant modes of the flow field. They contribute the vast majority of energy and play a more important role in the evolution of the flow.

Figure 18 presents the PSD of the temporal coefficients for the first three POD modes for the unsteady flow of the spiked-blunt body. Figure 18a represents the first mode, while Figure 18b represents the second and third modes. The PSD of the POD temporal coefficients exhibits multiple frequency peaks, whereas the PSD of the DMD temporal coefficients mentioned earlier has only one characteristic peak. This discrepancy arises because POD is based on the decomposition of second-order statistical data, which utilises time-averaged spatial tensors to obtain coherent structures at different energy levels. By contrast, DMD extracts the dominant coherent structures from a series of instantaneous velocity fields by approximating the linear projection between snapshots. In other words, POD focuses solely on a series of representative bases that are orthogonal in space, while DMD considers temporal orthogonality and spatial orthogonality.

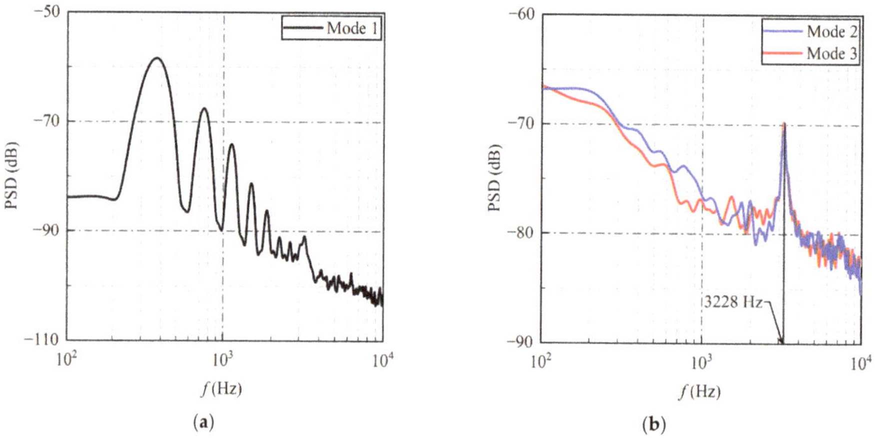

Figure 18. Power spectral density of the temporal coefficients amongst the first three modes of the POD for the unsteady flow of the spiked-blunt body. (**a**) Mode 1. (**b**) Modes 2 and 3.

The first POD mode represents the mean flow field of unsteady pulsations, which occupies the highest energy in the flow. Six distinct peaks are observed in the PSD of the temporal coefficients. The corresponding peak frequencies are 348, 750, 1154, 1547, 1833, and 3228 Hz. The peak frequencies at 750 and 3228 Hz align with the main frequencies of the dominant DMD modes obtained through the energy sorting criterion. For the other low-frequency peaks, they are speculated to be background noise from the experiment. They are likely attributed to alternating dark/bright changes between snapshots due to excessively moist air on the day of the experiment. The second and third modes can be regarded as perturbations to the mean flow field, given that they have the highest energy proportion after the mean flow field. The power spectra of temporal coefficients exhibit a similar trend, and both share the same peak frequency of 3228 Hz. This frequency is nearly identical to the principal frequency of modes 2 and 3 in the DMD using the energy sorting criterion. However, it notably differs from the main frequencies of the DMD modes that employ the two other sorting criteria.

An experiment was conducted to measure the dynamic pressure signals on the windward face of the cylindrical, blunt body. This approach enables a more comprehensive analysis of the spectral characteristics of the unsteady flow field around a spiked-blunt body and highlights the effectiveness of the energy sorting criterion in capturing these spectral features. Figure 19 presents the PSD of the dynamic pressure signals for the spiked-blunt body, with P1 to P4 representing four pressure measurement points distributed circumferentially on the windward face of the cylindrical blunt body. The dynamic pressure sensor used in the experiment operates at a sampling frequency of 10 kHz, which allows it to capture dynamic pressure signals up to 5 kHz according to the Nyquist sampling theorem. The spectra of dynamic pressure signals from the four channels exhibit a high degree of similarity around 3 kHz, with only very minor differences. These slight variations arise from the influence of the three-dimensional effects of the flow field. Simultaneously, this main frequency of 3 kHz closely matches the principal frequency of the dominant modes extracted by POD and DMD using the energy sorting criterion.

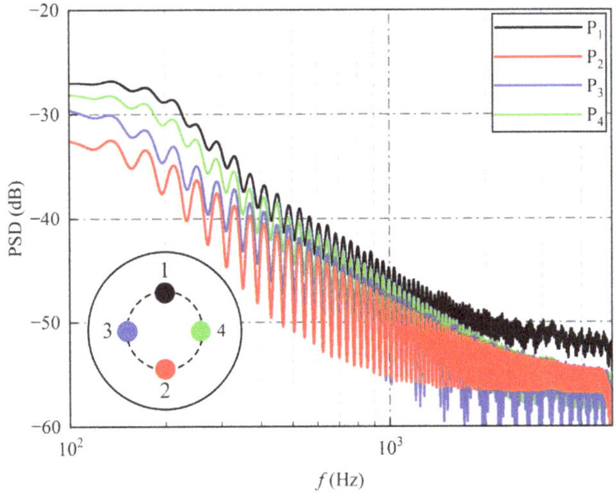

Figure 19. Power spectral density of the dynamic pressure signals for the spiked-blunt body. P_1 to P_4 represent the locations of dynamic pressure sensors.

The spatial structures of the dominant modes are extracted using POD to conduct a more in-depth comparison and discuss the similarities and differences in the spatial structures of modes extracted by DMD with three sorting criteria. Figure 20 illustrates the flow field structures of the first three modes of the POD for the spiked-blunt body, where Figure 20a–c corresponds to the first, second, and third modes, respectively. The first mode represents the mean flow field of unsteady pulsations, which accounts for 23.2% of the energy. The most prominent features in the spatial structure include the aftershock, recirculation zone, trace of ejected flow mass near the cylinder shoulder, and the interaction zone where the shear layer and aftershock interact. Notably, speckles and streaks are observed in mode 1, which validates that the multiple low-frequency signals present in the PSD of its temporal coefficients are indeed background noise in the flow field.

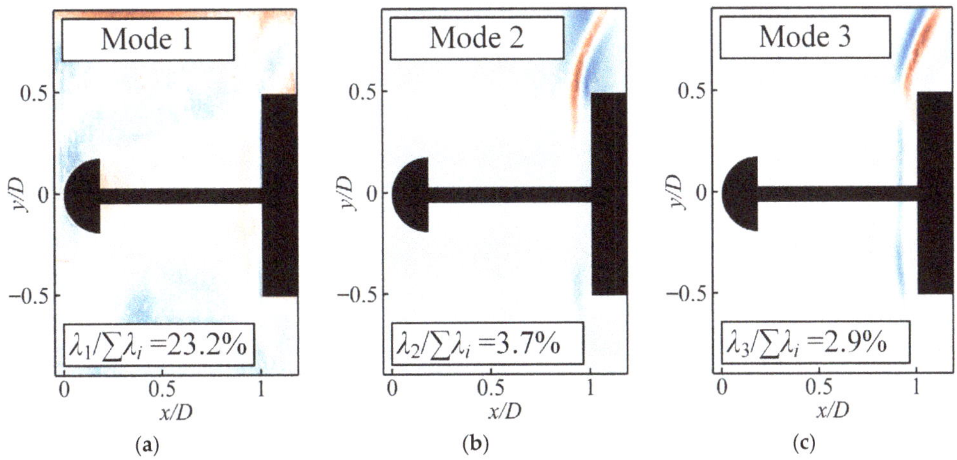

Figure 20. Flow field structures of the first three modes of the POD for the spiked-blunt body. Specifically, (**a**) represents the basic flow field (i.e., the mean flow field), while (**b**,**c**) depict perturbations to the basic flow field.

The second and third modes represent perturbations to the fundamental flow field, which account for 3.7% and 2.9% of the energy, respectively. The background noise of the flow field is no longer present, and their main features both include a single aftershock in front of the cylindrical, blunt body. However, differences are observed between them. In mode 2, the aftershock fully envelops the surface of the cylindrical, blunt body, with the pulsation intensity slightly higher at the shoulder than in the central area. It corresponds to the phase of withholding in the suppressed pulsation mode [53]. In mode 3, the structure of the aftershock is similar to that in the smash phase, but a shallow blue stripe covers the centre of the cylindrical blunt body. This phenomenon is speculated to be an intermittent weak shock wave, and this intermittent weak shock wave reduces the peak frequency in the dynamic pressure signal compared with the peak frequency of the temporal coefficients. Notably, the shock foot is located near the pressure measurement point, and the principal frequency is caused by the continuous pounding of the aftershock on the shoulder of the cylindrical, blunt body.

In summary, the results from the POD and dynamic pressure signals exhibit a high level of agreement with the outcomes of the DMD using the energy sorting criterion in terms of spectral characteristics and spatial structure. The POD and dynamic pressure signals serve as complementary tools that validate the results obtained through DMD using the energy sorting criterion, which further establishes the superiority of this criterion over the two other sorting criteria in the context of unsteady experimental flow fields.

In addition, it is crucial to note that despite the unique advantages of the energy sorting criterion, several limitations and boundaries should be kept in mind in practical applications, and these considerations are essential to ensuring that the results are correctly applied and interpreted.

1. Limited data length: The length of the data series used for DMD affects the accuracy of the decomposition. Insufficient data length may result in an incomplete representation of the dynamics, leading to inaccurate pattern recognition and ordering.
2. Truncation error: When truncating the SVD in DMD, a truncation error occurs, where modes with small singular values are removed. This error affects the accuracy of the decomposition, and the choice of truncation layer should be carefully considered to balance the simplicity and accuracy of the model.
3. Interpretation of modes: The interpretation of the modes obtained from the DMD should be performed with caution. Although the modes represent coherent spatio-temporal patterns, they may not always correspond to physically meaningful structures. The physical interpretation of the modes should be based on domain knowledge, complementary analysis, and validation.

4. Conclusions

This study applies the commonly used amplitude and frequency sorting criteria in conventional DMD, as well as the energy sorting criterion proposed in this study, to process and comparatively analyse the high-dimensional schlieren dataset of the unsteady flow field around a spiked-blunt body under Ma = 2.2 inflow conditions. Furthermore, the study employs POD and dynamic pressure signals to explore the effectiveness of the three criteria in capturing spectral characteristics and spatial structures. This study summarises the characteristics and application scenarios of DMD under each sorting criterion, which sheds light on the primary flow features of the unsteady flow field around the spiked-blunt body in supersonic conditions. The key conclusions are as follows:

1. DMD with the conventional amplitude–frequency sorting criterion presents substantial limitations. DMD using the amplitude sorting criterion can capture structures with large initial amplitudes from the flow field. However, these extracted modes may exhibit excessive decay rates, which make them unable to maintain stability in the flow field over extended periods. DMD using the frequency sorting criterion can extract high- and low-frequency structures from the flow field. However, this criterion has the drawback of limited differentiation amongst the extracted modes. The reason

is that they essentially represent similar types of flow field structures, which results in excessive loss of flow field information.

2. DMD, with the energy sorting criterion, can extract the predominant structures of unsteady pulsation in the flow field. This approach simultaneously considers spatial and temporal orthogonality, which effectively avoids the limitations of modes sorted by amplitude with high decay rates and modes sorted by frequency with low differentiation. Compared with the two other sorting criteria, the energy sorting criterion proves more suitable for the experimental dataset of unsteady flow fields.
3. POD can effectively capture dominant coherent structures in the flow field by determining spatially orthogonal bases. The results from POD, along with the spectral characteristics of experimentally measured dynamic pressure signals, exhibit a strong alignment with the DMD results obtained using the energy sorting criterion. This finding substantiates the superiority of the energy sorting criterion over the two other sorting criteria when applied to unsteady experimental flow fields.
4. The spatial composition of the flow field around a hemispherical aerodome and a cylindrical blunt body under supersonic inflow conditions primarily consists of several key elements: the aftershock in front of the cylindrical blunt body, the foreshock in front of the aerodome, and the shear layer and recirculation zone behind the aerodome. The unsteady flow field is predominantly influenced by the pulsation of the aftershock in front of the cylindrical, blunt body. This flow pattern exhibits multi-frequency coupling, with the primary frequency of 3.3 kHz originating from the periodic motion of the aftershock. This reciprocating motion continuously drives the compression and expansion of gas on the surface of the cylindrical, blunt body.

Author Contributions: Conceptualisation, Y.W. and R.G.; methodology, Y.W.; software, Y.W.; validation, J.X. and R.G.; formal analysis, Y.W.; investigation, Y.W.; resources, J.X.; data curation, Q.Q.; writing—original draft preparation, Y.W.; writing—review and editing, J.X. and R.G.; visualisation, Y.W. and L.C.; supervision, J.X. and R.G.; project administration, J.X.; funding acquisition, J.X. All authors have read and agreed to the published version of the manuscript.

Funding: This research was funded by the National Natural Science Foundation of China (Grant No. 12332018), the National Science and Technology Major Project (Grant No. 2019-II-0007-0027), the Defence Industrial Technology Development Programme (Grant No. JCKY2019605D001), and the Interdisciplinary Innovation Fund for Doctoral Students of Nanjing University of Aeronautics and Astronautics (Grant No. KXKCXJJ202309). The authors would like to express their sincere gratitude for the financial support.

Data Availability Statement: The data are not public due to privacy issues. The data presented in this study are available on request from the corresponding author.

Conflicts of Interest: The authors declare no conflicts of interest.

Nomenclature

English symbols:

A	system matrix of the high-dimensional flow field, (-)
$\widetilde{A}$	low-dimensional similar matrix, (-)
C	temporal coefficient of DMD mode, (-)
D	diameter of blunt body, (mm)
E	energy of DMD mode, (-)
I	identity matrix, (-)
L	length of blunt body, (mm)
Ma	Mach number, (-)
N	number of snapshots, (-)
$P_{1\sim4}$	pressure monitoring point, (-)
Re	Reynolds number, (-)
T	temperature, (K)
U, V	unitary matrix, (-)

$\boldsymbol{X}, \boldsymbol{Y}$	adjacent snapshot matrix, (-)
d	diameter of spike, (mm)
f	frequency, (Hz)
g	growth rate, (-)
l	length of spike, (mm)
p	pressure, (Pa)
t	time series, (ms)
x	column vector of single flow field snapshot, (-)
Greek symbols:	
$\boldsymbol{\Lambda}$	eigenvector, (-)
$\boldsymbol{\Phi}$	matrix of DMD mode, (-)
$\boldsymbol{\Sigma}$	diagonal matrix, (-)
α	amplitude of DMD mode, (-)
λ	eigenvalue of related DMD mode, (-)
ρ	density, (kg/m^3)
σ	standard deviation, (-)
ω	frequency of DMD mode, (Hz)
x	column vector of single flow field snapshot, (-)
Subscripts:	
$(\cdot)_A$	aerodome
$(\cdot)_b$	back pressure
$(\cdot)_e$	exit condition
$(\cdot)_F$	Frobenius norm
$(\cdot)_m$	test model
$(\cdot)_t$	total parameter

References

1. Stewartson, K. On the Linearized Potential Theory of Unsteady Supersonic Motion. *Q. J. Mech. Appl. Math.* **1950**, *3*, 182–199. [CrossRef]
2. Clemens, N.T.; Narayanaswamy, V. Low-Frequency Unsteadiness of Shock Wave/Turbulent Boundary Layer Interactions. *Annu. Rev. Fluid Mech.* **2014**, *46*, 469–492. [CrossRef]
3. Howlett, D.; Hunter, L. A Study of a Supersonic Axisymmetric Spiked Inlet at Angle of Attackusing the Navier-Stokes Equations. In Proceedings of the 24th Aerospace Sciences Meeting, Reno, NV, USA, 6–9 January 1986; American Institute of Aeronautics and Astronautics: Reston, VA, USA, 1986.
4. Sekar, K.R.; Karthick, S.K.; Jegadheeswaran, S.; Kannan, R. On the Unsteady Throttling Dynamics and Scaling Analysis in a Typical Hypersonic Inlet–Isolator Flow. *Phys. Fluids* **2020**, *32*, 126104. [CrossRef]
5. Jagadeesh, G.; Hashimoto, T.; Naitou, K.; Sun, M.; Takayama, K. Visualization of Unsteady Shock Oscillations in the High-Enthalpy Flow Field around Double Cones. *J. Vis.* **2003**, *6*, 195–203. [CrossRef]
6. Gomes-Fernandes, R. Flow Features around Double Cones at Hypersonic Speed. *Aeronaut. J.* **2013**, *117*, 741–748. [CrossRef]
7. Sudarshan, B.; Saravanan, S. Heat Flux Characteristics within and outside a forward Facing Cavity in a Hypersonic Flow. *Exp. Therm. Fluid Sci.* **2018**, *97*, 59–69. [CrossRef]
8. Charwat, A.F.; Roos, J.N.; Dewey, F.C.; Hitz, J.A. An Investigation of Separated Flows—Part I: The Pressure Field. *J. Aerosp. Sci.* **1961**, *28*, 457–470. [CrossRef]
9. Estruch-Samper, D.; Chandola, G. Separated Shear Layer Effect on Shock-Wave/Turbulent-Boundary-Layer Interaction Unsteadiness. *J. Fluid Mech.* **2018**, *848*, 154–192. [CrossRef]
10. Mair, W.A. Experiments on Separation of Boundary Layers on Probes in Front of Blunt-Nosed Bodies in a Supersonic Air Stream. *Philos. Mag. J. Sci.* **1952**, *43*, 695–716. [CrossRef]
11. Maull, D.J. Hypersonic Flow over Axially Symmetric Spiked Bodies. *J. Fluid Mech.* **1960**, *8*, 584–592. [CrossRef]
12. Panaras, A.G.; Drikakis, D. High-Speed Unsteady Flows around Spiked-Blunt Bodies. *J. Fluid Mech.* **2009**, *632*, 69–96. [CrossRef]
13. Guenther, R.A.; Reding, J.P. Fluctuating Pressure Environment of a Drag Reduction Spike. *J. Spacecr. Rockets* **1977**, *14*, 705–710. [CrossRef]
14. Ahmed, M.Y.M.; Qin, N. Drag Reduction Using Aerodisks for Hypersonic Hemispherical Bodies. *J. Spacecr. Rockets* **2010**, *47*, 62–80. [CrossRef]
15. Sahoo, D.; Das, S.; Kumar, P.; Prasad, J.K. Effect of Spike on Steady and Unsteady Flow over a Blunt Body at Supersonic Speed. *Acta Astronaut.* **2016**, *128*, 521–533. [CrossRef]
16. Motycka, D.L.; Murphy, J.B. *Experimental Investigation of Inlet-to-Inlet Shock Interference*; SAE International: Warrendale, PA, USA, 1965.
17. Farahani, M.; Daliri, A.; Younsi, J.S. Supersonic Inlet Buzz Detection Using Pressure Measurement on Wind Tunnel Wall. *Aerosp. Sci. Technol.* **2019**, *86*, 782–793. [CrossRef]

18. D'souza, N.; Molder, S.; Moretti, G. Numerical Method for Hypersonic Internal Flow over Blunt Leading Edges and Two Blunt Bodies. *AIAA J.* **1972**, *10*, 617–622. [CrossRef]
19. Ericsson, L.E. Aeroelastic Instability Caused by Slender Payloads. *J. Spacecr. Rockets* **1967**, *4*, 65–73. [CrossRef]
20. Tumuklu, O.; Theofilis, V.; Levin, D. On the Unsteadiness of Shock–Laminar Boundary Layer Interactions of Hypersonic Flows over a Double Cone. *Phys. Fluids* **2018**, *30*, 106111. [CrossRef]
21. Schuelein, E. Wave Drag Reduction Approach for Blunt Bodies at High Angles of Attack: Proof-of-Concept Experiments. In Proceedings of the 4th Flow Control Conference, Seattle, WA, USA, 23–26 June 2008; American Institute of Aeronautics and Astronautics: Reston, VA, USA, 2008.
22. Jones, J.J. *Flow Separation from Rods ahead of Blunt Noses at Mach Number 2.72*; Langley Aeronautical Laboratory: Langley Field, VA, USA, 1952.
23. Wood, C.J. Hypersonic Flow over Spiked Cones. *J. Fluid Mech.* **1962**, *12*, 614–624. [CrossRef]
24. Kabelitz, H.-P. *Zur Stabilität Geschlossener Grenzschichtablösegebiete an Konischen Drehkörpern bei Hyperschallanströmung*; Abt. Wissenschaftliches Berichtswesen der Deutschen Forschungs- und Versuchsanstalt für Luft- und Raumfahrt: Berlin, Germany, 1971.
25. Yamauchi, M.; Fujii, K.; Higashino, F. Numerical Investigation of Supersonic Flows around a Spiked Blunt Body. *J. Spacecr. Rockets* **1995**, *32*, 32–42. [CrossRef]
26. Crawford, D.H. *Investigation of the Flow over a Spiked-Nose Hemisphere-Cylinder at a Mach Number of 6.8*; NASA Technical Note; National Aeronautics and Space Administration: Washington, DC, USA, 1959.
27. Mehta, R.C. Numerical Analysis of Pressure Oscillations over Axisymmetric Spiked Blunt Bodies at Mach 6.80. *Shock Waves* **2002**, *11*, 431–440. [CrossRef]
28. Sahoo, D.; Karthick, S.K.; Das, S.; Cohen, J. Shock-Related Unsteadiness of Axisymmetric Spiked Bodies in Supersonic Flow. *Exp. Fluids* **2021**, *62*, 89. [CrossRef]
29. Schmid, P.J.; Sesterhenn, J. Dynamic Mode Decomposition of Numerical and Experimental Data. In Proceedings of the Sixty-First Annual Meeting of the APS Division of Fluid Dynamics, San Antonio, TX, USA, 23–25 November 2008.
30. Schmid, P.J. Dynamic Mode Decomposition of Numerical and Experimental Data. *J. Fluid Mech.* **2010**, *656*, 5–28. [CrossRef]
31. Seena, A.; Sung, H.J. Dynamic Mode Decomposition of Turbulent Cavity Flows for Self-Sustained Oscillations. *Int. J. Heat Fluid Flow* **2011**, *32*, 1098–1110. [CrossRef]
32. Duke, D.; Soria, J.; Honnery, D. An Error Analysis of the Dynamic Mode Decomposition. *Exp. Fluids* **2012**, *52*, 529–542. [CrossRef]
33. Muld, T.W.; Efraimsson, G.; Henningson, D.S. Flow Structures around a High-Speed Train Extracted Using Proper Orthogonal Decomposition and Dynamic Mode Decomposition. *Comput. Fluids* **2012**, *57*, 87–97. [CrossRef]
34. Sarmast, S.; Dadfar, R.; Mikkelsen, R.F.; Schlatter, P.; Ivanell, S.; Sørensen, J.N.; Henningson, D.S. Mutual Inductance Instability of the Tip Vortices behind a Wind Turbine. *J. Fluid Mech.* **2014**, *755*, 705–731. [CrossRef]
35. Jovanović, M.R.; Schmid, P.J.; Nichols, J.W. Sparsity-Promoting Dynamic Mode Decomposition. *Phys. Fluids* **2014**, *26*, 024103. [CrossRef]
36. Sayadi, T.; Schmid, P.J.; Richecoeur, F.; Durox, D. Parametrized Data-Driven Decomposition for Bifurcation Analysis, with Application to Thermo-Acoustically Unstable Systems. *Phys. Fluids* **2015**, *27*, 037102. [CrossRef]
37. Chen, K.K.; Tu, J.H.; Rowley, C.W. Variants of Dynamic Mode Decomposition: Boundary Condition, Koopman, and Fourier Analyses. *J. Nonlinear Sci.* **2012**, *22*, 887–915. [CrossRef]
38. Wynn, A.; Pearson, D.S.; Ganapathisubramani, B.; Goulart, P.J. Optimal Mode Decomposition for Unsteady Flows. *J. Fluid Mech.* **2013**, *733*, 473–503. [CrossRef]
39. Dawson, S.T.M.; Hemati, M.S.; Williams, M.O.; Rowley, C.W. Characterizing and Correcting for the Effect of Sensor Noise in the Dynamic Mode Decomposition. *Exp. Fluids* **2016**, *57*, 42. [CrossRef]
40. Leroux, R.; Cordier, L. Dynamic Mode Decomposition for Non-Uniformly Sampled Data. *Exp. Fluids* **2016**, *57*, 94. [CrossRef]
41. Bagheri, S. Effects of Weak Noise on Oscillating Flows: Linking Quality Factor, Floquet Modes, and Koopman Spectrum. *Phys. Fluids* **2014**, *26*, 094104. [CrossRef]
42. Hemati, M.S.; Williams, M.O.; Rowley, C.W. Dynamic Mode Decomposition for Large and Streaming Datasets. *Phys. Fluids* **2014**, *26*, 111701. [CrossRef]
43. Tu, J.H.; Rowley, C.W.; Kutz, J.N.; Shang, J.K. Spectral Analysis of Fluid Flows Using Sub-Nyquist-Rate PIV Data. *Exp. Fluids* **2014**, *55*, 1805. [CrossRef]
44. Gueniat, F.; Mathelin, L.; Pastur, L. A Dynamic Mode Decomposition Approach for Large and Arbitrarily Sampled Systems. *Phys. Fluids* **2015**, *27*, 025113. [CrossRef]
45. Proctor, J.L.; Brunton, S.L.; Kutz, J.N. Dynamic Mode Decomposition with Control. *SIAM J. Appl. Dyn. Syst.* **2016**, *15*, 142–161. [CrossRef]
46. Annoni, J.; Seiler, P. A Method to Construct Reduced-Order Parameter-Varying Models. *Int. J. Robust. Nonlinear Control* **2017**, *27*, 582–597. [CrossRef]
47. He, G.; Wang, J.; Pan, C. Initial Growth of a Disturbance in a Boundary Layer Influenced by a Circular Cylinder Wake. *J. Fluid. Mech.* **2013**, *718*, 116–130. [CrossRef]
48. Wan, Z.H.; Zhou, L.; Wang, B.F.; Sun, D.J. Dynamic Mode Decomposition of Forced Spatially Developed Transitional Jets. *Eur. J. Mech.-B Fluids* **2015**, *51*, 16–26. [CrossRef]

49. Ducoin, A.; Loiseau, J.C.; Robinet, J.C. Numerical Investigation of the Interaction between Laminar to Turbulent Transition and the Wake of an Airfoil. *Eur. J. Mech.-B Fluids* **2016**, *57*, 231–248. [CrossRef]
50. Statnikov, V.; Bolgar, I.; Scharnowski, S.; Meinke, M.; Kähler, C.J.; Schröder, W. Analysis of Characteristic Wake Flow Modes on a Generic Transonic Backward-Facing Step Configuration. *Eur. J. Mech.-B Fluids* **2016**, *59*, 124–134. [CrossRef]
51. Qin, Q.; Guan, R.; Ma, Z.; Xu, J. Experimental Verification of Pulsation Suppression over Spiked Cylinder Using Standard/Double/Inverse Aerodome. *Aerosp. Sci. Technol.* **2021**, *116*, 106848. [CrossRef]
52. Guan, R.; Qin, Q.; Wang, Y.; Xu, J. Experimental Investigation of the Pulsing Flow Field around Four Types of Spiked-Blunt Body at Mach 2.2. *Aerosp. Sci. Technol.* **2023**, *133*, 108129. [CrossRef]
53. Guan, R.; Qin, Q.; Wang, Y.; Xu, J. Experimental Investigation of Flow Unsteadiness over Four Types of Spiked-Blunt Body at Mach 2.2. *Aerosp. Sci. Technol.* **2023**, *140*, 108468. [CrossRef]

aerospace

MDPI

Article

Control of Cowl Shock/Boundary Layer Interaction in Supersonic Inlet Based on Dynamic Vortex Generator

Mengge Wang [1,2], Ziyun Wang [1,*], Yue Zhang [1,2,*], Daishu Cheng [3], Huijun Tan [1], Kun Wang [1] and Simin Gao [1]

1 College of Energy and Power Engineering, Nanjing University of Aeronautics and Astronautics, Nanjing 210016, China; wmg-sz2102122@nuaa.edu.cn (M.W.); thj@263.net (H.T.); kunwang@nuaa.edu.cn (K.W.)
2 Laboratory of Aerodynamics in Muitiple Flow Regimes, China Aerodynamics Research and Development Center, Mianyang 621000, China
3 Zhengde Polytechnic, Nanjing 211106, China; fzdso26@126.com
* Correspondence: ziyun_wang@nuaa.edu.cn (Z.W.); y.zhang@nuaa.edu.cn (Y.Z.)

Abstract: A shock wave/boundary layer interaction (SWBLI) is a common phenomenon in supersonic inlet flow, which can significantly degrade the aerodynamic performance of the inlet by inducing boundary layer separation. To address this issue, in this paper, we propose the use of a dynamic vortex generator to control the SWBLI in a typical supersonic inlet. The unsteady simulation method based on dynamic grid technology was employed to verify the effectiveness of the proposed method of control and investigate its mechanism. The results showed that, in a duct of finite width at the inlet, the SWBLI generated complex three-dimensional (3D) flow structures with remarkable swirling properties. At the same time, vortex pairs were generated close to the side wall as a result of its presence, and this led to the intensification of transverse flow and, in turn, the formation of a complex 3D structure of the flow of the separation bubble. The dynamic vortex generator induced oscillations of variable intensity in the vortex system in the supersonic boundary layer that enhanced the mixing between the boundary layer flow and the mainstream. Meanwhile, the unique effects of "extrusion" and "suction" in the oscillation process continued to charge the airflow, and the distribution of velocity in the boundary layer significantly improved. As the oscillation frequency of the vortex generator increased, its charging effect on low-velocity flow in the boundary layer increased, and its control effect on the flow field of the SWBLI became more pronounced. The proposed method of control reduced the length of the separation bubble by 31.76% and increased the total pressure recovery coefficient at the inlet by 6.4% compared to the values in the absence of control.

Keywords: dynamic micro-vortex generator; shock wave/boundary layer interaction; dynamic grid simulation; flow control; supersonic inlet

Citation: Wang, M.; Wang, Z.; Zhang, Y.; Cheng, D.; Tan, H.; Wang, K.; Gao, S. Control of Cowl Shock/Boundary Layer Interaction in Supersonic Inlet Based on Dynamic Vortex Generator. *Aerospace* **2023**, *10*, 729. https://doi.org/10.3390/aerospace10080729

Academic Editors: Kung-Ming Chung and Sebastian Karl

Received: 6 July 2023
Revised: 31 July 2023
Accepted: 15 August 2023
Published: 20 August 2023

1. Introduction

As an essential part of aspirated high-speed propulsion systems, the inlet has many functions, such as capturing flow, converting the energy of the incoming flow, regulating the velocity of flow at the outlet, and isolating the upstream and downstream disturbances. These functions profoundly influence the efficiency and working envelope of the engine [1]. The capture and deceleration of airflow in supersonic/hypersonic inlets primarily rely on the generation of shock waves and compression-induced wave systems. However, the inlet inevitably produces a boundary layer during its operation and may also ingest it from the forebody under conditions of integration to give rise to a shock wave/boundary layer interaction (SWBLI). This phenomenon is inherent to supersonic/hypersonic inlets and significantly impacts their performance. Specifically, the SWBLI at the entrance of the inlet is often intense at higher Mach numbers and may cause significant boundary layer separation. This is because the local adverse pressure gradient of the boundary layer introduced by the cowl shock exceeds its limit of separation and leads to apparent separation within the inlet, causing a compression–expansion–recompression flow process. SWBLIs, in general, lead to boundary layer separation, which causes a large decrease in total pressure, large distortion at the face of the engine, and potentially results in an inlet unstart [2]. In order to ensure the safe and efficient operation of an inlet within the entire envelope of operation of the engine, the complex phenomenon of the SWBLI in the inlet must be controlled [3,4].

Various methods of flow control have been proposed by researchers worldwide. Active methods of flow control primarily include boundary layer bleeding and blowing. The bleeding method removes low-energy fluid from the boundary layer before separation [5] and can suppress boundary layer separation. Blowing involves injecting high-pressure flow into the boundary layer to improve its ability to resist the adverse pressure gradients induced by the SWBLI [6]. However, both methods have drawbacks. Bleeding requires the removal of a portion of the flow captured by the inlet and complex suction structures, which incurs additional bleeding-induced resistance. Moreover, while blowing can weaken the effect of the SWBLI in a specific range, it requires a high-pressure source, which makes it less feasible.

Passive methods of flow control, such as vortex generators (VG) [5–8], have become a focus of research on controlling SWBLIs due to their simple construction, ease of installation, and greater degree of control. VGs introduce counter-rotating vortices to the flow without additional energy. These vortices transfer high-momentum flow to the wall, and this results in a fuller boundary layer profile. However, the additional shape-induced drag caused by traditional VGs creates strong disturbances in the mainstream that affect the overall efficiency of the aircraft. Consequently, researchers have shifted their focus to micro-vortex generators (MVG) [9], as they can significantly reduce the additional drag due to the components of control. MVGs have a height that is only 10–60% of the thickness of the boundary layer and can still achieve the desired control effects [9,10]. Significant progress has been made in the development of MVGs in recent years [11,12]. Blinde et al. [13] conducted detailed studies on the flow field in the SWBLI under MVG-based control. The results showed that the presence of the MVG reduced the region of separation from 20% to 30%, and the amplitude of the incident shock-induced oscillations was reduced by 20% compared with the condition without MVG-based control. Anderson et al. [14] demonstrated that triangular wedge- and blade-shaped MVG arrays can generate counter-rotating vortices and create alternating spatial distributions of the upwash and downwash in the spanwise direction that intensify the exchange of energy between the inner and outer fluids of the boundary layer to make it fuller. Zhang [15] proposed and assessed a highly swept ramp-type MVG that induced a "precompression effect," a "dividing effect," and a "mixing effect" due to the unique structure of the vortex generator. This phenomenon suppressed boundary layer separation and promoted the reattachment of the separated flow.

Moreover, researchers have explored combined methods of flow control. Anderson et al. [16] improved an active control system for an S-shaped inlet by adding an MVG to the rear of the jet to achieve the same control effect as that when the jet flow was reduced to one-tenth of its original rate of flow. Wagner et al. [17] found that placing an MVG at the inlet led to a 32% increase in maximum back-pressure and a 34% improvement in pressure stability within the isolation region when the inlet was in a stable start-up state.

In this paper, we propose a dynamic vortex generator (hereinafter referred to as the "dynamic MVG") for controlling the cowl-induced SWBLI in a typical supersonic inlet. The dynamic MVG can not only generate wake-induced vortices but also inject energy upstream of the SWBLI. Its influence on the SWBLI was studied using unsteady numerical simulations combined with the dynamic grid technique. The mechanism of control and effects of the MVG on supersonic inlets are also discussed in this paper.

2. Concept of Dynamic MVG Control

In order to enhance the ability of the supersonic inlet boundary layer to resist the adverse pressure gradient, a SWBLI control method based on an array of dynamic MVGs is proposed in this paper. Figure 1 shows that the array configuration of the vortex generators was obtained by lining them up along the spanwise direction. The array was composed of three dynamic MVGs to improve their range of control. This is also the minimum number of units that can accurately reflect the characteristics of interaction between the vortices generated by them. Its specific mode of operation is as follows: The vortex generator is placed at position X_p upstream of the point of incidence of the shock (X_p is the distance between the trailing edge of the vortex generator and the point of incidence of the shock). When the SWBLI does not cause significant boundary layer separation, the flow does not need to be controlled. The vortex generator is embedded into the wall of the inlet to keep it flat and avoid increasing the resistance in the duct. When large-scale boundary layer separation is induced by the SWBLI at the inlet, the dynamic MVG oscillates around its axis of rotation at high frequencies to induce dynamic vortices of higher intensity that yield stronger momentum mixing and lead to an increase in the suppression of the separation.

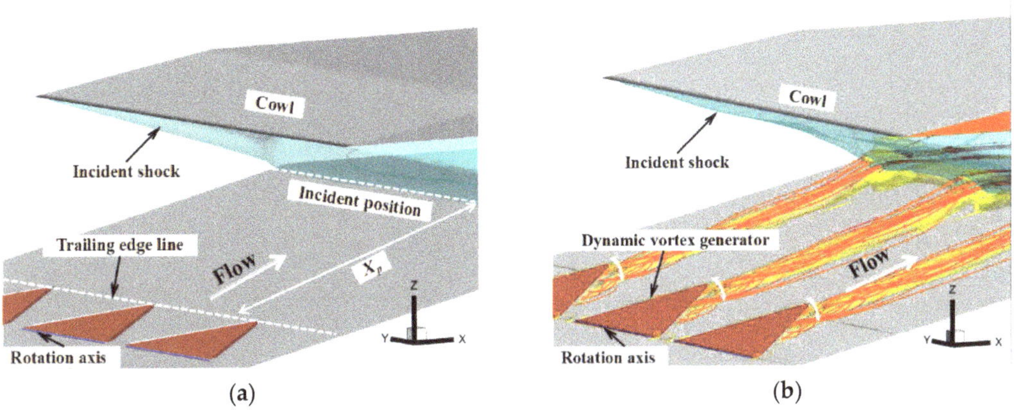

Figure 1. Design concept of the dynamic MVG. (**a**) Flat state of the dynamic MVG. (**b**) Oscillating state of the dynamic MVG.

3. Methodology

3.1. Description of the Model

The dynamic MVG was placed in a supersonic inlet with a working Mach number of 3.8 to verify its capability of control. Figure 2 shows the two-dimensional (2D) schematic diagram of the inlet. Its angle of compression α in the first stage was 12°, the height of the entrance to the duct (h_{in}) was 24.9 mm, and its height h was 14.4 mm. The red dotted box is used to highlight the vortex generator location. According to the design principle of the proposed dynamic MVG, the array of dynamic MVGs was arranged upstream of the shoulder of the inlet O to ensure that the point of incidence of shock was at a certain distance from the trailing edge of the vortex generator. According to the reference value of the position of shock incidence in [18,19], for this paper, we selected $10h_v$ as the distance X_p between the trailing edge of the dynamic MVG and the point of shock incidence. The basic configuration of the dynamic MVG proposed in this study was designed using the methods in [14,18], and its configuration is also the basic configuration commonly used by other researchers, and it can ensure the optimal control effect of the vortex generator [11,13,20]. The maximum height of oscillations of the dynamic MVG was h_v, an order of the thickness of the incoming boundary layer δ, and the spacing between the centerlines of two MVGs was $s = 7.5h_v$. The half-top angle AP of the dynamic MVG was 24°, and the chord length c was 7.2 times the dynamic MVG height h_v. The typical geometric parameters are listed in Table 1. Note that the dynamic MVG was a thin plate-type generator with a thickness d of 0.5 mm; the leading edge was chamfered, and there was a 0.1 mm margin between the perimeter of the vortex generator and the bottom wall. The MVG (three in a group) oscillated periodically around the axis of oscillation of the leading edge of the vortex generator at a frequency of 100 Hz.

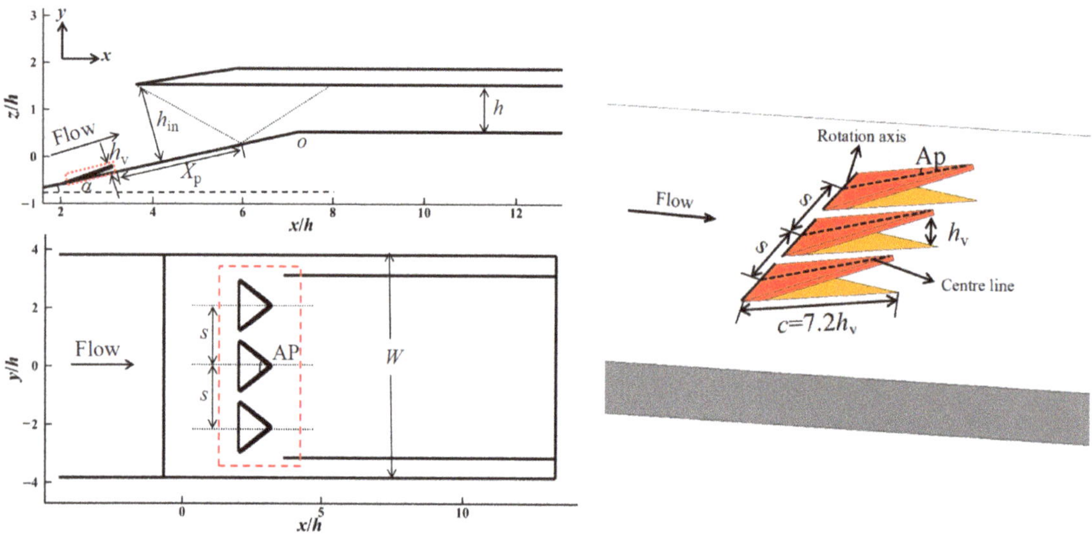

Figure 2. Schematic diagram of the proposed model. (**a**) Schematic diagram of the configuration of the inlet. (**b**) Structure of the dynamic MVG.

Table 1. Main design parameters of the model and the dynamic MVG.

Design Parameter	Value
h (mm)	14.4
h_v (mm)	4.0
h_{in} (mm)	24.9
W (mm)	110.0
AP	24.0
s (mm)	30.0
c (mm)	28.8
α (°)	12.0
X_p (mm)	40

This paper establishes a coordinate system, with the x-axis direction representing the flow direction, the y-axis direction representing the extension of the flow field, and the z-axis direction representing the perpendicular wall normal. Figure 3 shows the variations in the flow field of the array of vortex generators along the directions of flow (x–z plane) and the extension (y–z plane); the flow characters on different flow direction positions and extension directions are discussed in this paper. The distribution of pressure and velocity on these cross sections were analyzed; x was defined as the distance between the measurement position and the trailing edge of the vortex generator, and y represented the distance to the vortex generator symmetric plane $y = 0h_v$. The static pressure and temperature of the incoming flow were given according to atmospheric conditions at a flight altitude of 20 km. For a freestream Mach number Ma_0 of 3.8, the freestream flow pressure p_0 is 5529 Pa, and the static temperature T_0 is 216 K; therefore, the total pressure P^* is 640 kPa, and the total temperature T^* is 842 K. Due to the forebody compression surface, an oblique shock wave is generated at the front edge of the compression surface, resulting in a deceleration of airflow after passing through this shock wave. As a result, upstream of the vortex generator, there is a decrease in Mach number to 3.0. The velocity u and wall-normal direction distance z are dimensionless treated with freestream streamwise velocity u_0 and vortex generator height h_v, respectively, and the static pressure p is dimensionless treated with the freestream flow pressure p_0.

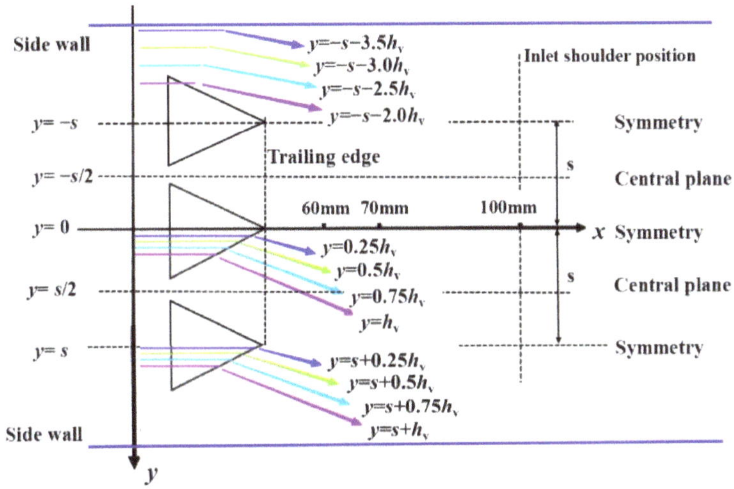

Figure 3. Relative positions of the array of dynamic MVGs.

3.2. Numerical Approach

3.2.1. Numerical Method

The dynamic mesh method in ANSYS FLUENT can be used to model flows where the shape of the domain changes with time due to motion in the domain boundaries. The MVGs were dynamically adjusted through unsteady numerical simulations, and spring smoothing and local reconstruction were used in the algorithm for the dynamic mesh. The update of the volume mesh is handled automatically at each time step based on the new positions of the boundaries. To use the dynamic mesh model, it is necessary to provide a starting volume mesh and a description of the motion of any moving zones in the model. The motion of these zones can be defined using a user-defined function (UDF) in ANSYS FLUENT, and the moving part was set to the motion of a rigid body rotating about a fixed point. Because the model contains moving and non-moving regions, it is important to identify these regions by grouping them into their respective cell zones in the starting volume mesh. Furthermore, regions deforming as a result of motion on their adjacent regions must also be grouped into separate zones in the starting volume mesh.

The grid topology of the model of the inlet and the array of dynamic MVGs are shown in Figure 4. The interface divided the entire computational domain into a non-moving domain (on the left) and a moving inner domain (on the right). Mixed grids were used for the calculations to satisfy the requirements of the method to update the dynamic grid. Unstructured grids were used near the vortex generator (right figure) to ensure that the grid could be updated and reconstructed during the motion of the vortex generator, while structural grids were used in the other areas (left figure) to ensure the quality of the grid. Maintain consistent surface mesh distribution at the interface between structured and unstructured grids, and set the boundary condition to the interface.

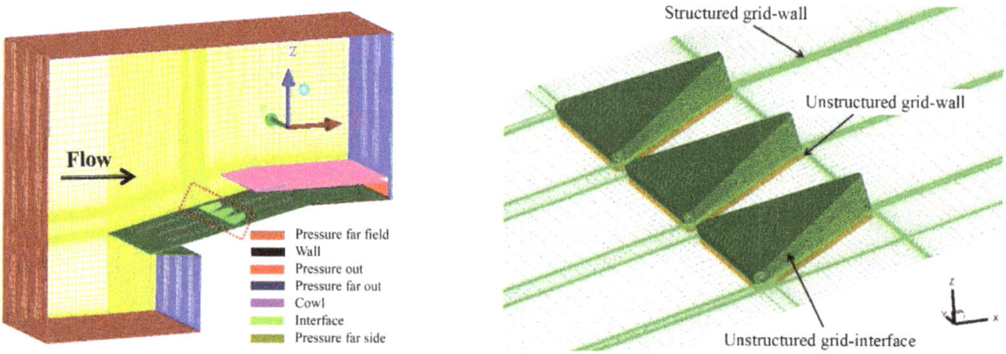

Figure 4. Grid topology of the array of dynamic MVGs.

The total number of grid nodes was about 15.9 million, and the y+ of the near-wall grid was guaranteed to be less than 1. In addition, the grid downstream of the vortex generator was encrypted to capture the structure of the flow of the vortex generator. The boundary conditions of the flow field were set as follows: The bottom and side walls were set to adiabatic solid-wall boundaries. The left and right sides were set as the side walls (hidden in order to better show the internal grid); the boundaries of pressure at the outlet were imposed on all outlets, and the other surfaces were subjected to the far-field boundary conditions of pressure.

To balance the efficiency of the search for appropriate values of the parameters with the accuracy of the calculations, calculations based on the 3D unsteady Reynolds-averaged Navier–Stokes (URANS) equations were carried out using the Computational Fluid Dynamics (CFD) software ANSYS FLUENT. This was carried out as follows: The inviscid convection flux is discretized by Roe's approximated Riemann method [21], and the interface flux is interpolated using Monotone Upwind Scheme for Conservation Laws (MUSCL) to obtain the second-order difference scheme of the inviscid convection flux. At the same time, the viscous flux is discretized through the second-order upwind scheme. The ideal gas model was used for the calculations, and the Shear Stress Transport (SST) k-ω model was used to model the turbulence [22]. In a previous study in the literature, a RANS solver, based on the SST turbulence model in FLUENT, was successfully applied to the flow with SWBLI, and the results of this study are in good agreement with experimental results [23–26]. The viscosity coefficient follows the Sutherland formula [27]. Moreover, the Mach number at the exit of the duct and mass flow through the exit were monitored in the simulations, and it was ensured that the residual convergence and monitored parameters were unchanged in each time step.

The equations to be solved are as follows [28]:

Continuity equation:

$$\frac{\partial(\rho u_i)}{\partial x_i} = 0 \tag{1}$$

Momentum equation:

$$\frac{\partial(\rho u_i u_j)}{\partial x_j} = -\frac{\partial p}{\partial x_i} + \frac{\partial \tau_{ij}}{\partial x_j} \tag{2}$$

$$\tau_{ij} = \mu\left(\frac{\partial u_i}{\partial x_j} + \frac{\partial u_j}{\partial x_i}\right) - \frac{2}{3}\mu\frac{\partial u_i}{\partial x_i}\delta_{ij}. \tag{3}$$

Energy equation:

$$\frac{\partial[u_j(\rho E+p)]}{\partial x_j}=\frac{\partial\left(k_t\frac{\partial T}{\partial x_j}\right)}{\partial x_j}+\frac{\partial(\tau_{ij}u_i)}{\partial x_j} \tag{4}$$

where ρ is the density; u_i and u_j are velocity components; p is the pressure; T is the temperature; E is the total energy per unit mass of fluid; τ_{ij} is the viscous stress tensor; and k_t is the heat conduction coefficient. Einstein summation marker and Kronecker operator δ_{ij} are used in the above equations.

The k–ω SST model was used to simulate the turbulence characteristics [29]. Many scholars [30–32] have shown that SST models can simulate SWBLI flows well.

Equation of turbulent kinetic energy k:

$$\frac{\partial(\rho u_j k)}{\partial x_j}=\frac{\partial}{\partial x_j}\left[\Gamma_k\frac{\partial k}{\partial x_j}\right]+G_k-\beta^*\rho f_{\beta^*}\omega k. \tag{5}$$

Equation of the specific dissipation rate of turbulence ω:

$$\frac{\partial(\rho u_j\omega)}{\partial x_j}=\frac{\partial}{\partial x_j}\left[\Gamma_\omega\frac{\partial\omega}{\partial x_j}\right]+D_w+G_\omega-\beta\rho f_\beta\omega^2, \tag{6}$$

where G_k represents the turbulent kinetic energy produced by the averaged velocity gradient, Γ_k and Γ_k represent the effective diffusivity of k and ω, G_ω represents the generation of ω, and the last terms in Equations (5) and (6) represent the dissipation of k and ω due to turbulence. D_ω represents the cross-diffusion term, and D_ω is defined as

$$D_w=2(1-F_1)\rho\frac{1}{\omega\sigma_{\omega 2}}\frac{\partial k}{\partial x_j}\frac{\partial\omega}{\partial x_j} \tag{7}$$

Effective diffusivity:

$$\Gamma_k=\mu+\frac{\mu_t}{\sigma_k}\Gamma_\omega=\mu+\frac{\mu_t}{\sigma_\omega}, \tag{8}$$

where σ_k and σ_ω are turbulent Prandtl numbers corresponding to k and ω, respectively, and they are defined as follows:

$$\sigma_k=\frac{1}{F_1/\sigma_{k1}+(1-F_1)/\sigma_{k2}},$$

$$\sigma_\omega=\frac{1}{F_1/\sigma_{\omega 1}+(1-F_1)/\sigma_{\omega 2}}, \tag{9}$$

The blending function F_1 is given by

$$F_1=\tanh\left(\Phi_1^4\right),$$

$$\Phi_1=\min\left[\max\left(\frac{\sqrt{k}}{0.09\omega y},\frac{500\mu}{\rho y^2\omega}\right),\frac{4\rho k}{\sigma_{\omega 2}D_w^+y^2}\right], \tag{10}$$

$$D_w^+=\max\left[2\rho\frac{1}{\sigma_{\omega 2}}\frac{1}{w}\frac{\partial k}{\partial x_j}\frac{\partial\omega}{\partial x_j},10^{-10}\right]$$

The turbulent viscosity μ_t is calculated by k and ω. The SST model can obtain a proper transport behavior by a limiter to the formulation of the eddy-viscosity:

$$\mu_t = \frac{\rho k}{\omega}\frac{1}{\max\left[\frac{1}{a^*}, \frac{SF_2}{a_1\omega}\right]},$$

$$F_2 = \tanh\left(\Phi_2^2\right),$$

$$\Phi_2 = \max\left[2\frac{\sqrt{k}}{0.09\omega y}, \frac{500\mu}{\rho y^2\omega}\right]. \tag{11}$$

Here, S is the magnitude of the vorticity. The corresponding generations of k and ω are

$$G_k = \mu_t S^2 G_\omega = \frac{aa^*}{\nu_t}G_k,$$

$$a_\infty = F_1 a_{\infty,1} + (1 - F_1)a_{\infty,2},$$

$$a_{\infty,1} = \frac{\beta_{i,1}}{\beta^*_\infty} - \frac{\kappa^2}{\sigma_{\omega 1}\sqrt{\beta^*_\infty}}, \tag{12}$$

$$a_{\infty,2} = \frac{\beta_{i,2}}{\beta^*_\infty} - \frac{\kappa^2}{\sigma_{\omega 2}\sqrt{\beta^*_\infty}}.$$

In the dissipative terms of k and ω,

$$f_{\beta^*} = \begin{cases} 1, & \chi_k \leq 0, \\ \frac{1 + 680\chi_k^2}{1 + 400\chi_k^2}, & \chi_k > 0, \end{cases}$$

$$\chi_k \equiv \frac{1}{\omega^3}\frac{\partial k}{\partial x_j}\frac{\partial \omega}{\partial x_j}, \tag{13}$$

$$f_\beta = \frac{1 + 70\chi_\omega}{1 + 80\chi_\omega},$$

$$\chi_\omega = \left|\frac{\Omega_{ij}}{(0.09\omega)^3}\right|,$$

$$\Omega_{ij} = \frac{1}{2}\left(\frac{\partial u_i}{\partial x_j} - \frac{\partial u_j}{\partial x_i}\right), \tag{14}$$

$$\beta^* = \beta^*_i[1 + 1.5F(M_t)],$$

$$\beta^*_i = \beta^*_\infty\frac{\frac{4}{15} + (Re_t/8)^4}{1 + (Re_t/8)^4},$$

$$Re_t = \frac{\rho k}{\mu\omega}, \tag{15}$$

$$\beta = \beta_i\left[1 - 1.5\frac{\beta^*_i}{\beta_i}F(M_t)\right],$$

$$\beta_i = F_1\beta_{i,1} + (1 - F_1)\beta_{i,2}.$$

The compressibility function $F(M_t)$ here improves the applicability of the model in free-shear flow at high Mach numbers, and the expression is as follows:

$$F(M_t) = \begin{cases} 0, & M_t > M_{t0}, \\ M_t^2 - M_{t0}^2, & M_t < M_{t0}, \end{cases} M_t^2 \equiv \frac{2k}{\alpha^2} \tag{16}$$

$$\alpha = \sqrt{\gamma RT}.$$

The following are some constants in the above expression: $a_1 = 0.34$, $M_{t0} = 0.3$, $\beta_{i,1} = 0.075$, $\beta_{i,2} = 0.0828$, $\beta^*_\infty = 0.09$, $\kappa = 0.41$, $a = a^* = 1$, $\sigma_{k1} = 1.176$, $\sigma_{\omega 1} = 2.0$, $\sigma_{k2} = 1.0$, and $\sigma_{\omega 2} = 1.168$.

Following the above, unsteady time step (δt) sensitivity tests were performed to determine the appropriate time step size to achieve efficient and accurate unsteady computation. According to the method in Ref. [33], the preliminary value of the time step is set as 1×10^{-5}. At a Mach number of 3.0, the characteristic velocity was defined as $u = Mac$ (c is the speed of sound), and the dimensionless time step ($\delta t^* = \delta t \times u/h_{\text{in}}$) was 3.5×10^{-3}, indicating that the choice of time step was sufficiently accurate. To further validate the accuracy of the specified time step, three time steps (1×10^{-4}, 1×10^{-5}, and 1×10^{-6}) were chosen for the unsteady numerical simulation verification. Considering the time cost of the calculations, the curve of the Mach number at the exit of the duct at 0–3 ms was calculated for comparative analysis. Figure 5 shows the curves of the Mach number of the exit of the duct at different time steps. It is clear that at time steps of 1×10^{-4} and 1×10^{-5}, a specific deviation occurred between the curves. However, the curves of the Mach number with time steps of 1×10^{-5} and 1×10^{-6} coincided with each other. After calculation, it was found that when δt equals 1.0×10^{-4}, 1.0×10^{-5}, and 1.0×10^{-6}, the CFL values are found to be 9.8, 0.98, and 0.098, respectively. Hulme et al. [34] pointed out that low CFL numbers can reduce vibration, reduce numerical diffusion, and improve accuracy. Generally, CFL < 1 and near 1 can be taken so that the accuracy of the numerical solution can be guaranteed while considering the calculation speed and convergence. When $\delta t = 1.0 \times 10^{-5}$, CFL < 1 and is near 1, thus meeting the requirements. Therefore, it is reasonable to conclude that a time step of 1×10^{-5} met the needs of the unsteady numerical simulations at the lowest cost.

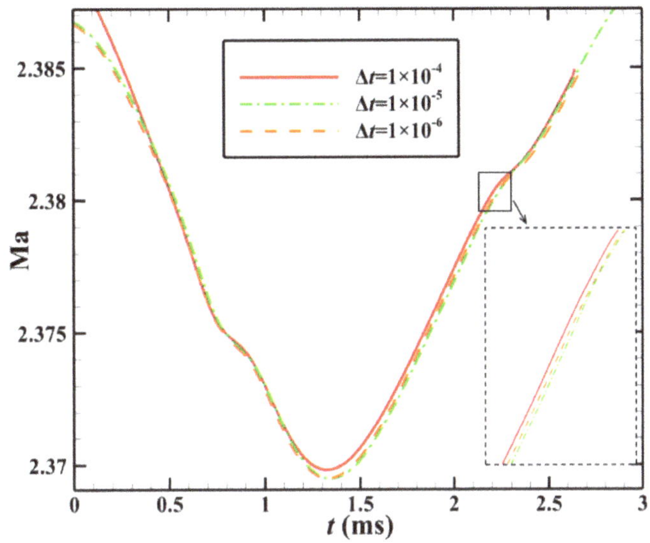

Figure 5. Curves of the Mach number at the exit of the duct with timesteps of different sizes.

To measure the impact of the number of nodes of the grid on the results of the calculations, three different sets of grids are used in this paper for numerical simulations under the same conditions of incoming flow: the coarse grid, fine grid, and dense grid contained 8.90×10^6, 1.59×10^7, and 2.39×10^7 cells, respectively. The distributions of velocity in the boundary layer at $x/h = 4.1$ and the curves of the wall pressure obtained by the simulations with the three sets of grids are shown in Figure 6. Although the densities of the dense and fine grids were different, the simulation results obtained when using them were in good agreement with those obtained with the use of the coarse grid. The fine grid was eventually chosen for subsequent simulations to save time and decrease the cost of calculation.

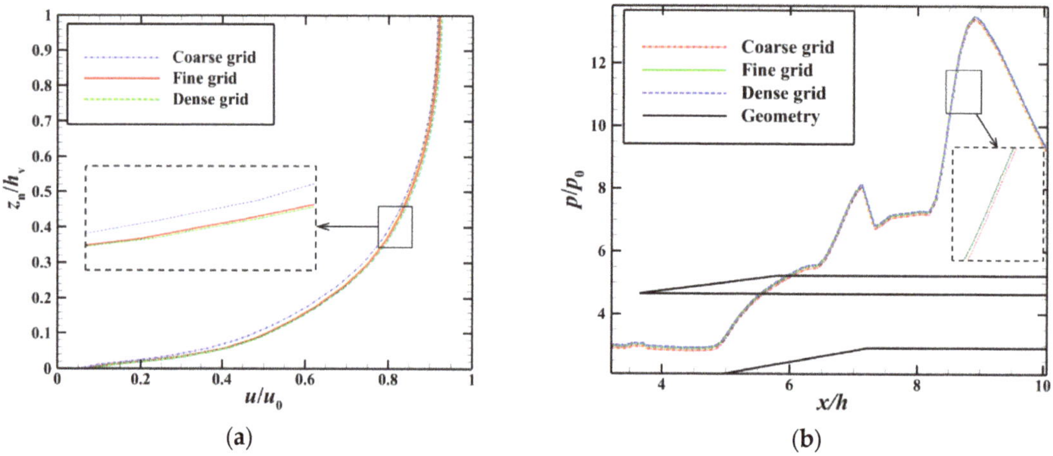

Figure 6. Profile of the boundary layer and distribution of the wall pressure with grids of different sizes. (**a**) Distributions of velocity at the boundary layer ($x/h = 4.1$) with grids of different sizes. (**b**) Wall pressure with grids of different sizes.

3.2.2. Numerical Validation

Because the process of the oscillation of dynamic MVGs was simulated based on a dynamic grid, the accuracy of the unsteady numerical simulations combined with the dynamic grid was further verified. The transient numerical simulation of the oscillations and transient pitch of the NACA0012 airfoil were carried out, and the numerical results were compared with those from the experiments [35]. In the experiments, the reference point of the pitch of oscillations of the wing was 0.25 times the chord length, and the oscillation motion is defined by: $a = a_0 + a_m \sin(\omega t + \varepsilon)$, where α and ε are the angle of attack and the phase angle depending on the time reference, respectively; α_0 is the initial angle of attack; and α_m is the amplitude of the angle of attack. The values of α_m, α_0, and ω were 2.51°, 0.016°, and 392.5, respectively. Figure 7 shows the hysteretical curve of the coefficient of the pitching moment C_m with the angle of attack α, as obtained from numerical simulations of the NACA0012 airfoil. In the figure, it is clear to see that the numerical result was close to the experimental data, and the maximum error between them was no more than ±5%. Therefore, the unsteady numerical simulations with the dynamic grid technique were determined to be adequately accurate.

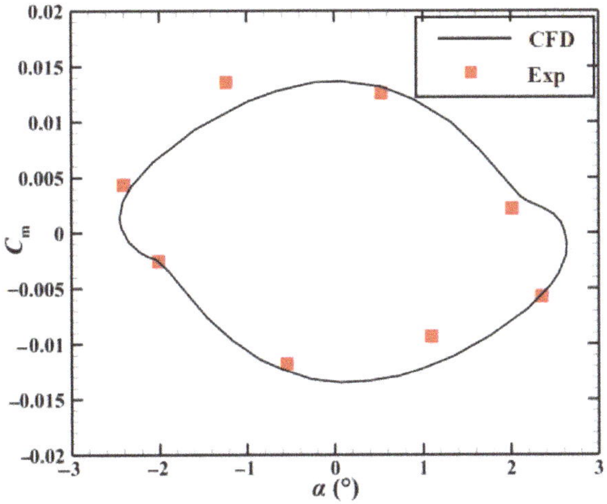

Figure 7. Comparison between the coefficients of the pitching moment obtained from experiments and simulations.

To validate the predictive capability of the model of turbulence used for vortices induced by the vortex generator, the experimental results of Giepman [36] were selected for testing. The parameters set in the simulation were consistent with the experimental conditions. The Mach number of the incoming flow was $Ma = 2.0$, and the height of the vortex generator was $h_v = 8.0$ mm. According to the results of 3D particle image velocimetry (PIV) experiments in Ref. [36], the nominal thickness of the boundary layer was $\delta = 17.5$ mm, and the shape factor was $H = 2.85$ when the vortex generator was not installed. The time-averaged velocity distribution of the boundary layer at this location was obtained from experiments and simulations, as shown in Figure 8a. The characteristics of the distribution of velocity in the downstream wake of the vortex generator need to be measured to assess the control effect. The normal positional distributions of the maximum velocity V_{max} and the minimum wake-induced velocity V_{min} on the symmetric surface at different flow stations shown in Figure 8b were obtained through the simulations. Figure 8 also shows that the results of Giepman's experiments and simulations were in good agreement with that of our experiments and simulations. The position of V_{max} on the symmetric surface was consistent with the experimental measurements along the direction of the flow, and the maximum difference between the positions was only $0.2h_v$ in the range of $h_v < x < 27.5h_v$. The value of V_{min} of the wake was within the scope of $h_v < x < 27.5h_v$, and the maximum difference between its positions in the experiment and the simulation was only $0.1h_v$. This shows that the unsteady simulations used in this paper can be used to obtain the complex vortex-induced flow fields within the specific range of the accuracy requirements.

3.2.3. Quasi-Steady Verification

The above method was selected to conduct an unsteady numerical simulation of the flow field induced by a dynamic MVG array to validate the correlation of the flow field between adjacent periods. The frequency of oscillation of the vortex generator was 100 Hz, and the period T is 0.001 s, as shown in Figure 9 (the relative positions of measuring points near the vortex generator are shown in the upper-right part of the figure). The changes in the wall pressure at different monitoring points of the vortex generator over time yielded an interesting phenomenon when $t = nT$ and $t = (n + 1)\ T$, whereby the curves of the wall pressure were similar at different measuring points. It can be inferred that the flow field induced by the vortex generator had quasi-stable characteristics; that is, the adjacent periods of oscillation were identical. Thus, the flow in a single period was considered to be representative. To prevent the first period from being affected by the reference flow field, the results of the second period were chosen for data processing.

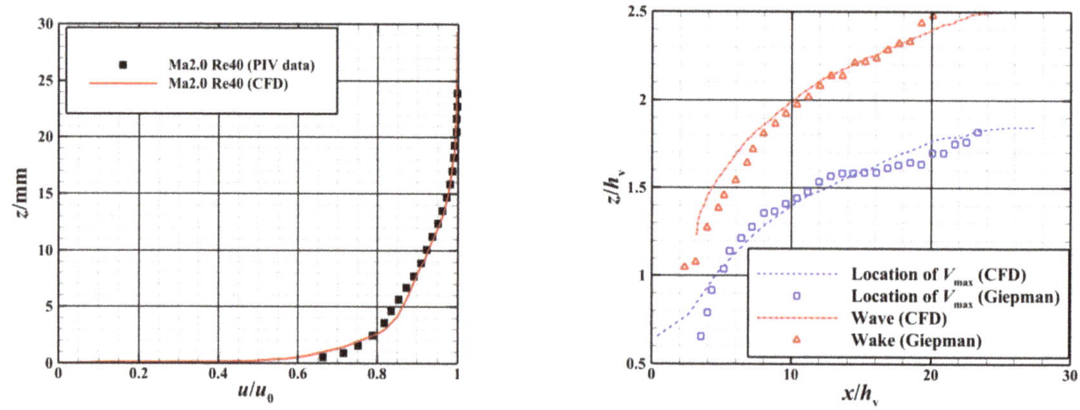

(**a**) (**b**)

Figure 8. Comparison between the results of experiments and simulations. (**a**) Comparison of the velocity distribution of the boundary layer between the experiment and the simulation [36]. (**b**) The points of the highest and lowest velocities of the wake on a plane of symmetry [36].

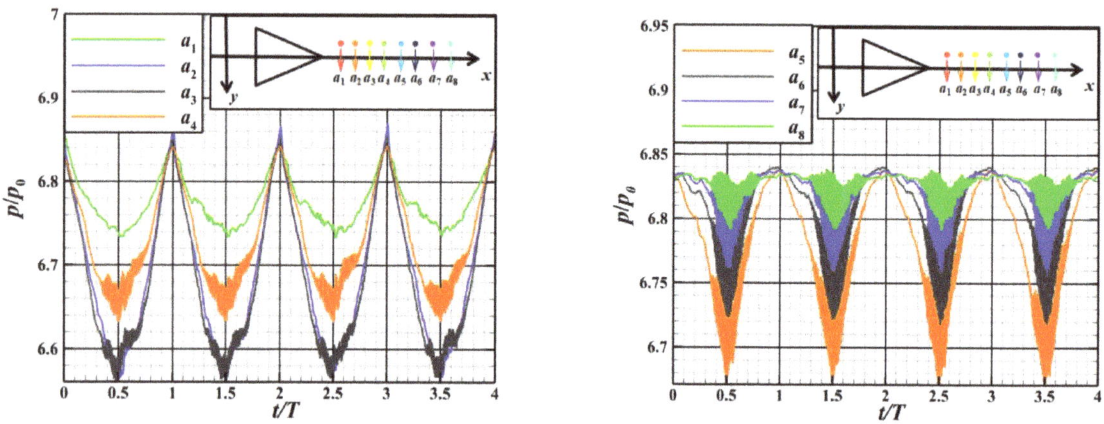

Figure 9. Variations in pressure at typical measuring points near the dynamic MVG over time.

4. Results and Discussion

4.1. Flow Structure of SWBLI in Supersonic Inlet without Control

Firstly, the 3D flow inside the inlet without dynamic MVG control was investigated. Figure 10 shows contours of the Mach number on the plane of symmetry of the inlet and the iso-surface of zero velocity at it (the blue curved surface in the figure). The figure shows that the SWBLI led to substantial boundary layer separation at the inlet under this condition, and an extensive range of low-momentum flows were accumulated along the direction of the flow near the side walls, which caused the flow field to exhibit strong three-dimensionality.

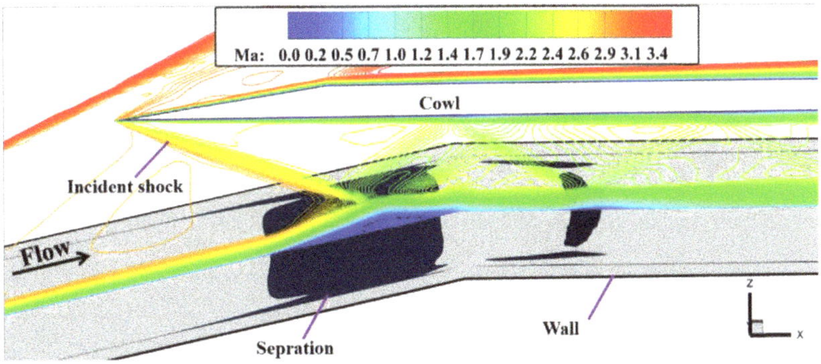

Figure 10. Contours of the Mach number on the planes of symmetry and zero velocity of the inlet without flow control.

Figure 11 shows the boundary layer at different spanwise positions at $x/h = 4.7$ without flow control (the relative positions are shown in Figure 3). The fullness of the near-wall boundary layer gradually decreased as the distance between it and the side wall decreased. Separation had already occurred in the near-wall boundary layer ($y = s + 3.5h_v$) at $x/h = 4.7$, indicating that flow was unstable there, and the zone of separation near the side wall was longer than that of the mainstream flow. Figure 12 shows the streamlines emitted $0.05h_v$ from the bottom and side walls of the inlet. Under the incident shock, the vortices generated near the side wall forced the flow toward the center of the channel, and this enhanced the transverse flow near the bottom wall of the inlet.

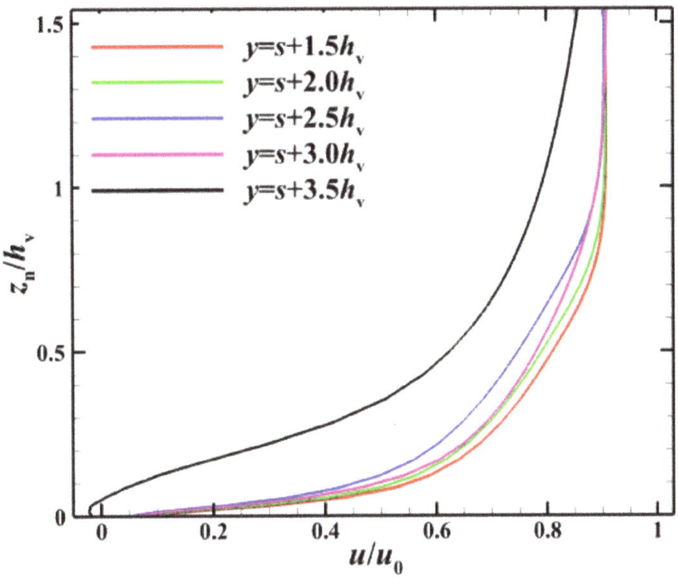

Figure 11. Boundary layer at different spanwise positions at $x/h = 4.7$ without control.

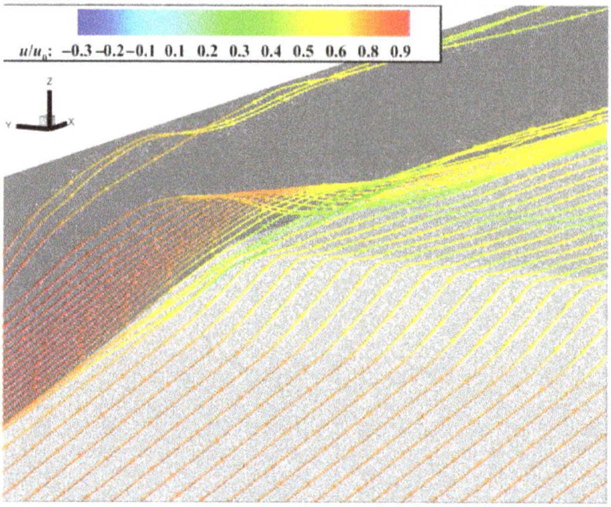

Figure 12. Streamlines emanating from the bottom and side walls at $0.05h_v$.

The spreading slices at different spanwise positions were used for further analysis. The distributions of the Mach number on different slices in the uncontrolled case in Figure 13 shows that the zone of separation gradually became shorter from the plane of symmetry ($y = s$) to the slice near the side wall. The shape of the separation near the side wall ($y = s + 2h_v$) changed considerably because the glancing interaction between the shock and the boundary layer of the side wall dominated the flow near the latter. It is clear that there was a strong transverse flow at the bottom wall of the inlet. Under its influence, the separated boundary layer was squeezed by the flow of the side wall so that the separation was more significant near the plane of symmetry ($y = s$).

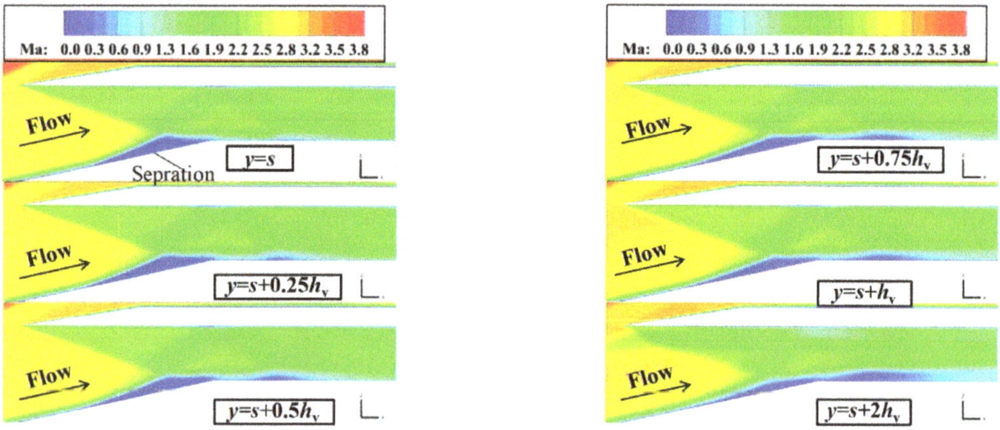

Figure 13. Distributions of the Mach number in different *xoz* planes without control.

Figure 14 shows the distributions of the wall limit streamlines on the bottom and side walls of the inlet without flow control. The wall limit streamlines distributions can represent the direction of flow near the wall and reflect the size of the separation area. A pair of prominent corner vortices were generated near each side wall under the action of the incident shock. The size of the separation near the side wall was influenced by this, and the flow field was clearly different from that under the 2D condition. In addition, the boundary layer of the side wall separated upstream of the incident shock. Figure 14b shows that wall limit streamlines also appeared on the side wall, indicating that the interaction between its boundary layer and the incident shock induced large-scale 3D flow separation, leading to the local accumulation of low-energy flow, which had adverse effects on the aerodynamic performance and structural strength of the inlet. Such significant boundary layer separation with complex 3D flow structures and swirling properties negatively impacted the aerodynamic performance of the inlet. Therefore, an effective method of controlling flow is needed to suppress the unfavorable flow induced by the SWBLI and improve the aerodynamic performance of the supersonic inlet.

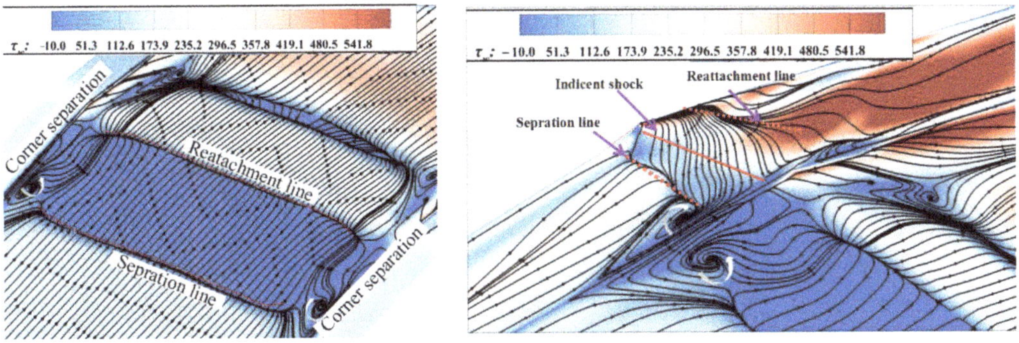

(**a**) (**b**)

Figure 14. Distributions of wall limit streamlines on the bottom and side walls of the duct without control. (**a**) Streamlines emitted $0.05h_v$ from the bottom and side walls of the duct. (**b**) Distributions of the wall limit streamlines on the side walls.

4.2. *Capability of Dynamic MVG Array to Control SWBLI in Supersonic Inlet*

4.2.1. Effect of Dynamic MVG Array on Disturbance in Supersonic Boundary Layer

The dynamic array of MVGs was installed to control the SWBLI in the supersonic inlet. The evolution of spatial vortices induced by dynamic vortex generator arrays was investigated first. The vortex structure was represented by a Liutex-Omega isosurface, and the Liutex-*R* contours of vortex intensity at the five flow stations away from the trailing edge of the vortex generator at different moments were extracted [37]. Figure 15 shows that, at different times, the vortex structures had similar patterns of distribution. Two pairs of vortices were located downstream of the three vortex generators: a main vortex with a more significant range of influence and a secondary vortex below it. The solid black line represents the main vortex's core line, and the dashed black line represents that of the secondary vortex. The main vortex played a dominant role in the evolution of the downstream flow field, while the secondary vortex had a smaller range of influence and was restricted near the plane of symmetry. According to the contours of the intensities of the vortex at five flow stations at different times, it can be seen that the *R* values of the Liutex-*R* of the core positions of the main vortex were relatively high, gradually decreasing around its core. The contours at different times show that each pair of vortices exhibited different changes with the oscillations of the dynamic MVG. As the vortex generator oscillated upward, the radius and intensity of the vortices increased and reached their maximum values at the highest point ($t = 1/2T$). When the vortex generator swung downward, the radius and intensity of the vortices tended to decrease. Ultimately, a controllable vortex structure was obtained within the supersonic boundary layer.

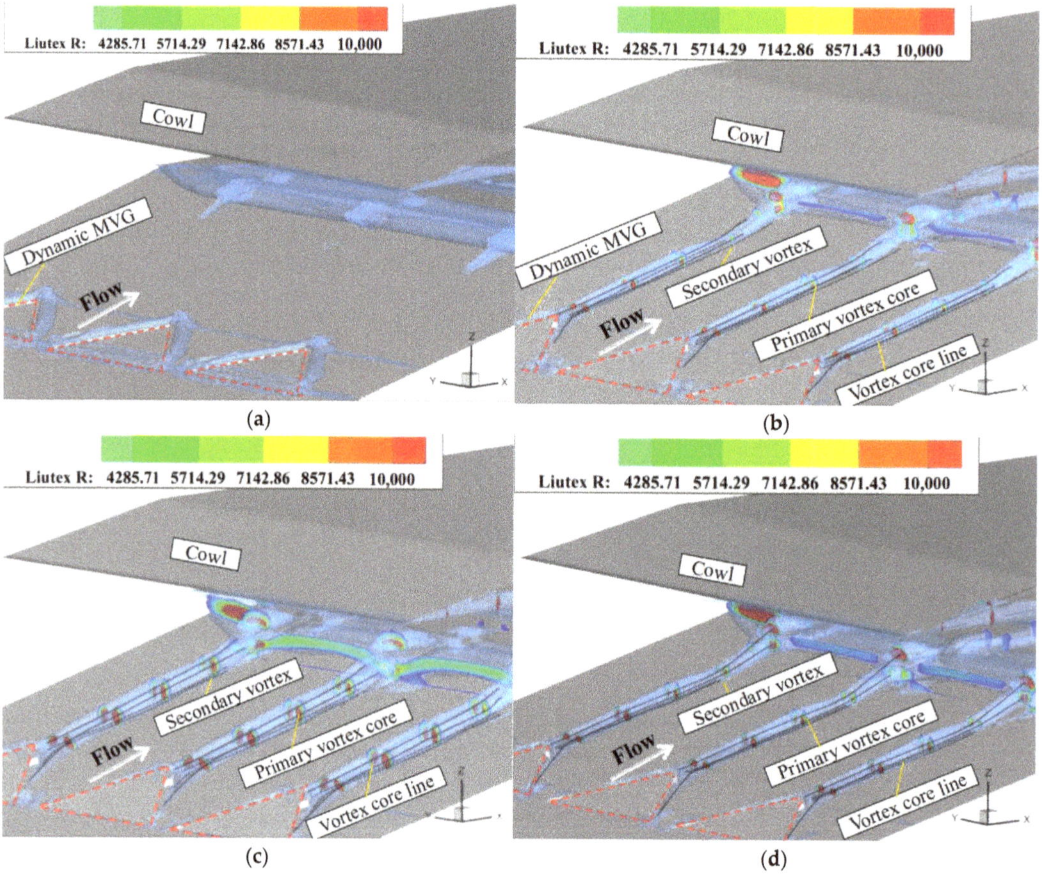

Figure 15. Evolution of the vortex along the direction of flow of the array of dynamic MVG. (**a**) $t = 0T$. (**b**) $t = 1/4T$. (**c**) $t = 2/4T$. (**d**) $t = 3/4T$.

Moreover, the presence of the SWBLI led to the formation of a strong separation induced vortex in the flow field, as shown in Figure 15. As the vortex generator oscillated upward, the wake-induced vortex penetrated the vortical structure of the SWBLI and weakened it. As the dynamic MVG continued to oscillate, the vortex acting on the region of the SWBLI was further enhanced to make the energy transport more prominent.

Figure 16 shows the time-averaged velocity profiles at the boundary layer at different streamwise positions after the array of dynamic MVGs and includes a schematic diagram of different spanwise positions (shown in the lower-left corner). The velocity u and the distance z_n (i.e., the normal distance to the bottom wall of the inlet) were non-dimensionalized by the mainstream velocity u_0 and the height of the vortex generator h_v. To illustrate the influence of the vortex generator, the velocity distribution at the boundary layer without the vortex generator control is also shown in Figure 16. At the location of $x/h = 4.7$, the boundary layer velocity distribution at different spanwise positions throughout one cycle is presented. The boundary layer had the same distribution at different spanwise positions except at the plane of symmetry ($y = 0h_v$). The profile of the near-wall boundary layer was relatively thin when the vortex generator was completely embedded into the wall ($t = 0T$). The profile of the boundary layer became fuller as the vortex generator slowly oscillated upward and became full when it swung to the highest point ($t = 1/2T$). As the spanwise position gradually moved away from the plane of symmetry $y = 0h_v$, the velocity of the near-wall boundary layer increased on the spanwise plane of $y = 0.5h_v$ and further increased at spanwise planes of $y = 0.75h_v$ and $y = h_v$. Therefore, the best position for controlling the flow was not on the plane of symmetry $y = 0h_v$ but at positions at a certain distance from it. In addition, the near-wall boundary layer at any streamwise position under dynamic MVG control was fuller than that under the uncontrolled condition. This indicates that the array of dynamic MVGs could increase the momentum in the near-wall region and enable the boundary layer to overcome the strong adverse pressure gradient induced by the cowl shock.

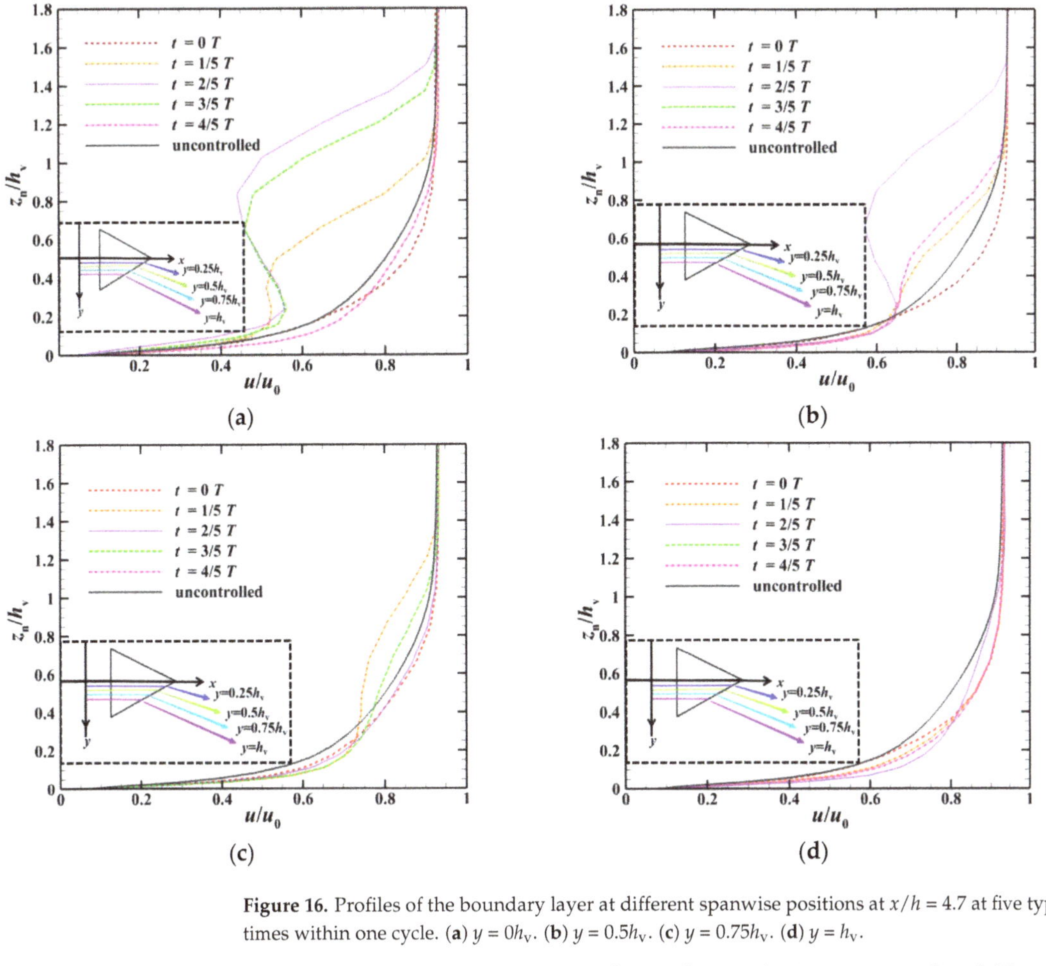

Figure 16. Profiles of the boundary layer at different spanwise positions at $x/h = 4.7$ at five typical times within one cycle. (**a**) $y = 0h_v$. (**b**) $y = 0.5h_v$. (**c**) $y = 0.75h_v$. (**d**) $y = h_v$.

Figure 17 shows the distribution of streamlines in the instantaneous flow field around the dynamic MVG to better understand the origin of the wake of the vortex generator. The starting points of the streamlines were located at the leading edge of the dynamic MVG, and the normal heights z_n were $10\%h_v$, $2.5\%h_v$, $1\%h_v$, and $0.5\%h_v$. The typical feature of the instantaneous structure of the wake of the dynamic MVGs was a pair of counter-rotating vortices, and this is consistent with the analysis in the previous section. Streamlines from the upstream boundary layer fell off along the side edge of the dynamic MVG and rolled up to form streamwise vortices. Streamwise vortices from both sides eventually merged into the streamwise counter-rotating vortex pair at the rear of the dynamic MVG. Some of the streamlines emitted from the upstream boundary layer at the normal height of $z_n = 10\%h_v$ bypassed the flow of the dynamic MVG and did not directly enter the wake, indicating that a lower volume of the fluid was entrained inside the vortex at this height. The streamlines emitted below $z_n = 10\%h_v$ constituted the wake of the dynamic MVG, and their final position was close to the core of the vortex, especially near the wall. In summary, the dynamic MVG induced a pair of streamwise vortices in supersonic flow that gathered low-momentum flow from the bottom layer of the boundary layer on both sides toward the centerline and lifted it upward. Meanwhile, the high-momentum flow above the boundary layer was entrained by the streamwise vortices and mixed with the flow inside the boundary layer to help achieve the objective of control.

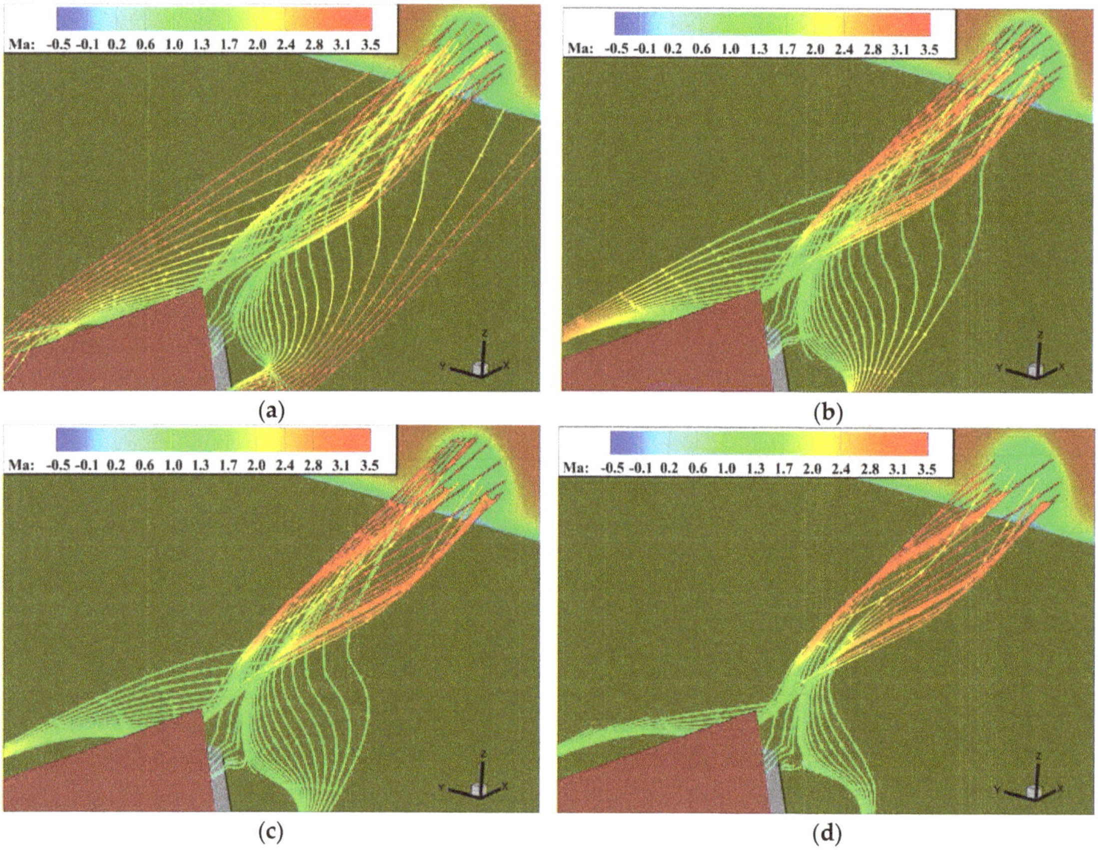

Figure 17. Distribution of streamlines around the array of dynamic MVGs in an instantaneous flow field. (**a**) $z_n = 10\%h_v$. (**b**) $z_n = 2.5\%h_v$. (**c**) $z_n = 1.0\%h_v$. (**d**) $z_n = 0.5\%h_v$.

The above analysis indicates that the vortex generator has a unique collection function for the near-wall boundary layer. It is clear from the distribution of streamlines around the dynamic MVG shown in Figure 18 that although, when it swung down to $t = 3/4T$, the height of the vortex generator was the same as that at $t = 1/4T$, the airflow below it was extruded by its trailing edge at $t = 3/4T$, while only prominent suction was observed at $t = 1/4T$. When the vortex generator swung upward, air flowed into the cavity below it from both sides. The typical profile of the velocity of the boundary layer at the position shown in Figure 19 indicates a significant increase in the near-wall velocity at $t = 3/4T$ compared with that at $t = 1/4T$, indicating that the unique "extrusion" and "suction" functions of the dynamic MVG continued to charge the airflow.

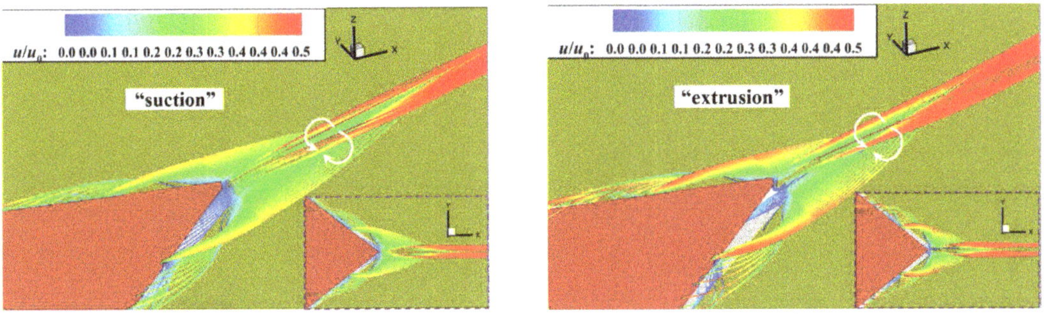

Figure 18. Distribution of streamlines around the dynamic MVG. (**a**) $t = 1/4T$. (**b**) $t = 3/4T$.

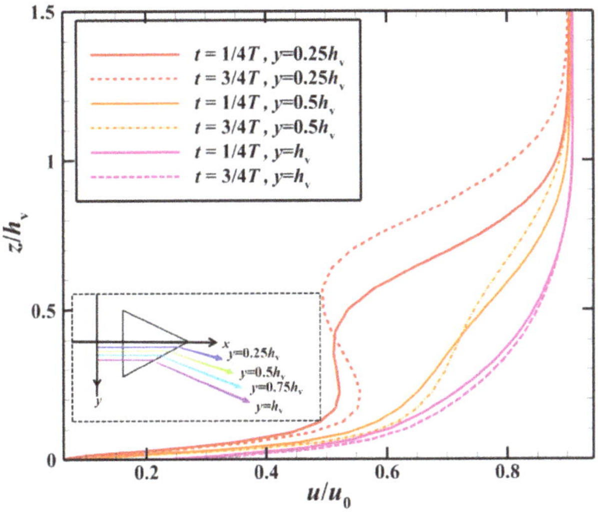

Figure 19. Profiles of velocity at $x/h = 4.7$ during "suction" and "extrusion" (the solid lines represent "suction", and the dashed lines represent "extrusion").

4.2.2. Effect of Dynamic MVG Array on SWBLI

The above analysis indicates that the dynamic MVG can induce vortices of controllable strength and significantly change the fullness of the boundary layer under conditions of supersonic inflow. Subsequently, a control characteristics analysis was conducted to examine its influence on the SWBLI region of the supersonic inlet. Figure 20 shows the time-averaged contours of the Mach number on different spanwise planes under the control of the array of MVGs. Compared with the case without control, the low-momentum region near the wall decreased at all spanwise positions under dynamic MVG-based control. The separation bubble gradually decreased in size from the plane of symmetry of the vortex generator to the other spanwise positions because the planes $y = 0h_V$ and $y = s$ were in the middle of the streamwise vortices induced by the vortex generator, and mixing had a minor effect on it. The planes of $y = 0.5h_V$, $y = 0.75h_V$, $y = s + 0.5h_V$, and $y = s + 0.75h_V$ were located in the streamwise vortices induced by the trailing edge of the vortex generator. The mixing resulting from the streamwise vortices was strong and reduced the size of the separation bubble.

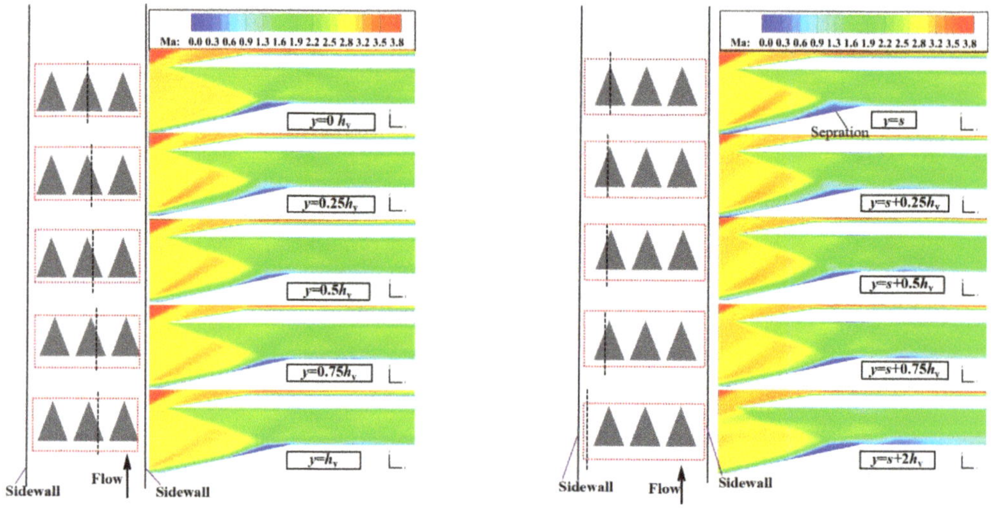

Figure 20. Time-averaged Mach number of the vortex generator at different spanwise positions.

Figure 21 shows the distributions of the time-averaged static pressure and wall friction coefficient c_f along the wall. The method proposed by Kendall and Koochesfahani [38] was used to estimate the friction velocity u_τ. The wall shear stress τ_w can be calculated using the relation $\tau_w = \rho_w u_\tau^2$. The wall friction coefficient was $c_f = 2\tau_w/\rho_0 u_0^2$. Increasing the local c_f value can delay the separation and reduce the disturbance area at the same time. Therefore, the c_f value reflects the flow field characteristics and control effect of the SWBLI [39]. The solid lines represent the time-averaged wall static pressure, and the dashed lines represent the distribution of c_f on the wall. The static pressure on the wall at different spanwise positions in the separation when the array of dynamic MVGs was used was significantly lower than in the uncontrolled flow field, indicating that the array could reduce local high pressure near the separation. The distribution of c_f of the wall on the corresponding *xoz* plane showed that the uncontrolled state had a shear stress on the wall opposite to the flow direction in the separation (x/h = 5.30~x/h = 7.00). The velocity gradient was negative, which is highly unfavorable for the efficiency of the inlet. However, the reverse shear stress at each spanwise position was reduced in the case involving control. The effects on the shear stress in the separation at $y = 0.5h_v$ (x/h = 5.65~x/h = 6.81) and $y = 0.75h_v$ (x/h = 5.61~x/h = 6.92) were better with control than without it. The length of the separation could be reduced by up to 31.76% compared with the uncontrolled case. However, the plane of symmetry $y = 0h_v$ was located in an area in which the two inner vortices interfered with each other, and the control effect was not as good as that at the other spanwise planes, which is consistent with the above discussion on the distribution of the time-averaged Mach number. In summary, the array of dynamic MVGs was able to reduce local high pressure near the separation and improve the downstream c_f of the wall.

To verify the control characteristics of the vortex generator at different times, an instantaneous Mach number contour was extracted at the position of $y = 0.25h_v$ at $t = 0T$, $t = 1/4T$, $t = 1/2T$, and $t = 3/4T$. Figure 22 shows that the size of the separation bubble decreased as the vortex generator swung upward. When it swung to the highest point at $t = 1/2T$, the separated boundary layer quickly reattached, and the control effect was better than at other times. This is because the scale and strength of the vortex peaked when the vortex generator swung to the highest point ($t = 1/2T$), as discussed above. When it swung downward, the separation bubble continued to decrease in size due to the "extrusion" effect of the dynamic MVGs.

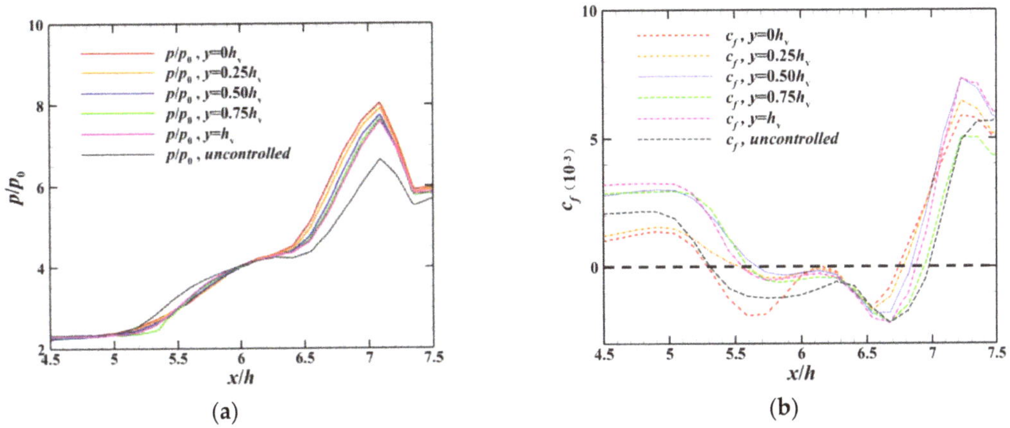

Figure 21. Distributions of the time−averaged parameters of the array of dynamic MVGs in different *xoz* planes. (**a**) Distribution of the time-averaged static pressure along the wall. (**b**) Distribution of the time−averaged wall friction coefficient along the wall.

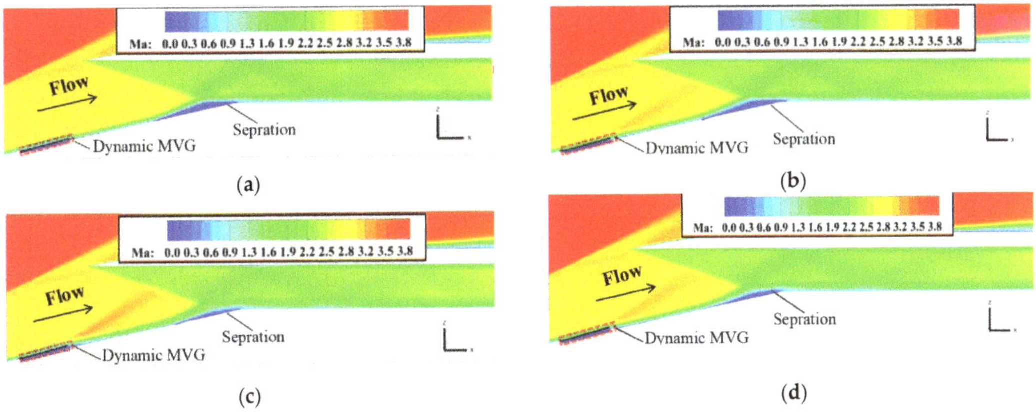

Figure 22. Contours of the Mach number of the array of dynamic MVGs at different times: (**a**) $t = 0T$, (**b**) $t = 1/4T$, (**c**) $t = 1/2T$, (**d**) $t = 3/4T$.

Figure 23 shows the distributions of the shear stress and limit streamlines on the wall at different times when controlled by the array of MVGs. Figure 23 shows that as the array of dynamic MVGs oscillated, the size of the separation bubble varied. The improvement in the shear stress on the wall on the plane of symmetry was better at $t = 1/2T$ and $t = 3/4T$. Under the control of the array of MVGs, the corner vortices disappeared, and the corner separation decreased. The separation line gradually formed a sawtooth shape, leading to a significant difference in the length of the separation at different spanwise positions. The separation near the $y = 0h_{\mathrm{v}}$ plane had the shortest flow, and the control effect at the $y = \pm s/2$ planes was not as good as the other spanwise planes because they were located between the trailing vortices of the array of vortex generators. In conclusion, the dynamic MVGs could satisfactorily control the interaction between the shock and the boundary layer. Compared with that in the absence of control, the length of flow in the separation region was significantly reduced, and the total pressure recovery coefficient increased by 6.4%.

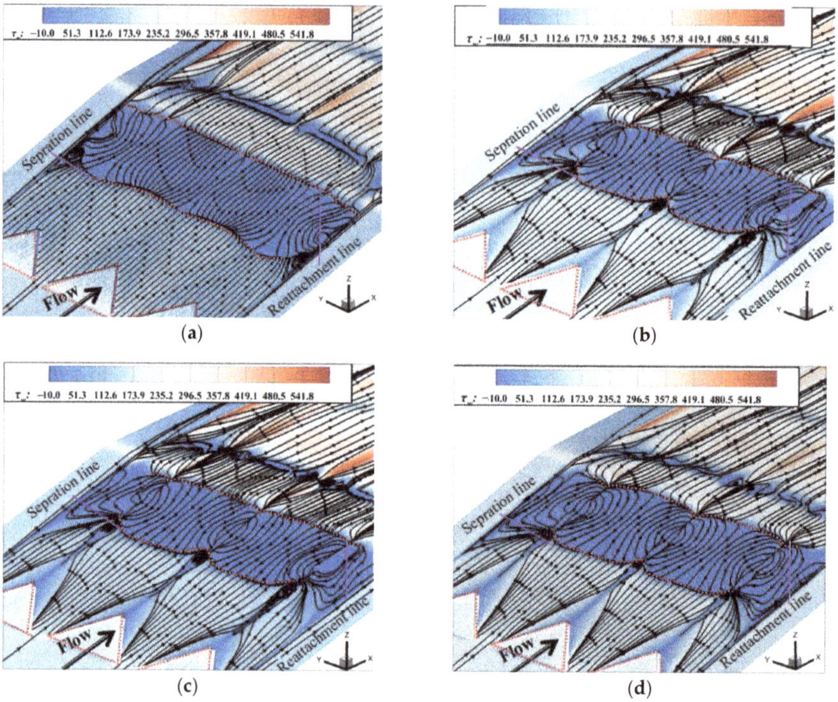

Figure 23. The distributions of the shear stress and limit streamlines on the wall at different times: (**a**) $t = 0T$, (**b**) $t = 1/4T$, (**c**) $t = 2/4T$, (**d**) $t = 3/4T$.

4.3. Influence of Dynamic Frequency of MVG Array on its Control Effect

Figure 24 shows variations in the shape factor and contours of the time-averaged Mach number on the plane $y = 0.25h_v$ under different frequencies. The variations in the shape factor H are shown in Figure 24a. The kinetic energy loss caused by viscosity is closely related to the fullness of the velocity profile in the boundary layer; the shape factor was introduced to express the relative kinetic energy loss in the boundary layer. The smaller the H, the fuller the boundary layer [40]. The monitoring position (dashed black line on the plane of symmetry $y = 0.25h_v$ in Figure 24b) was located upstream of the separation bubble at $x/h = 5.4$. Considering the limits imposed by the speed of the motor and the feasibility of the structure, the frequency of oscillations f of the vortex generator was controlled to within 300 Hz, and three frequencies (f_1 = 50 Hz, f_2 = 150 Hz, and f_3 = 300 Hz) were selected for analysis. Figure 24 shows that H decreased for $3/4T$ as the duration of oscillations increased, indicating that the vortex generator had enhanced the exchange of energy between the boundary layer and the mainstream, increased the fullness of the near-wall boundary layer, and was expected to achieve flow control. When the vortex generator oscillated at f_3, the value of the shape factor H at each instance was smaller than those at f_1 and f_2, and the maximum reduction reached 30%. This indicates that as the oscillation frequency of the dynamic MVGs increased, the frequencies of "suction" and "extrusion" of the airflow increased, leading to a higher intensity of energy transfer to airflow and a more significant effect in flow control. As can be seen from the time-averaged Mach number contour in Figure 24b, the height and length of the separation bubble decrease with the increase in the vortex generator's oscillation frequency, and this is because a higher oscillation frequency leads to more momentum injection into the near-wall region, which helps to overcome the adverse pressure gradient caused by SWBLI. This reduces the size of the separation bubble, which is consistent with the control effect of the shape factor change mentioned earlier.

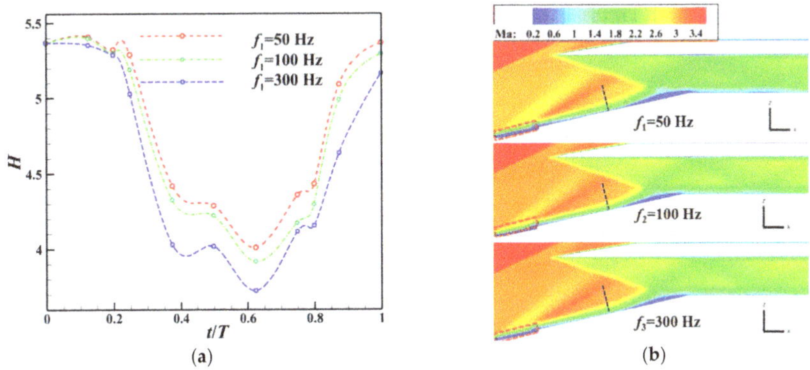

Figure 24. Contours of the mean Mach number and changes in the shape factor at different frequencies of oscillation at $y = 0.25h_v$. (**a**) Changes in the H at different times. (**b**) Changes in the contour of the mean Mach number.

Figure 25a shows the profiles of velocity near the wall ($z_n/h_v < 0.2$) at the streamwise position $x/h = 4.7$ ($x = 70$ mm) and the spanwise positions $y = 0.25h_v$ and $y = 0.75h_v$. As the frequency of oscillation increased, the profile of the boundary layer near the wall at both spanwise positions gradually became fuller. However, compared with those at $y = 0.25h_v$, the changes in the near-wall velocity were more pronounced at the spanwise position of $y = 0.75h_v$. When the MVG oscillated at a frequency of 300 Hz, the profile of the boundary layer was the fullest, and the near-wall velocity could be increased by up to 8.8% compared with that at $f_1 = 50$ Hz. The vortex generator had the most significant control effect on the flow field at a frequency of 300 Hz. In addition, Figure 25b shows the profiles of velocity near the wall at the streamwise positions of $x/h = 4.1$ and $x/h = 4.7$ (spanwise position, $y = 0.5h_v$). When the dynamic MVG oscillated at a frequency of 300 Hz, the near-wall velocity increased to varying degrees at both streamwise positions, with more significant changes observed at $x/h = 4.7$. In summary, when the frequency of oscillations of the dynamic MVG increased, the near-wall boundary layer at each station obtained more kinetic energy, and this led to a more pronounced control effect.

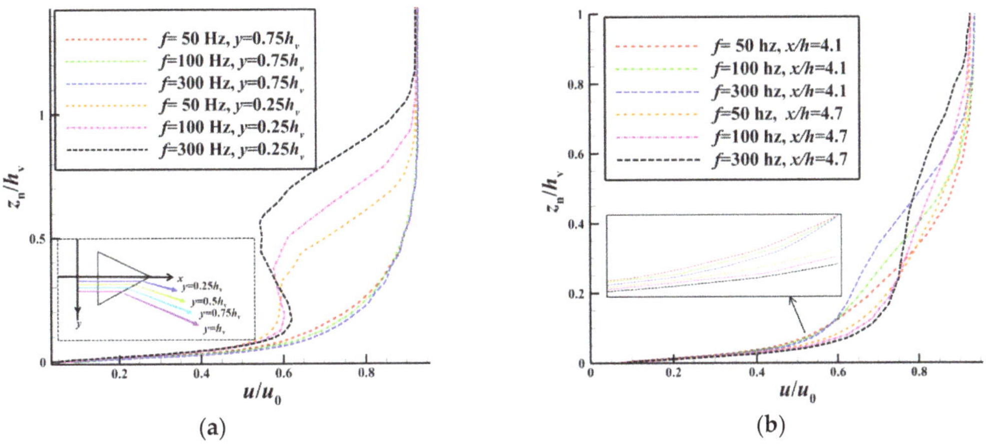

Figure 25. Changes in velocity at different frequencies of oscillation. (**a**) Velocity distributions at different spanwise positions ($x/h = 4.7$). (**b**) Velocity distributions at different streamwise positions ($y = 0.5h_v$).

5. Conclusions

In this study, we proposed a method for controlling the SWBLI in a supersonic inlet by using an array of dynamic MVGs. The aerodynamic feasibility of this design concept was preliminarily verified, and the mechanism of flow and the law of influence of the relevant parameters were analyzed. Our main conclusions are as follows:

(1) The incident shock in the supersonic inlet imposed a strong adverse pressure gradient on the boundary layer, which led to its local thickening and separation. Due to the presence of the side wall of the inlet, vortices that intensified transverse flow were generated near the side wall, leading to a complex 3D structure of flow of the separation bubble. A large separation was formed in the middle of the bottom wall.

(2) The dynamic MVGs induced a vortex structure with variable intensity in the supersonic boundary layer due to their oscillation. This enhanced the mixing of the flow of the boundary layer with the high-speed mainstream flow and caused the profile of the velocity of the separation to become fuller while enhancing the stability of the boundary layer. At the same time, the unique effects of "extrusion" and "suction" of the vortex generators during their oscillation continued to charge the airflow, further enhancing its ability to suppress the separation.

(3) When flow was controlled by the array of dynamic MVGs, the height of the separation bubble in the supersonic inlet decreased more significantly than that in the absence of control. Compared with the case without control, the length of the separation in the streamwise direction decreased by up to 31.76%, and the coefficient of total pressure recovery increased by 6.4%.

(4) When the frequency of the dynamic MVG was in the range of 50–300 Hz, the effect of charging the low-speed airflow near the boundary layer was enhanced as the frequency of oscillations of the vortex generators increased, and the shape factor of the boundary layer decreased by up to 30% at a frequency of 300 Hz compared with that in the absence of control.

Author Contributions: Conceptualization, Y.Z. and M.W.; methodology, M.W., Z.W. and H.T.; validation, M.W. and Y.Z.; formal analysis, M.W.; investigation, M.W.; resources, Y.Z.; data curation, Y.Z. and Z.W.; writing—original draft preparation, M.W.; writing—review and editing, Y.Z., Z.W., D.C., K.W. and S.G.; visualization, K.W. and D.C.; supervision, Y.Z. and Z.W.; project administration, Y.Z.; funding acquisition, Y.Z. All authors have read and agreed to the published version of the manuscript.

Funding: This work was funded by the National Natural Science Foundation of China (Grant Nos. 12025202, 12172175), the Postgraduate Research & Practice Innovation Program of NUAA (Grant No. xcxjh20220203), the Open Fund from Laboratory of Aerodynamics in Multiple Flow Regimes (Grant No. KLYSYS-KFJJ-ZD-2022-02), and the Young Scientific and technological Talents Project of Jiangsu Association for Science and Technology (Grant No. TJ-2021-052).

Institutional Review Board Statement: Not applicable.

Informed Consent Statement: Written informed consent has been obtained from the patient(s) to publish this paper.

Data Availability Statement: The data within the article can be made available upon request.

Conflicts of Interest: The authors declare that they have no known competing financial interests or personal relationships that could have appeared to influence the work reported in this paper.

References

1. Seddon, J.; Goldsmith, E.L. *Intake Aerodynamics*; AIAA Education Series; American Institute of Aeronautics and Astronautics: New York, NY, USA, 1985; pp. 1–30.
2. Holden, M. Historical review of experimental studies and prediction methods to describe laminar and turbulent shock wave/boundary layer interactions in hypersonic flows. In Proceedings of the 44th AIAA Aerospace Sciences Meeting and Exhibit, Reno, NV, USA, 9–12 January 2006.

3. Tan, H.-J.; Li, C.-H.; Zhang, Y.; Li, G.-S. Review of progress in shock control technology with fixed geometry. *J. Propul. Technol.* **2016**, *37*, 2001–2008.
4. Van Wie, D.M. Scramjet inlets. In *Scramjet Propulsion*; Curran, E.T., Murthy, S.N.B., Eds.; American Institute of Aeronautics and Astronautics: Reston, VA, USA, 2000; Volume 189, pp. 447–511.
5. Pan, H.-L.; Li, J.-H.; Shen, Q. Studies of turbulence/shock interaction in a scramjet inlet. *J. Propul. Technol.* **2013**, *34*, 1172–1178.
6. Yuan, H.C.; Liang, D.W. Effect of suction on starting of hypersonic inlet. *J. Propul. Technol.* **2006**, *27*, 525–528.
7. Yan, H.-M.; Zhong, J.-J.; Han, J.-A.; Feng, Z.-M. Research on boundary layer suction in the throat of supersonic inlet. *J. Propul. Technol.* **2009**, *30*, 175–181.
8. Zhao, J.; Fan, X.-Q.; Wang, Y.; Tao, J. Classification of flow field in supersonic boundary layer bleed slot. *J. Propul. Technol.* **2017**, *38*, 2463–2470.
9. Shi, X.; Lyu, M.; Zhao, Y.; Tao, S.; Hao, L.; Yuan, X. Flow control technique for shock wave/turbulent boundary layer interactions. *Acta Aeronaut. Astronaut. Sin.* **2022**, *43*, 625929.
10. Wu, H.; Wang, J.; Huang, W.; Du, Z.; Yan, L. Research progress on shock wave/boundary layer interactions and flow controls induced by micro vortex generators. *Acta Aeronaut. Astronaut. Sin.* **2021**, *42*, 25371.
11. Babinsky, H.; Li, Y.; Pitt Ford, C.W. Microramp control of supersonic oblique shock-wave/boundary-layer interactions. *AIAA J.* **2009**, *47*, 668–675. [CrossRef]
12. Wang, B.; Liu, W.; Zhao, Y.; Fan, X.; Wang, C. Experimental investigation of the microramp based shock wave and turbulent boundary layer interaction control. *Phys. Fluids* **2012**, *24*, 055110.
13. Blinde, P.L.; Humble, R.A.; van Oudheusden, B.W.; Scarano, F. Effects of micro-ramps on a shock wave/turbulent boundary layer interaction. *Shock Waves* **2009**, *19*, 507–520. [CrossRef]
14. Anderson, B.H.; Tinapple, J.; Surber, L. Optimal control of shock wave turbulent boundary layer interactions using micro-array actuation. In Proceedings of the 3rd AIAA Flow Control Conference, San Francisco, CA, USA, 5–8 June 2006.
15. Zhang, Y.; Tan, H.-J.; Du, M.-C.; Wang, D.-P. Control of shock/boundary-layer interaction for hypersonic inlets by highly swept microramps. *J. Propul. Power* **2015**, *31*, 133–143. [CrossRef]
16. Anderson, B.H.; Mace, J.L.; Mani, M. Active "fail safe" micro-array flow control for advanced embedded propulsion systems. In Proceedings of the 47th Aerospace Sciences Meeting, sponsored by the American Institute of Aeronautics and Astronautics, Orlando, FL, USA, 5–8 January 2009.
17. Wagner, J.L. Experimental Studies of Unstart Dynamics in Inlet/Isolator Configurations in a Mach 5 Flow. Ph.D. Thesis, The University of Texas at Austin, Austin, TX, USA, 2009.
18. Giepman, R.H.M.; Schrijer, F.F.J.; Van Oudheusden, B.W. Flow control of an oblique shock wave reflection with micro-ramp vortex generators: Effects of location and size. *Phys. Fluids* **2014**, *26*, 066101. [CrossRef]
19. Wang, M.G.; He, X.M.; Wang, J.J.; Zhang, Y.; Wang, K.; Tan, H.J.; Li, L.G. Shock wave/boundary layer interaction control method based on oscillating vortex generator. *J. Aviat.* **2023**, 1–16. Available online: http://kns.cnki.net/kcms/detail/11.1929.v.20230421.1348.010.html (accessed on 14 August 2023).
20. Lee, S.; Goettke, M.K.; Loth, E.; Tinapple, J.; Benek, J. Microramps upstream of an oblique-shock/boundary-layer interaction. *AIAA J.* **2010**, *48*, 104–118. [CrossRef]
21. Van Leer, B. Towards the ultimate conservative difference scheme. V. A second-order sequel to Godunov's method. *J. Comput. Phys.* **1979**, *32*, 101–136. [CrossRef]
22. White, F.M. *Viscous Fluid Flow*; McGraw-Hill: New York, NY, USA, 1991.
23. Jiao, X.; Chang, J.; Wang, Z.; Yu, D. Numerical study on hypersonic nozzleinlet starting characteristics in a shock tunnel. *Acta Astronaut.* **2017**, *130*, 167–179. [CrossRef]
24. Sekar, K.R.; Karthick, S.K.; Jegadheeswaran, S.; Kannan, R. On the unsteady throttling dynamics and scaling analysis in a typical hypersonic inlet–isolator flow. *Phys. Fluids* **2020**, *32*, 126104. [CrossRef]
25. Huang, H.-X.; Tan, H.-J.; Sun, S.; Ling, Y. Evolution of supersonic corner vortex in a hypersonic inlet/isolator model. *Phys. Fluids* **2016**, *28*, 126101. [CrossRef]
26. Jiao, X.; Chang, J.; Wang, Z.; Yu, D. Mechanism study on local unstart of hypersonic inlet at high Mach number. *AIAA J.* **2015**, *53*, 3102–3112. [CrossRef]
27. Menter, F.R. Two-equation eddy-viscosity turbulence models for engineering applications. *AIAA J.* **1994**, *32*, 1598–1605. [CrossRef]
28. ANSYS Inc. *Fluent Theory Guide*; ANSYS Inc.: Canonsburg, PA, USA, 2018.
29. Wilcox, D.C. *Turbulence Modeling for CFD*; DCW Industries: La Cañada, CA, USA, 1993.
30. Li, N.; Chang, J.T.; Xu, K.J.; Yu, D.R.; Bao, W.; Song, Y.P. Prediction dynamic model of shock train with complex background waves. *Phys. Fluids* **2017**, *29*, 116103. [CrossRef]
31. Jordan, C.; Edwards, J.R.; Stefanski, D.L. Evaluation of RANS closure models using LES datasets for hypersonic shock boundary layer interactions. In Proceedings of the AIAA Scitech 2021 Forum, Virtual, 11–15 January 2021.
32. Bonne, N.; Brion, V.; Garnier, E.; Bur, R.; Molton, P.; Sipp, D.; Jacquin, L. Analysis of the two-dimensional dynamics of a Mach 1.6 shock wave/transitional boundary layer interaction using a RANS based resolvent approach. *J. Fluid Mech.* **2019**, *862*, 1166–1202. [CrossRef]
33. Hong, W.; Kim, C. Computational study on hysteretic inlet buzz characteristics under varying mass flow conditions. *AIAA J.* **2014**, *52*, 1357–1373. [CrossRef]

34. Hulme, I.; Clavelle, E.; Van der Lee, L.; Kantzas, A. CFD Modeling and Validation of Bubble Properties for a Bubbling Fluidized Bed. *Ind. Eng. Chem. Res.* **2005**, *44*, 4254–4266. [CrossRef]
35. Lambourne, N.C.; Landon, R.H.; Zwaan, R.J. *Compendium of Unsteady Aerodynamics Measurements*; AGARD-R-702; Technical Editing and Reproduction Ltd.: London, UK, 1982.
36. Giepman, R.; Srivastava, A.; Schrijer, F.; van Oudheusden, B. The effects of Mach and Reynolds number on the flow mixing properties of micro-ramp vortex generators in a supersonic boundary layer. In Proceedings of the 45th AIAA Fluid Dynamics Conference, Dallas, TX, USA, 22–26 June 2015.
37. Liu, C.Q. Liutex-vortex definition and the third generation vortex recognition method. *Acta Aerodyn. Sin.* **2020**, *38*, 413–431.
38. Kendall, A.; Koochesfahani, M. A method for estimating wall friction in turbulent wall-bounded flows. *Exp. Fluids* **2008**, *44*, 773–780. [CrossRef]
39. Babinsky, H.; Titchener, N. *Shock Boundary Layer Interaction Flow Control with Micro Vortex Generators*; European Office of Aerospace Research and Development: London, UK, 2011.
40. Babinsky, H.; Harvey, J.K. *Shock Wave–Boundary-Layer Interactions*; Cambridge University Press: Cambridge, UK, 2011.

aerospace

MDPI

Article

Mitigation of Shock-Induced Separation Using Square-Shaped Micro-Serrations—A Preliminary Study

Fangyou Yu [1,2], Zhanbiao Gao [1], Qifan Zhang [1,*], Lianjie Yue [1,2] and Hao Chen [1,*]

1 State Key Laboratory of High Temperature Gas Dynamics, Institute of Mechanics, Chinese Academy of Sciences, Beijing 100190, China; yufangyou16@mails.ucas.ac.cn (F.Y.); gaozb@imech.ac.cn (Z.G.); yuelj@imech.ac.cn (L.Y.)

2 School of Engineering Science, University of Chinese Academy of Sciences, Beijing 100049, China

* Correspondence: zhangqifan@imech.ac.cn (Q.Z.); chenhao@imech.ac.cn (H.C.)

Abstract: Suppressing shock-induced flow separation has been a long-standing problem in the design of supersonic vehicles. To reduce the structural and design complexity of control devices, a passive control technique based on micro-serrations is proposed and its controlling effects are preliminarily investigated under test conditions in which the Mach number is 2.5 and the ramp creating an incident shock is 15 deg. Meanwhile, a vorticity-based criterion for assessing separation scales is developed to resolve the inapplicability of the zero skin friction criterion caused by wall unevenness. The simulations demonstrate that the height of the first stair significantly influences the separation length. Generally, the separation length is shorter at higher stairs, but when the height is greater than half of the thickness of the incoming boundary layer, the corresponding separation point moves upstream. A stair with a height of only 0.4 times the thickness of the boundary layer reduces the separation length by 2.69%. Further parametric analysis reveals that while the remaining serrations have limited effects on the flow separation, an optimization of their shape (depth and width) can create more favorable spanwise vortices and offer a modest improvement of the overall controlling performance. Compared to the plate case, a 9.13% reduction in the separation length can be achieved using a slightly serrated design in which the leading stair is 0.1 high and the subsequent serrations are 0.2 deep and 0.05 wide (nondimensionalized, with the thickness of the incoming boundary layer). Meanwhile, the micro-serration structure even brings less drag. Considering the minor modification to the structure, the proposed method has the potential for use in conjunction with other techniques to exert enhanced control on separations.

Keywords: micro-serration; separation control; shock wave/boundary layer interaction

Citation: Yu, F.; Gao, Z.; Zhang, Q.; Yue, L.; Chen, H. Mitigation of Shock-Induced Separation Using Square-Shaped Micro-Serrations—A Preliminary Study. *Aerospace* **2024**, *11*, 148. https://doi.org/10.3390/aerospace11020148

Academic Editor: Sergey Leonov

Received: 17 January 2024
Revised: 7 February 2024
Accepted: 10 February 2024
Published: 12 February 2024

1. Introduction

The shock wave/boundary layer interaction (SWBLI) is a fundamental flow phenomenon involving the complex shock–shock and shock–boundary layer interactions that are common in both external and internal flows, such as flows at external surfaces, the intake and isolators of supersonic and hypersonic vehicles, transonic airfoil surfaces, and other positions. With a strong adverse pressure gradient, the boundary layer in the interaction region will inevitably separate, resulting in a significant energy loss and a lower total pressure recovery [1–3]. At the same time, separation can cause flow oscillations [4,5], excessive thermal loads [6–8], and other dangerous situations. Furthermore, SWBLI can also lead to a thickening of the boundary layer and even prevent the start in the intake of a scramjet engine [9,10]. SWBLI must be considered in aircraft design, and control of SWBLI is conducive to the safe and stable operation of aircraft.

Although significant progress has been achieved in SWBLIs in recent decades, since Ferri [11] first discovered this phenomenon experimentally, SWBLI control, particularly the control of separation, is still a hot topic in the field of aerodynamics. Several techniques have been used for separation control [12]: decrease in the imposed adverse pressure gradient, removal of the low-momentum near-wall flow, or addition of the momentum of the near-wall flow. The specific flow control methods can be categorized into passive control methods and active control methods. Passive control includes the use of a micro-vortex generator (MVG) [13–15], a local wall modification in the form of a bump [16,17], and a backward facing step [18,19]. Active control includes the application of boundary layer bleed/suction ahead of the shock-induced interaction [20,21], steady microjets [22,23], plasma jets [24,25], and spark jets [26]. Active control has a range of advantages, and researchers also prefer complex flow control systems that can be actively controlled with feedback. For example, boundary layer bleed/suction is an effective method of suppressing the separation in both laminar and turbulent boundary layers. However, the bleed system simultaneously dumps considerable amounts of captured airflow to obtain an acceptable control effect, which may lead to the poor aerodynamic performance and reduced propulsion efficiency of the aircraft. For example, engines equipped with active control may result in an intake with increased weight and aerodynamic drag [27] that cannot compensate for the induced loss. Thus, the current economical, structurally simple, and safe method for practical use is still based on passive control.

To date, the micro-vortex generator (MVG) has been widely used for passive control; it can reduce the size of the separation zone by 10~30% after design optimization [28,29]. The height of the MVG is approximately 10~70% of the boundary layer. The streamwise vortex pair generated by the wake of the MVG transports the high-energy airflow in the upper boundary layer into the bottom layer and mixes it with the low-energy airflow to increase the momentum of the low-velocity region near the wall, enhancing its resistance to the adverse pressure gradient and realizing the control of the separation of the boundary layer. Even though the vortices induced by the MVG can reconfigure the downstream boundary layer and the downstream shock wave shows clear deformation or even degeneration under the action of the vortices, the separation of the boundary layer remains severe at other locations in the spanwise direction due to the limited influence range of the vortices induced by MVGs. In addition, the MVG is far from capable of controlling the strong SWBLIs that occur in a finite-width channel [30]. To maximize the performance of the MVG, it is usually necessary to combine and rearrange a series of MVGs [29,31] and place them in suitable positions [32]. Because MVGs must be designed carefully for different configurations with different sizes, arrangements, and mounting positions, an alternative passive control method should be developed.

In this paper, a new technique using micro-surface serrations is developed to weaken shock-induced separations. This approach uses a simple structure and does not make any special demands on installation space. The primary objective of this work is to examine the effectiveness of this concept and to determine the basic effects of the key geometric parameters. The rest of the paper is organized as follows. First, the numerical method and its validation are introduced in Section 2. Then, some details about the control effect of the micro-serrations are described, including an alternative method to determine the location of the separation zone, the influence of the first windward stair, and the influence of the subsequent micro-serrations. A concise conclusion is provided in the final section.

2. Methodology

2.1. Description of the Computational Domain and Boundary Conditions

In this study, the wall micro-serration configuration is composed of a windward stair and subsequent micro-serrations. To capture the main flow characteristics, the computational domain for the numerical simulation is simplified to a two-dimensional case, as shown in Figure 1. The entire domain is filled with a structured mesh. The micro-serrations start at $x = 0$ and are arranged in a rectangular concave–convex shape with a length of 100, as illustrated in Figure 2. In addition, to eliminate the influence of the first windward stair on the incoming flow, a smooth plate with a length of 10 is set before the initial position. The height of the computational domain is 25. All the variables related to length are dimensionless and are given in terms of the thickness of the incoming boundary layer, except where specifically noted otherwise. The thickness of boundary layer used in the study is 2 mm.

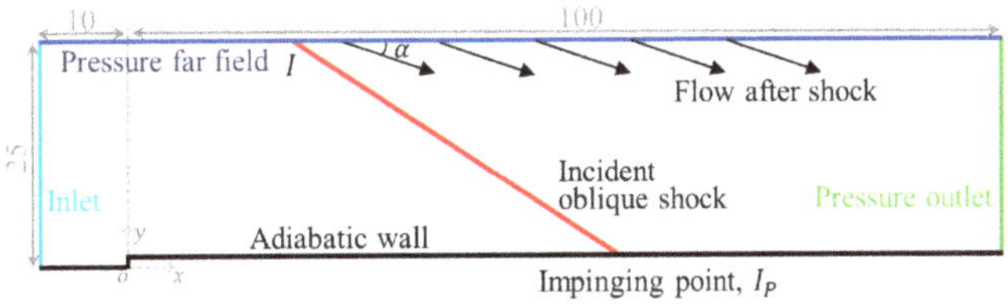

Figure 1. Computational domain and boundary conditions.

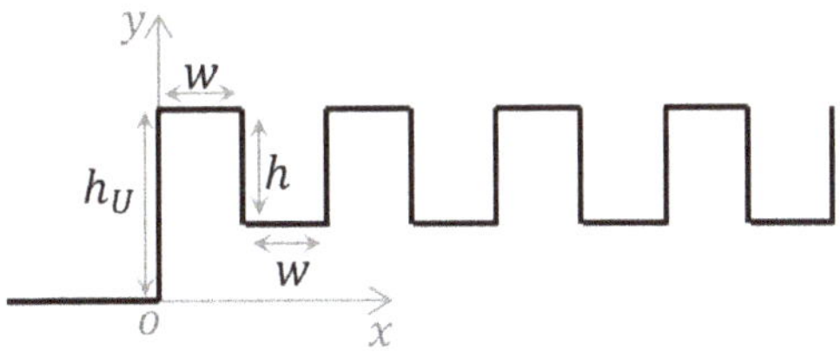

Figure 2. Schematic diagram of micro-serrations.

As micro-serrations are described by many parameters, Table 1 provides a summary description of the relevant symbols. For example, h_U0 represents a smooth plate, and $h_U0.1$ is a small stair of height 0.1. Without loss of generality, if either w or h are equal to zero, these parameters are not labeled when naming the simulation case.

Table 1. Some parameters related to SWBLIs in this study.

Parameters	Explanations
M_0	Incoming Mach number
h_U	Dimensionless height of the first windward stair at $x = 0$
w	Dimensionless width of the micro-serration
h	Dimensionless depth of the micro-serration
α	The deflection angle of incident shock (deg.)
I_P	The impinging point of the inviscid incident shock at the bottom wall ($x = I_P, y = 0$)

Prior to the simulation, the flow field of a smooth plate is calculated with the same settings, and the corresponding boundary layer thickness section is taken from the plate case as the inlet boundary condition of the subsequent cases so that the initial flow field is obtained. The outlet is set as the pressure outlet condition, the upper boundary is set as the pressure far field condition, and the lower boundary is set as an adiabatic no-slip wall. For all cases, the incoming stagnation pressure p_{t0} is 101,325 Pa, and the stagnation temperature T_{t0} is 300 K.

Instead of a traditional geometric shock generator, the incident shock is generated by setting discontinuous aerodynamics parameters upstream and downstream of point I on the upper pressure far field boundary [33–35]. Specifically, the conditions upstream of point I are set as the incoming flow parameters, i.e., M_0, p_0, and T_0, and the conditions downstream of point I are set as the post-shock parameters with a deflection angle α. In this way, based on the initial flow field, the airflow can be deflected from the upper boundary, thus achieving the purpose, which is to generate an incident shock. Compared to the physical geometric shock generator, the use of this pneumatic shock generator prevents the separation shock of the large-scale separation zone from impinging on the wedge surface of the geometric generator and reflecting. This reflected shock may act on the separation zone on the lower wall again, thus destroying the main flow structure of the SWBLIs [35]. In addition, it can also reduce the complexity of the mesh and improve the computational efficiency.

2.2. Numerical Method

All the calculations in this study are based on the two-dimensional Reynolds-averaged Navier–Stokes (RANS) solver. For the calculation of the flow field, the turbulence model is k-ω shear stress transport (SST), which has been successfully applied to supersonic flows [36–39]. The fluid is an ideal gas model and is processed as calorically perfect air. Meanwhile, the viscosity coefficient is calculated according to the Sutherland formula. The Roe FDS scheme is utilized for vector flux splitting. As for the spatial discretization, a second-order upwind scheme is used for the gradient term, the flow term, the turbulent kinetic energy term, and the specific dissipation rate term.

The equations to be solved are as follows:

Continuity equation:

$$\frac{\partial(\rho u_i)}{\partial x_i} = 0. \tag{1}$$

Momentum equation:

$$\frac{\partial(\rho u_i u_j)}{\partial x_j} = -\frac{\partial p}{\partial x_i} + \frac{\partial \tau_{ij}}{\partial x_j}, \tag{2}$$

$$\tau_{ij} = \mu\left(\frac{\partial u_i}{\partial x_j} + \frac{\partial u_j}{\partial x_i}\right) - \frac{2}{3}\mu\frac{\partial u_i}{\partial x_i}\delta_{ij}. \tag{3}$$

Energy equation:

$$\frac{\partial\left[u_j(\rho E + p)\right]}{\partial x_j} = \frac{\partial\left(k_t \frac{\partial T}{\partial x_j}\right)}{\partial x_j} + \frac{\partial(\tau_{ij} u_i)}{\partial x_j}, \tag{4}$$

where ρ is the density, u_i and u_j are velocity components, p is the pressure, T is the temperature, E is the total energy per unit mass of fluid, τ_{ij} is the viscous stress tensor, and k_t is the heat conduction coefficient. The Einstein summation convention and Kronecker operator δ_{ij} are used in the above equations.

Menter improved the standard k-ω model and first proposed the SST turbulence model [40]. The SST model combines the k-ω turbulence model and the k-ε turbulence model, which are suitable for solving the turbulence near the wall and the free-shear turbulence far away from the wall, respectively. SST realizes a more accurate prediction of the separated flow and the boundary layer flow under the adverse pressure gradient. The SST turbulence model is given by

$$\frac{\partial(\rho u_j k)}{\partial x_j} = \frac{\partial}{\partial x_j}\left(\Gamma_k \frac{\partial k}{\partial x_j}\right) + G_k - \rho\beta^* f_{\beta^*} k\omega, \tag{5}$$

and

$$\frac{\partial(\rho u_j \omega)}{\partial x_j} = \frac{\partial}{\partial x_j}\left(\Gamma_\omega \frac{\partial \omega}{\partial x_j}\right) + D_\omega + G_\omega - \rho\beta f_\beta \omega^2. \tag{6}$$

In these equations, G_k represents the generated turbulence kinetic energy due to mean velocity gradients, G_ω represents the generation of ω, and Γ_k and Γ_ω represent the effective diffusivity of k and ω, respectively. The last terms in Equations (5) and (6) represent the dissipation of k and ω due to turbulence. D_ω represents the cross-diffusion term which is defined as

$$D_\omega = 2(1 - F_1)\rho\sigma_{\omega 2}\frac{1}{\omega}\frac{\partial k}{\partial x_j}\frac{\partial \omega}{\partial x_j}. \tag{7}$$

The effective diffusivities for the k-ω model are given by

$$\left.\begin{array}{l}\Gamma_k = \mu + \frac{\mu_t}{\sigma_k}, \\ \Gamma_\omega = \mu + \frac{\mu_t}{\sigma_\omega},\end{array}\right\} \tag{8}$$

where σ_k and σ_ω are the turbulent Prandtl numbers for k and ω, respectively, which are defined as follows:

$$\left.\begin{array}{l}\sigma_k = \frac{1}{F_1/\sigma_{k1}+(1-F_1)/\sigma_{k2}}, \\ \sigma_\omega = \frac{1}{F_1/\sigma_{\omega 1}+(1-F_1)/\sigma_{\omega 2}},\end{array}\right\} \tag{9}$$

where

$$F_1 = \tanh\left(\Phi_1^4\right), \tag{10}$$

$$\Phi_1 = \min\left[\max\left(\frac{\sqrt{k}}{0.09\omega y}, \frac{500\mu}{\rho y^2 \omega}\right), \frac{4\rho\sigma_{\omega 2}k}{D_w^+ y^2}\right], \tag{11}$$

$$D_w^+ = \max\left(2\rho\sigma_{\omega 2}\frac{1}{\omega}\frac{\partial k}{\partial x_j}\frac{\partial \omega}{\partial x_j}, 10^{-20}\right). \tag{12}$$

The turbulent viscosity μ_t is computed using k and ω as follows:

$$\mu_t = \frac{\rho k}{\omega}\frac{1}{\max\left(\frac{1}{a^*}, \frac{SF_2}{a_1\omega}\right)}, \tag{13}$$

$$F_2 = \tanh\left(\Phi_2^2\right), \tag{14}$$

$$\Phi_2 = \max\left(2\frac{\sqrt{k}}{0.09\omega y}, \frac{500\mu}{\rho y^2\omega}\right), \tag{15}$$

where S is the vorticity magnitude.

The corresponding generation of k and ω is described by

$$\left.\begin{array}{l}G_k = \mu_t S^2, \\ G_\omega = a\frac{\omega}{k}G_k.\end{array}\right\} \tag{16}$$

The coefficient a is given by

$$a = \frac{a_\infty}{a^*}\left(\frac{a_0 + Re_t/R_\omega}{1 + Re_t/R_\omega}\right), \tag{17}$$

where

$$Re_t = \frac{\rho k}{\mu\omega}. \tag{18}$$

In the dissipative terms of k and ω,

$$f_{\beta^*} = \begin{cases} 1, & \chi_k \leq 0, \\ \frac{1+680\chi_k^2}{1+400\chi_k^2}, & \chi_k \geq 0, \end{cases} \tag{19}$$

$$f_\beta = \frac{1+70\chi_\omega}{1+80\chi_\omega}, \tag{20}$$

$$\left.\begin{array}{l}\chi_k \equiv \frac{1}{\omega^3}\frac{\partial k}{\partial x_j}\frac{\partial \omega}{\partial x_j},\\ \chi_\omega = \left|\frac{\Omega_{ij}}{(0.09\omega)^3}\right|,\end{array}\right\} \tag{21}$$

$$\Omega_{ij} = \frac{1}{2}\left(\frac{\partial u_i}{\partial x_j} - \frac{\partial u_j}{\partial x_i}\right), \tag{22}$$

$$\left.\begin{array}{l}\beta^* = \beta_i^*[1 + 1.5F(M_t)],\\ \beta_i^* = \beta_\infty^* \frac{\frac{4}{15}+(Re_t/8)^4}{1+(Re_t/8)^4},\end{array}\right\} \tag{23}$$

$$\left.\begin{array}{l}\beta = \beta_i\left[1 - 1.5\frac{\beta_i^*}{\beta_i}F(M_t)\right],\\ \beta_i = F_1\beta_{i,1} + (1 - F_1)\beta_{i,2}.\end{array}\right\} \tag{24}$$

The compressibility function $F(M_t)$ improves the applicability of the model in free-shear flow at high Mach numbers, and the expressions are given by

$$F(M_t) = \begin{cases} 0, & M_t \leq M_{t0}, \\ M_t^2 - M_{t0}^2, & M_t \geq M_{t0}, \end{cases} \tag{25}$$

$$\left.\begin{array}{l}M_t^2 \equiv \frac{2k}{c^2},\\ c = \sqrt{\gamma RT}.\end{array}\right\} \tag{26}$$

The constants in the above expressions are [40,41]: $a^* = 1$, $a_\infty = 0.52$, $a_0 = 1/9$, $a_1 = 0.34$, $R_\omega = 2.95$, $M_{t0} = 0.3$, $\beta_{i,1} = 0.075$, $\beta_{i,2} = 0.0828$, $\beta_\infty^* = 0.09$, $\sigma_{k1} = 1.176$, $\sigma_{\omega 1} = 2.0$, $\sigma_{k2} = 1.0$, and $\sigma_{\omega 2} = 1.168$. More details can be found in References [40,41].

2.3. Code Validation

Code validation is conducted by comparison with the experimental results obtained by Grossman and Bruce [42]. The simulations match the experimental free-stream conditions with an incoming Mach number of 2.0 and a unit Reynolds number of approximately 2.0×10^7 m^{-1}. The deflection angle of the oblique incident shock is 12°. Figure 3 presents the comparison between the experimental and numerical surface pressure distributions, where the zero of the x-axis, named "CoordinateX", is the impinging point of the inviscid shock wave and the y-axis is the surface pressure ratio based on the incoming static pressure. The curve obtained by simulation is essentially in agreement with the experimental results, including the starting point of the pressure jump, which suggests the initial position of the separation. It should be noted that the pneumatic shock generator, as mentioned above, is used to generate the incident shock. Therefore, the expansion wave emitting from the end of the wedge shock generator is not considered in this validation; it is also possible that, due to this reason, the pressure obtained from the simulation differs from the experimental results near the reattachment point.

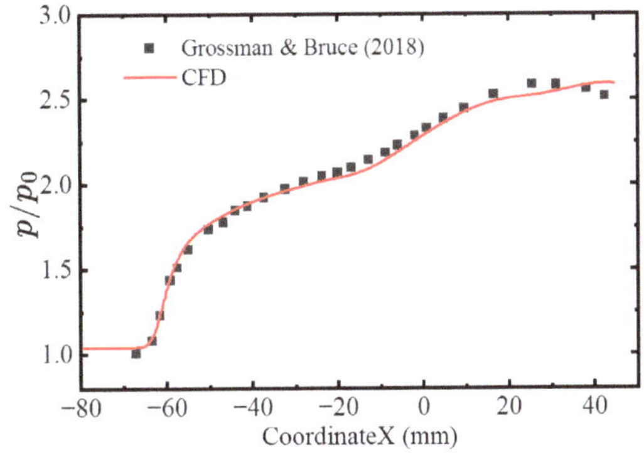

Figure 3. Comparison between experimental [42] and numerical surface pressure distributions.

Due to the complexity of the micro-serrations of the lower wall of the computational domain and to save computing resources, we adopted automatic mesh adaptation to capture as much of the flow information as possible. After a certain number of iterations, the cells, whose density gradient is larger than $d\rho_0$, will be refined once. In our study, $d\rho_0$ is 0.002 with an incoming Mach number of 2.5. Figure 4 shows the surface pressure distributions near the separation point with different refinement levels. The corresponding refinement levels are 0, 2, 4, and 5 and are called the original grid, level 2, level 4, and level 5, respectively. A total of 1000×400 grids are used originally. The curves from level 4 and level 5 overlap, suggesting that the simulation converges after four refinements. Under these circumstances, the streamwise size of the refined cells is less than 1.5×10^{-4} m. All of the other cases meet this cell size limitation after refining, and the final number of grid points is approximately 2×10^6. In addition, a slat with a length of 0.1 mm has 7 grid points, and there are more grid points when the slat size is larger.

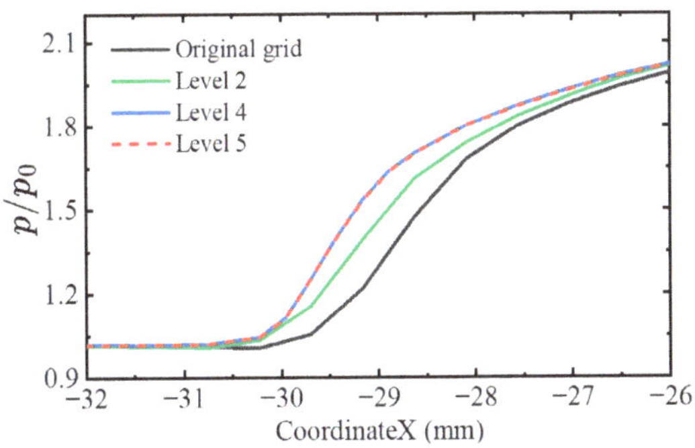

Figure 4. Surface pressure distributions with different refinement levels (M_0 = 2.5, h_U = 0, α = 15°, I_P = 65).

3. Results and Discussion

3.1. Vorticity-Based Criterion for Separation Assessment

For conventional configurations, i.e., plate or curved surfaces, the skin friction coefficient C_f can be utilized to determine the separation and reattachment point of the separation zone [43–45]. However, the wall surface with micro-serrations is not geometrically continuous; therefore, the method based on C_f is no longer applicable, and a new basis for determining the separation zone must be considered. The analysis of the flow field of SWBLI with micro-serrations found that a sign inversion of the vorticity magnitude gradient appears near the separation point and reattachment point, and this parameter does not depend on the wall profile. Therefore, it can be inferred that the location of the separation zone can be determined by the vorticity magnitude along a line close to the wall.

To test the applicability of vorticity magnitude in determining the location of the separation, the obtained results are compared to the numerical results for a typical smooth plate (M_0 = 2.5, h_U = 0, α = 15°, I_P = 65). It is observed from the vorticity magnitude contours shown in Figure 5 that when the near-wall streamline is deflected, particularly in the case when the streamline is perpendicular to the wall, the vorticity magnitude is exactly near the extreme point, and this trend is synchronized, providing a preliminary verification of the feasibility of the new method. For a further quantitative evaluation, the C_f along the lower wall is compared in Figure 6 with the vorticity magnitude distribution curve. It is easy to find that the vorticity magnitude takes the minimum value which has an essentially one-to-one correspondence with the location of the zero position of C_f; the resulting difference in the length of the separation zone is only approximately 1.14%, as shown in Table 2. This indicates that the minimum value of the vorticity magnitude can be used as an alternative method to determine the location of the separation zone in this study.

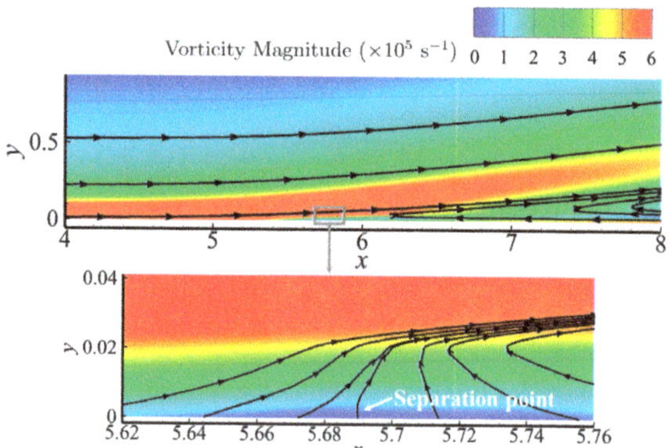

Figure 5. Computed contour plots of vorticity magnitude and streamline distributions near the separation point (M_0 = 2.5, h_U = 0, α = 15°, I_P = 65).

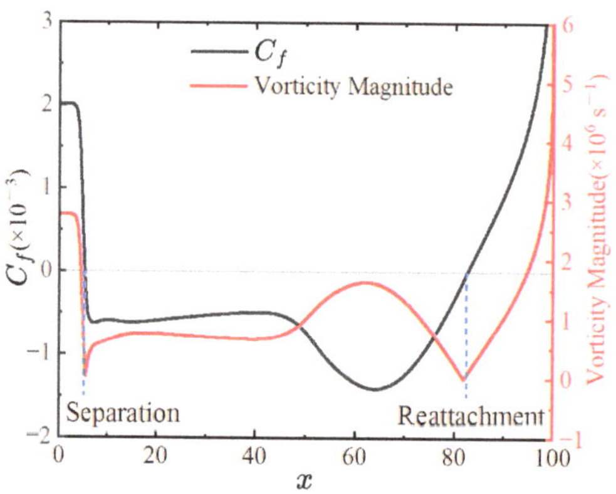

Figure 6. Comparison of surface C_f and vorticity magnitude distribution (M_0 = 2.5, h_U = 0, α = 15°, I_P = 65).

Table 2. Comparison of the separation zone determined by C_f and vorticity magnitude.

	Separation Point	Reattachment Point	Separation Length
C_f	5.32	82.53	77.21
Vorticity magnitude	5.69	82.02	76.33

3.2. *Effects of a Single Stair on Shock-Induced Separation*

The wall profile shown in Figure 2 contains a series of convex structures, of which the first stair has the most direct effect on the flow. Because several geometrical parameters describe the micro-serration, the purely stair configurations, i.e., the cases where both w and h are 0, are first studied in detail to obtain preliminary information about the effect of the micro-serration on the separation.

Figure 7 illustrates the numerical results for several stair configurations, with the h_U values of 0, 0.1, 0.2, 0.4, 0.7, 1.0, 2.0, and 4.0. The left y-axis is the separation length (L_s), and the right y-axis represents the position of the separation point compared to that in the plate case (Δx_s). It should be noted that L_s and Δx_s are given relative to the thickness of the incoming boundary layer and are therefore dimensionless. It is observed from Figure 7 that when the height of the small stair is $h_U < 0.4$, the separation length decreases rapidly and reaches a minimum at $h_U = 0.4$, which is approximately half of the thickness of the incoming boundary layer, and the separation point moves downstream in this range. Then, with increasing h_U, the separation length first increases slightly and then decreases. At the same time, the separation point starts to move upstream. Interestingly, the separation length does not vary much when the height of the first stair is close to the thickness of the incoming boundary layer, that is, $0.5 < h_U < 2.0$.

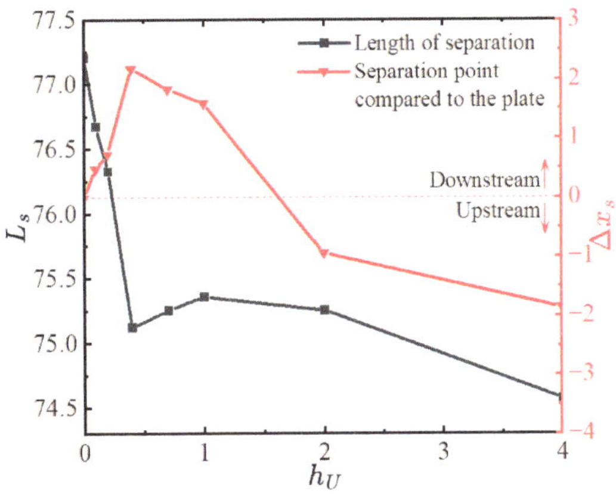

Figure 7. Separation length and location of the separation point ($M_0 = 2.5$, $\alpha = 15°$, $I_P = 65$).

Figure 8a,b show the typical flow field of a pure stair configuration. A small separation that is similar to a pneumatic wedge is generated at the front of the stair. Due to the compression of this small pneumatic wedge, a weak shock wave is formed at the stair and intersects with the incident oblique shock, weakening the intensity of the incident shock to some extent. Additionally, it is also observed from Figure 8c that the local adverse pressure gradient near the separation point is reduced due to the pre-pressurization effect of the compression surface on the windward side. These two effects result in a reduction in the separation length under the influence of the stair. The height of the initial point of the streamlines in Figure 8a,b is equal to the thickness of the incoming boundary layer. After passing through the stair, the height of this streamline in $h_U1.0$ increases by 9.5% compared to $h_U0.4$, which can also be considered as an increase in the thickness of the boundary layer in front of the separation point. The thicker boundary layer attenuates the above two beneficial effects of separation control to some extent. Therefore, the separation length increases from h_U = 0.4 to h_U = 1.0 and the separation point moves upstream. However, the length of the separation zone of the stair configuration is generally smaller than that of the smooth plate.

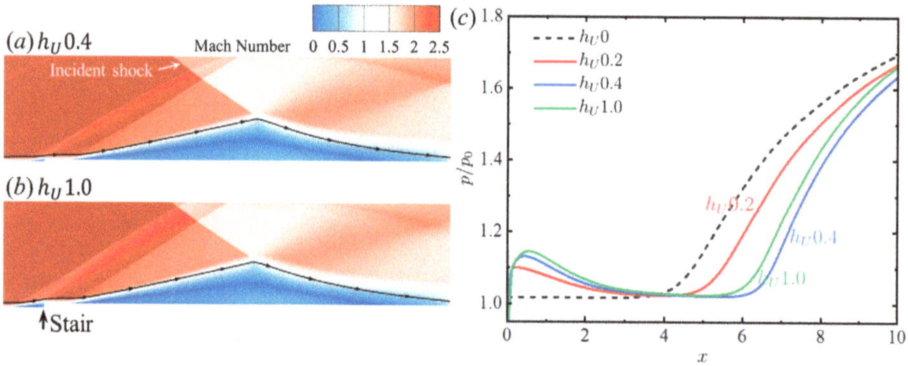

Figure 8. Contours of the Mach number distributions with stair heights of (**a**) 0.4 and (**b**) 1.0 and (**c**) surface pressure distributions near the separation point (M_0 = 2.5, α = 15°, I_P = 65).

3.3. *Effects of Serration Size on Shock-Induced Separation*

The above results indicate that the windward stair has a nonnegligible influence on the separation length, particularly for relatively large stair heights. Thereafter, based on the different effects of the size of h_U on the separation length, the control effect of the wall micro-serration on the separation is investigated by taking h_U equal to 0.1, 0.4, 1.0, and 4.0 as examples for the three typical cases in which h_U is less than, equal to, and greater than the thickness of the incoming boundary layer, respectively.

Figure 9 shows the pressure distributions for the different heights of the windward stair and the corresponding micro-serration configurations of equal size under an incoming Mach number of 2.5. The pressure value is taken from the line that is adjacent to the upper surface of the micro-serration, i.e., $y = h_U$. For comparison, the pressure of the plate is also shown in the figure. It is observed that the onset of the pressure disturbance in the pressure distribution curve of the micro-serration configuration at h_U = 0.1 clearly lags behind those of the plate and the pure stair. For h_U = 0.4, 1.0, or 4.0, if the pressure fluctuation generated by the successive serrations of the wall microstructure is ignored, the pressure distribution curves basically coincide with that of the pure stair, and the pressure is slightly higher than that of the plate. In addition, a significant drop in the pressure for the micro-serration configuration at x = 4 is observed in Figure 9d, which is due to the end of the plateau at x = 4 and the subsequent larger notch. Furthermore, Figure 10 presents the pressure distributions of the micro-serration configurations with different depths and widths at h_U = 0.4. The pressure profiles of the micro-serration configurations strongly resemble the pressure profile of the pure stair, indicating the negligible influence from subsequent micro-serrations on the separation length when the height of the first stair is approximately half of the thickness of the incoming boundary layer.

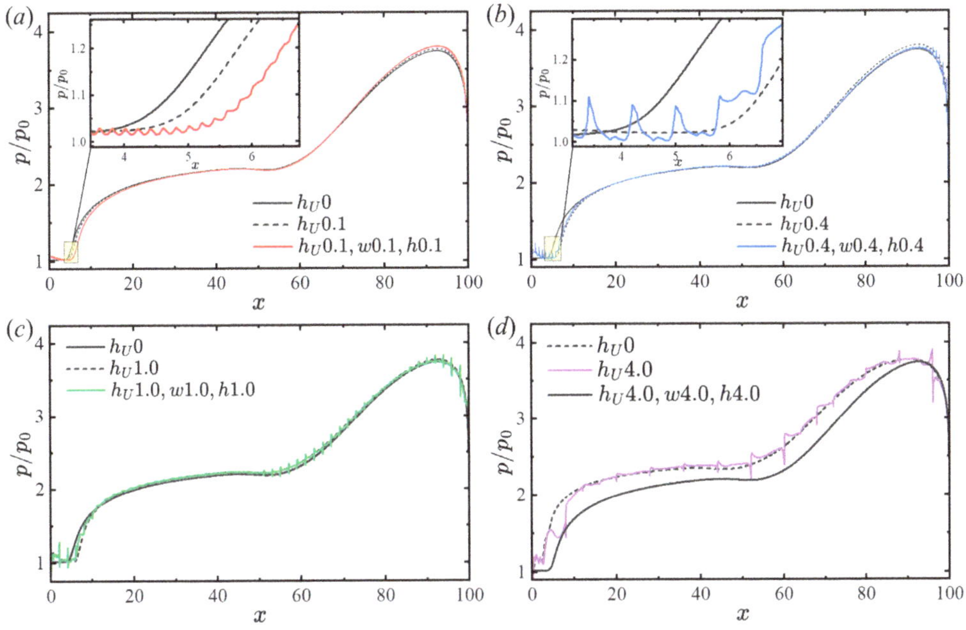

Figure 9. Comparison of the pressure distributions for the stair and micro-serration configurations at typical sizes (**a**) 0.1, (**b**) 0.4,(**c**) 1.0 and (**d**) 4.0 (M_0 = 2.5, α = 15°, I_P = 65).

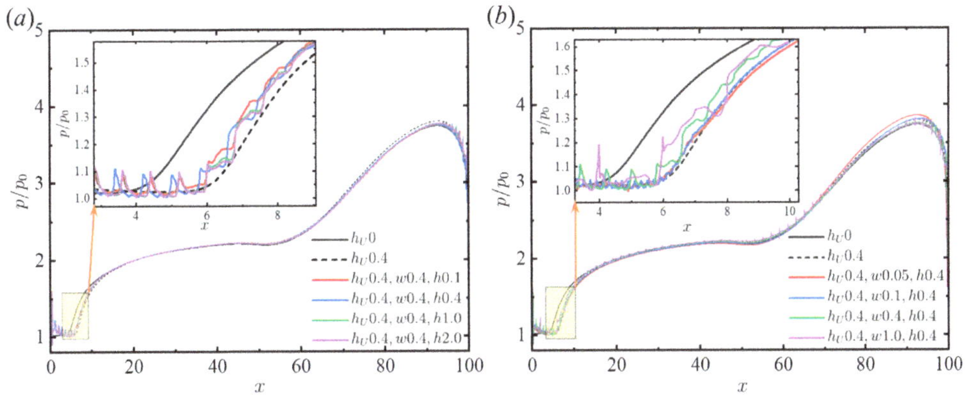

Figure 10. Pressure distributions of micro-serration configurations with different (**a**) depths and (**b**) widths at h_U = 0.4 (M_0 = 2.5, α = 15°, I_P = 65).

The above results suggest that micro-serration plays a certain role in the control of separation length. On the one hand, from the point of view of the pressure distribution, the large-scale micro-serration configuration is indistinguishable from that of the pure stair. On the other hand, the small-scale micro-serration configuration, although close to the effect of the pure stair, still produces a visible improvement in the control effect, indicating that, with the exception of the first stair, the subsequent micro-serrations still contribute to the flow control. Therefore, the influence laws of other parameters of micro-serrations are further investigated for small scales.

First, the influence of the depth of the micro-serrations in small-scale configurations is examined. Taking h_U = 0.1 as an example, Figure 7 shows that this height of the stair has a slight influence on the separation length and the incoming boundary layer. The pressure distributions are shown in Figure 11a, with w = 0.1 unchanged and h = 0.1, 0.2, 0.3, and 1.0. As shown in the locally enlarged figure, the separation length, L_s, is the smallest at h = 0.2, which is reduced by 4.56% compared with the smooth plate. As h continues to increase, the onset of pressure disturbance moves upstream, and L_s increases concurrently. The comparison of this series of cases, which is summarized in Figure 11b, verifies that L_s decreases as h increases within a specific range, beyond which the micro-serrations become less effective in controlling the separation length.

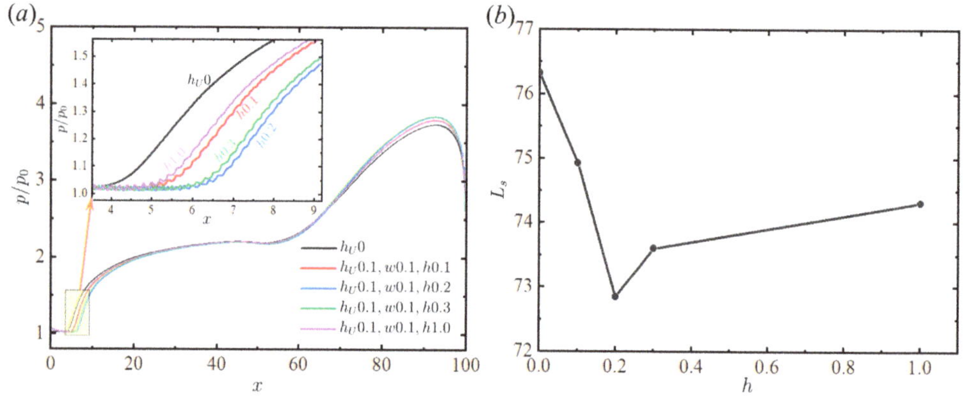

Figure 11. (**a**) Pressure distributions and (**b**) separation length of micro-serration configurations with different depths at h_U = 0.1 (M_0 = 2.5, α = 15°, I_P = 65).

Next, based on the results of Figure 11, the pressure distributions for the depth h = 0.2 and the widths w of 0.025, 0.05, 0.1, and 0.2, are shown in Figure 12a. For comparison, the pressure curve of h_U0.1 is also shown in Figure 12a. The locally enlarged figure demonstrates that the onset of pressure disturbance for w = 0.05 is clearly located further downstream, where it exhibits better control of L_s than w = 0.025, 0.1, and 0.2, resulting in a reduction of 9.13% compared to the smooth plate. Even taking into account the 2.69% error mentioned above, this reduction is still appreciable when compared to the original scale of the entire separation zone. To summarize, there exists an optimal width of micro-serrations that yields the best effect on the control of the separation length, and deviations from this optimal width that make it either too large or too small will tend towards the effect of the windward stair of the corresponding size.

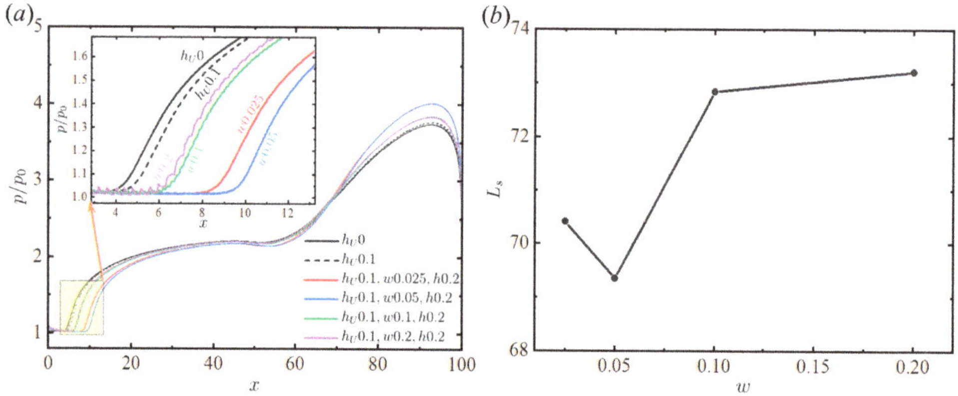

Figure 12. (**a**) Pressure distributions and (**b**) separation length of micro-serration configurations with different widths at h_U = 0.1 (M_0 = 2.5, α = 15°, I_P = 65).

The drag is also evaluated for the two-dimensional (2D) cases with a default spanwise distance of 1 m. Taking the plate and h_U0.1-w0.05-h0.2 as examples, it is assumed that there is an incoming flow condition on the other side of the wall (plate or micro-serration configuration), which is closer to the actual situation. Compared with 1.859 N for the plate, the drag of the micro-serration is reduced to 1.503 N. It follows that the micro-serration configuration does not introduce additional resistance.

To further explore the role of micro-serrations in separation control, some typical flow fields from the above cases are investigated. Figure 13a displays the flow image near the first stair, revealing a series of vortices formed within the micro-serration. The presence of these vortices entrains high-momentum fluid, locally increasing the velocity near the wall, which is advantageous for separation control, as shown in Figure 13b. Interestingly, the vortices detach from the micro-serration before the separation point (Figure 14). These detaching vortices may increase the capability of the configuration to resist an adverse pressure gradient, as demonstrated in Figure 13c. In addition, as mentioned in the previous section, the effect of these micro-serrations is limited, and the thickening of the boundary layer brought about by the increase in h_U will counteract the factors that are beneficial for separation control.

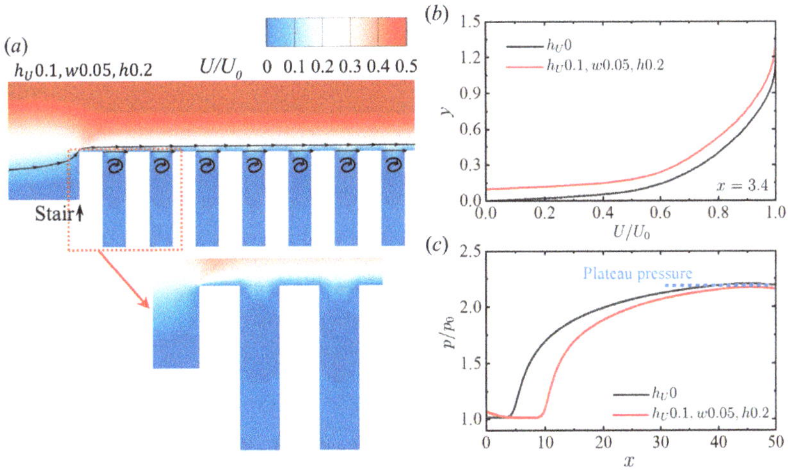

Figure 13. Role of vortices in separation control. (**a**) The vortices in the micro-serrations. (**b**) Comparison of velocity profiles upstream of the separation point. (**c**) Pressure distributions before the first pressure plateau (M_0 = 2.5, α = 15°, I_P = 65).

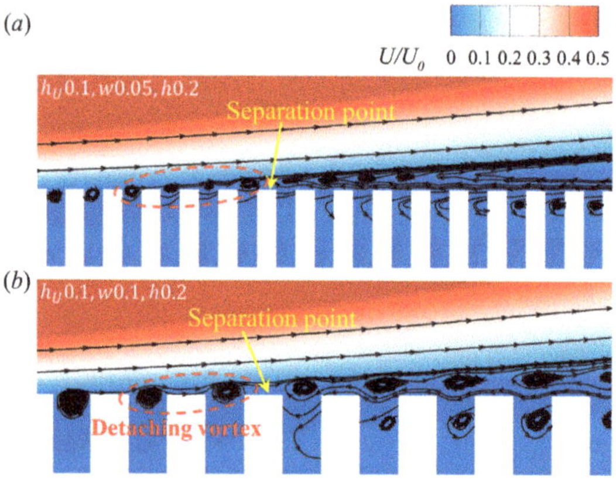

Figure 14. Vortices upstream of the separation point with different widths of the micro-serration: (**a**) $w = 0.05$, (**b**) $w = 0.1$ ($M_0 = 2.5$, $\alpha = 15°$, $I_P = 65$).

Finally, some speculations are made regarding the origins of the influence of the depth and width of the micro-serration. The impact of the depth of the micro-serration is mainly reflected in the position of the vortex within the micro-serration, as shown in Figure 15. The vortex develops in different locations for different h. A shallower position of the vortex, that is, a vortex that is in closer proximity to the upper surface of the micro-serration, is detached more easily. However, when the vortex is in a deep position, it cannot directly act on the boundary to bring high-momentum fluid. The development in both directions of h results in an optimal depth of the micro-serration. The width of the micro-serration exhibits a similar behavior, as depicted in Figure 14. The width directly determines the number of detaching vortices in front of the separation point, which to some extent represents the resistance that these vortices can provide to the adverse pressure gradient. The vortices in the micro-serration cannot be fully developed when w is too small, and the wall profile tends towards that of the plate when w is too large, which is also reflected in the pressure distributions in Figure 12a.

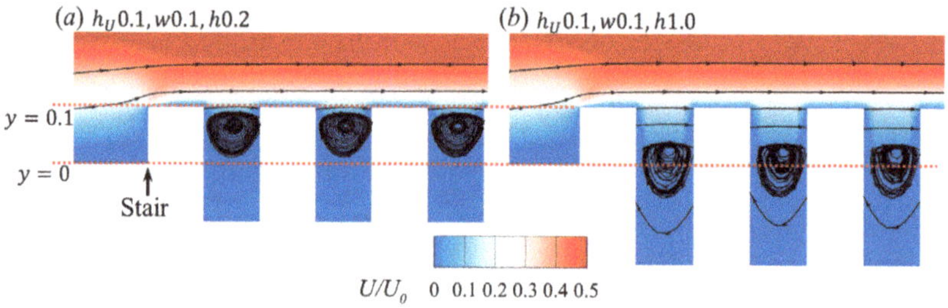

Figure 15. Location of vortices in the micro-serration with different depths: (**a**) $h = 0.2$ and (**b**) $h = 1.0$; the micro-serration is not fully displayed ($M_0 = 2.5$, $\alpha = 15°$, $I_P = 65$).

4. Conclusions

A new separation control method is proposed based on a square-shaped micro-serration configuration, and its control effects are examined in a preliminary investigation. Due to the geometrical discontinuity of the micro-serration, it is imperative to establish a new criterion for the determination of the separation zone. The differences between the separation zone determined by C_f and the vorticity magnitude in the plate case are compared, and it is observed that both the separation point and the reattachment point correspond to the minimum vorticity magnitude. The separation length, which is determined by the vorticity magnitude, aligns closely with the separation location found by $C_f = 0$ and can thus serve as a reliable basis for the subsequent quantification of the separation zone. First, the impact of the windward stair on the separation length is investigated; then, three representative sizes are chosen based on the findings from the small stairs to examine the controlling effects of micro-serration with varying sizes. The results show that the height of the stair significantly affects the separation length due to the pre-pressurization effect and the weak shock caused by the windward stair. Generally, the separation length is shorter with high stairs, whereas when the height of the stair exceeds a specific range, the separation point moves upstream, which is unfavorable for separation control. Additionally, when the height of the stair is less than half of the thickness of the incoming boundary layer, the micro-serration plays the dominant role in determining the separation control. However, when the height of the stair is large, the micro-serration becomes essentially ineffective. It is observed from the pressure distributions that the pressure curves of the micro-serration configurations coincide with that of the stair with the corresponding size. Based on this premise, a further investigation is conducted into the potential impacts of other parameters associated with micro-serrations in small-scale configurations. It is found that both the depth and width of the micro-serration exhibit significant effects arising from variations in the vortex positioning and the number of detached vortices. Relative to the plate, the separation length can be reduced by 9.13% using a slightly serrated design with less influence on the incoming boundary layer. In future applications that consider the minor modification of the structure, the micro-serration can be employed together with other techniques, such as MVG, to achieve better control.

Author Contributions: Conceptualization, Q.Z. and H.C.; methodology, F.Y. and H.C.; software, H.C.; validation, Z.G.; formal analysis, F.Y.; investigation, F.Y.; resources, H.C.; data curation, Q.Z. and Z.G.; writing—original draft preparation, F.Y.; writing—review and editing, H.C.; visualization, L.Y.; supervision, H.C.; project administration, L.Y.; funding acquisition, Q.Z. All authors have read and agreed to the published version of the manuscript.

Funding: This research was funded by National Natural Science Foundation of China, grant numbers 12102440, U2141220, and 11902325.

Data Availability Statement: The data supporting the findings of this study are available from the author upon reasonable request.

Conflicts of Interest: The authors declare no conflicts of interest.

References

1. Huang, W.; Du, Z.B.; Yan, L.; Moradi, R. Flame propagation and stabilization in dual-mode scramjet combustors: A survey. *Prog. Aerosp. Sci.* **2018**, *101*, 13–30. [CrossRef]
2. Pasquariello, V.; Grilli, M.; Hickel, S.; Adams, N.A. Large-eddy simulation of passive shock-wave/boundary-layer interaction control. *Int. J. Heat Fluid Flow* **2014**, *49*, 116–127. [CrossRef]
3. Zhang, Y.; Tan, H.J.; Wang, Z.Y.; Li, X.; Guo, Y.J. Progress of shockwave/boundary layer interaction and its control in inlet. *J. Propuls. Technol.* **2020**, *41*, 241–259. (In Chinese) [CrossRef]
4. Hadjadj, A.; Dussauge, J.P. Shock wave boundary layer interaction. *Shock Waves* **2009**, *19*, 449–452. [CrossRef]
5. Meier, G.; Szumowski, A.P.; Selerowicz, W.C. Self-excited oscillations in internal transonic flows. *Prog. Aerosp. Sci.* **1990**, *27*, 145–200. [CrossRef]
6. Spaid, F.W.; Frishett, J.C. Incipient separation of a supersonic, turbulent boundary layer, including effects of heat transfer. *AIAA J.* **1972**, *10*, 915–922. [CrossRef]
7. Hou, W.T.; Qiao, W.Y.; Luo, H.L. Numerical simulation of effects of film-cooling on interaction between shock wave and boundary layer. *J. Propuls. Technol.* **2009**, *30*, 555–560. (In Chinese) [CrossRef]
8. Huang, W.; Chen, Z.; Yan, L.; Yan, B.B.; Du, Z.B. Drag and heat flux reduction mechanism induced by the spike and its combinations in supersonic flows: A review. *Prog. Aerosp. Sci.* **2019**, *105*, 31–39. [CrossRef]
9. Zhang, Q.F.; Tan, H.J.; Chen, H. Experimental study of unstart oscillatory flow control of hypersonic inlet with movable slot-plate. *J. Propuls. Technol.* **2017**, *38*, 1450–1458. (In Chinese) [CrossRef]
10. He, L.H.; Chen, H.; Yue, L.J.; Zhang, Q.F. Interaction of a shock train with inherent isentropic waves in a curved isolator. *Phys. Fluids* **2022**, *34*, 066106. [CrossRef]
11. Ferri, A. Experimental Results with Airfoils Tested in the High-Speed Tunnel at Guidonia. 1940, NACA-TM-946. Available online: https://ntrs.nasa.gov/citations/19930094471 (accessed on 6 September 2013).

12. Panaras, A.G.; Lu, F.K. Micro-vortex generators for shock wave/boundary layer interactions. *Prog. Aerosp. Sci.* **2015**, *74*, 16–47. [CrossRef]
13. Giepman, R.H.M.; Schrijer, F.F.J.; van Oudheusden, B.W. Flow control of an oblique shock wave reflection with micro-ramp vortex generators: Effects of location and size. *Phys. Fluids* **2014**, *26*, 066101. [CrossRef]
14. Martis, R.R.; Misra, A. Separation attenuation in swept shock wave–boundary-layer interactions using different microvortex generator geometries. *Shock Waves* **2017**, *27*, 747–760. [CrossRef]
15. Rybalko, M.; Babinsky, H.; Loth, E. Vortex generators for a normal shock/boundary layer interaction with a downstream diffuser. *J. Propul. Power* **2012**, *28*, 71–82. [CrossRef]
16. Colliss, S.P.; Babinsky, H.; Nübler, K.; Lutz, T. Joint experimental and numerical approach to three-dimensional shock control bump research. *AIAA J.* **2014**, *52*, 436–446. [CrossRef]
17. Bruce, P.J.K.; Colliss, S.P. Review of research into shock control bumps. *Shock Waves* **2015**, *25*, 451–471. [CrossRef]
18. Li, W.P.; Liu, H. Large-eddy simulation of shock-wave/boundary-layer interaction control using a backward facing step. *Aerosp. Sci. Technol.* **2019**, *84*, 1011–1019. [CrossRef]
19. Murugan, J.N.; Govardhan, R.N. Shock wave–boundary layer interaction in supersonic flow over a forward-facing step. *J. Fluid Mech.* **2016**, *807*, 258–302. [CrossRef]
20. Slater, J.W. Improvements in modeling 90-degree bleed holes for supersonic inlets. *J. Propul. Power* **2012**, *28*, 773–781. [CrossRef]
21. Li, J.; Tian, Y.; Zhong, F.Y.; Yang, S.H. Effects of boundary-layer bleeding on flow field in scramjet combustor. *J. Propuls. Technol.* **2019**, *40*, 2700–2707. (In Chinese) [CrossRef]
22. Szwaba, R. Influence of air-jet vortex generator diameter on separation region. *J. Therm. Sci.* **2013**, *22*, 294–303. [CrossRef]
23. Reese, B.M.; Collins, E.G., Jr.; Fernandez, E.; Alvi, F.S. Nonlinear adaptive approach to microjet-based flow separation control. *AIAA J.* **2016**, *54*, 3002–3014. [CrossRef]
24. Kinefuchi, K.; Starikovskiy, A.Y.; Miles, R.B. Numerical investigation of nanosecond pulsed plasma actuators for control of shock-wave/boundary-layer separation. *Phys. Fluids* **2018**, *30*, 106105. [CrossRef]
25. Bisek, N.J.; Rizzetta, D.P.; Poggie, J. Plasma control of a turbulent shock boundary-layer interaction. *AIAA J.* **2013**, *51*, 1789–1804. [CrossRef]
26. Yang, G.; Yao, Y.F.; Fang, J.; Gan, T.; Li, Q.S.; Lu, L.P. Large-eddy simulation of shock-wave/turbulent boundary layer interaction with and without SparkJet control. *Chin. J. Aeronaut.* **2016**, *29*, 617–629. [CrossRef]
27. Herrmann, D.; Blem, S.; Gulhan, A. Experimental study of boundary-layer bleed impact on ramjet inlet performance. *J. Propul. Power* **2011**, *27*, 1186–1195. [CrossRef]
28. Wang, B.; Liu, W.D.; Zhao, Y.X.; Fan, X.Q.; Wang, C. Experimental investigation of the micro-ramp based shock wave and turbulent boundary layer interaction control. *Phys. Fluids* **2012**, *24*, 055110. [CrossRef]
29. Blinde, P.L.; Humble, R.A.; van Oudheusden, B.W.; Scarano, F. Effects of micro-ramps on a shock wave/turbulent boundary layer interaction. *Shock Waves* **2009**, *19*, 507–520. [CrossRef]
30. Zhang, Y. Flow Control Methods Based on Shape Memory Alloy for Hypersonic Inlets. Ph.D. Thesis, Nanjing University of Aeronautics and Astronautics, Nanjing, China, 2015. (In Chinese).
31. Lu, F.K.; Li, Q.; Liu, C.Q. Microvortex generators in high-speed flow. *Prog. Aerosp. Sci.* **2012**, *53*, 30–45. [CrossRef]
32. Pitt Ford, C.; Babinsky, H. Micro-ramp control for oblique shock wave/boundary layer interactions. In Proceedings of the 37th AIAA Fluid Dynamics Conference and Exhibit, Miami, FL, USA, 25–28 June 2007; p. 4115. [CrossRef]
33. Benek, J.A.; Suchyta, C.J.; Babinsky, H. Simulations of incident shock boundary layer interactions. In Proceedings of the 54th AIAA Aerospace Sciences Meeting, San Diego, CA, USA, 4–8 January 2016; p. 0352. [CrossRef]
34. Benhachmi, D.; Greber, I.; Hingst, W.R. Experimental and numerical investigation of an oblique shock wave/turbulent boundary layer interaction with continuous suction. In Proceedings of the 27th Aerospace Sciences Meeting, Reno, NV, USA, 9–12 January 1989; p. 357. [CrossRef]
35. Xie, W.Z.; Yang, S.Z.; Zeng, C.; Liao, K.; Ding, R.H.; Zhang, L.; Guo, S.M. Improvement of the free-interaction theory for shock wave/turbulent boundary layer interactions. *Phys. Fluids* **2021**, *33*, 075104. [CrossRef]
36. Huang, H.X.; Tan, H.J.; Sun, S.; Ling, Y. Evolution of supersonic corner vortex in a hypersonic inlet/isolator model. *Phys. Fluids* **2016**, *28*, 126101. [CrossRef]
37. Li, N.; Chang, J.T.; Xu, K.J.; Yu, D.R.; Bao, W.; Song, Y.P. Prediction dynamic model of shock train with complex background waves. *Phys. Fluids* **2017**, *29*, 116103. [CrossRef]
38. Sekar, K.R.; Karthick, S.; Jegadheeswaran, S., Kannan, R. On the unsteady throttling dynamics and scaling analysis in a typical hypersonic inlet–isolator flow. *Phys. Fluids* **2020**, *32*, 126104. [CrossRef]
39. Tang, D.G.; Li, J.P.; Zeng, F.Z.; Li, Y.; Yan, C. Bayesian parameter estimation of SST model for shock wave-boundary layer interaction flows with different strengths. *Chin. J. Aeronaut.* **2023**, *36*, 217–236. [CrossRef]
40. Menter, F.R. Two-equation eddy-viscosity turbulence models for engineering applications. *AIAA J.* **1994**, *32*, 1598–1605. [CrossRef]
41. ANSYS Inc. *Fluent Theory Guide*; ANSYS Inc.: Canonsburg, PA, USA, 2022.
42. Grossman, I.J.; Bruce, P.J. Confinement effects on regular–irregular transition in shock-wave–boundary-layer interactions. *J. Fluid Mech.* **2018**, *853*, 171–204. [CrossRef]
43. Matheis, J.; Hickel, S. On the transition between regular and irregular shock patterns of shock-wave/boundary-layer interactions. *J. Fluid Mech.* **2015**, *776*, 200–234. [CrossRef]

44. Teramoto, S. Large eddy simulation of shock wave/boundary layer interaction. *Trans. Jpn. Soc. Aeronaut. Space Sci.* **2005**, *47*, 268–275. [CrossRef]
45. Wollblad, C.; Davidson, L.; Eriksson, L.-E. Large eddy simulation of transonic flow with shock wave/turbulent boundary layer interaction. *AIAA J.* **2006**, *44*, 2340–2353. [CrossRef]

aerospace

Article

Experimental Investigation on the Control of Hypersonic Shock Wave/Boundary Layer Interaction Using Surface Arc Plasma Actuators at Double Compression Corner

Bo Yang [1,2], Hesen Yang [3,*], Chuanbiao Zhang [3], Ning Zhao [1], Hua Liang [3,*] and Dongsheng Zhang [3]

[1] College of Aerospace Engineering, Nanjing University of Aeronautics and Astronautics, Nanjing 210016, China; yandr_2002@139.com (B.Y.); zhaoam@nuaa.edu.cn (N.Z.)
[2] Hypervelocity Aerodynamics Institute of China Aerodynamics Research and Development Center, Mianyang 621000, China
[3] National Key Lab of Aerospace Power System and Plasma Technology, Air Force Engineering University, Xi'an 710038, China; zhangcb2020@126.com (C.Z.); zds13475138164@163.com (D.Z.)
* Correspondence: yanghesen96@126.com (H.Y.); lianghua82702@163.com (H.L.)

Abstract: Compression corner shock wave/boundary layer interaction (SWBLI) is a typical shock wave/boundary layer interaction (SWBLI) problem in supersonic/hypersonic flows. In previous studies, the separation flow is usually caused by a single shock wave. However, in the actual aircraft surface configuration, two-stage compression or even multistage compression will produce more complex SWBLI problems. The multi-channel shock structure makes the flow field structure more complicated and also puts forward higher requirements for the flow control scheme. In order to explore a flow control method for the double compression corner shock wave/boundary layer interaction problem, an experimental study is carried out to control the double compression corner shock wave/boundary layer interaction with a high-energy flow pulsed arc discharge array under the condition that the incoming flow velocity *Ma* 6.0 has both noise flow fields and quiet flow fields. The results show that when U_{DC} = 0.5 kV actuation is applied, the influence range of the hot gas mass flow direction is about 65 mm, which can weaken the shock wave intensity to a certain extent. When U_{DC} = 1 kV actuation is applied, the influence range of the hot gas mass flow direction extends to 85 mm, and the actuation has a significant control effect on the flow field. Through spatio-temporal evolution analysis and spatial gradient threshold processing of high-speed schlieren images of actuated flow fields, the feasibility of controlling the hypersonic double compression corner shock wave/boundary layer interaction by using a high-energy flow pulsed arc discharge array is verified. The control law of a high-energy flow pulsed arc discharge array acting on the double compression corner shock wave/boundary layer interaction is revealed.

Keywords: hypersonic; double compression corner; shock wave/boundary interaction; plasma actuation; flow control

Citation: Yang, B.; Yang, H.; Zhang, C.; Zhao, N.; Liang, H.; Zhang, D. Experimental Investigation on the Control of Hypersonic Shock Wave/Boundary Layer Interaction Using Surface Arc Plasma Actuators at Double Compression Corner. *Aerospace* **2023**, *10*, 1016. https://doi.org/10.3390/aerospace10121016

Academic Editor: Sergey Leonov

Received: 19 September 2023
Revised: 2 December 2023
Accepted: 4 December 2023
Published: 6 December 2023

1. Introduction

In a supersonic flow, the disturbance of the air stream cannot propagate in the direction of the flow, so shock waves will inevitably exist in the process of deceleration and shock wave/boundary layer interaction (SWBLI) will inevitably occur in the interaction of the boundary layer on the surface of the aircraft body [1–3]. When the inverse pressure gradient induced by shock waves is large enough, the boundary layer will become thicker or even return inside, resulting in the boundary layer separating from the object surface, which will directly damage the aerodynamic performance of the aircraft. There are various forms of shock wave/boundary layer interaction in the outflow of the aircraft, including oblique shock wave/boundary layer interaction, normal shock wave/boundary layer interaction, and three-dimensional shock wave/boundary layer interaction, among which Compression

Ramp-SWBLI is the most typical form of oblique shock wave/boundary layer interaction. It mainly occurs at the rudder surface of the aircraft [4,5].

At present, studies on CR-SWBLI control at a single compression corner have been extensive. For example, Verma and Chidambaranathan [6] adopted a stable microjet array for a single compression corner of 24° and found that jet spacing of 13 d has a better effect on reducing the size of pressure in the separation region. The 135° jet achieves a better control effect on the instability of separated shock waves. Verma compared the vortex generators of two positions and two configurations and found that the closer the interaction region is, the better the effect of the vortex generator on weakening the instability of the separation shock wave and reducing the intensity of the separation shock wave. The vortex generator with the configuration of s = 0 significantly reduces the root-mean-square value of the separation region. There is relatively little research on double compression Ramp-SWBLI and it mainly focuses on numerical simulation [7,8]. Durna et al. [9] studied the influence of the second order wedge angle on the flow field structure under Mach 7 conditions through numerical calculation. The results show that when the first-order wedge angle = 30° remains unchanged, the vortex structure has a more and more significant effect on the flow with the increase in the second-order wedge angle. Estruch-Samper et al. [10] conducted detailed experimental studies on Micro-Vortex generators of different heights, and the results show that when the height of the Micro-Vortex Generator is relatively small, it can inhibit flow separation. When the ratio of the height of the Micro-Vortex Generator to the boundary layer is about 0.3, the separation is effectively suppressed and the large-scale unsteady characteristics of SWBLI are significantly reduced. Babinsky et al. [11] also conducted an experimental study on the control details of the separation region through micro-ramps and found that when micro-ramps were used to disturb the flow field in supersonic inflow with Ma = 2.5, the flow separation caused by SWBLI could not be completely suppressed under the control of micro-ramps of all sizes but that micro-ramps could break the separation region. The space size is reduced to a plurality of broken small separation regions. Tong et al. [12] conducted a direct numerical simulation of a double compression corner with Mach number 2.9, the angle of the two stages was fixed at 12° and 24°, respectively, and studied the influence of the flow direction length L_c of the first stage on the flow field structure. It was found that with the increase in L_c, the size of the separation region is significantly reduced, the width of the Görtler vortex is reduced, and the directional coherence is enhanced [13–15].

As a passive flow control method without moving parts [16–18], plasma actuation can effectively avoid the increase in aerodynamic drag, flow loss, and other problems caused by the actuation itself, compared with micro-vortex generators, boundary layer venting, and other control methods. In recent years, due to the breakthrough of high repetition, frequency, energy, and array plasma actuation technology, its great potential in the field of supersonic flow control has been reflected. Preliminary studies have been conducted on dielectric barrier discharge, plasma synthetic jet, and surface arc discharge and other forms [19–27], particularly surface arc discharge [20]. In recent years, in the field of supersonic/hypersonic flow control, surface arc discharge plasma actuation (especially array-type surface arc discharge plasma actuation) has highlighted the advantages of strong energy injection, high frequency response characteristics, and flush with the wall surfaces [28,29]. Breakthrough progress has been made in shock wave control and shock wave/boundary layer interaction control [30,31].

Watanabe Y. et al. [32] examined the effect of the Reynolds number on plasma-assisted flow control. A linear dependency was found between the ramp pressure change per averaged plasma power and the Reynolds number. In addition, the effect of near-surface discharge on supersonic flow near 15° compression surface is studied by experiments and simulations [33]. Further simulations attempted to find an optimal range of plasma power and position in terms of achievable effect, effectiveness of the method, and response time of the system to the plasma actuation. Tang et al. [34] studied the evolution characteristics of high energy arc discharge actuation under a low pressure environment and used high-energy plasma actuation to effectively weaken the shock intensity of bow shock in front of a cylinder. In the incident shock wave/boundary layer interaction control, Luo et al. [35] found that both 5 kHz and 10 kHz high-frequency arc discharge actuation can weaken the intensity of separated shock waves and 5 kHz actuation can suppress the low-frequency motion of separated shock waves, while 10 kHz actuation can increase the low-frequency motion energy. Tang et al. [36] used spanwise array pulsed arc discharge actuation to achieve forced boundary layer transition and used high-frequency flow pulsed arc discharge array to achieve effective control of the shock wave/boundary layer interaction induced by the compression corner and proposed the corresponding conceptual model. Gan et al. [30] also realized the weakening of the intensity of separated shock waves in the shock wave/boundary layer interaction induced by the compression corner through high-energy and low-frequency array pulsed arc actuation.

In this paper, experimental studies on the control of double-compression corner shock/boundary layer interaction flow fields by array plasma actuation under Mach 6 conditions are carried out, and the coupling evolution characteristics of the array surface arc actuation and hypersonic double-compression corner shock/boundary layer interaction flow fields are discussed and analyzed. The ability of plasma actuation to regulate the shock wave/boundary layer interaction is verified under Mach 6 conditions, and the corresponding control rules are finally summarized.

2. Experimental System and Model

2.1. Hypersonic Quiet Wind Tunnel and High-Speed Schlieren System

The experiment is conducted in the Φ300 mm hypersonic quiet wind tunnel of the Hypervelocity Aerodynamics Institute of China Aerodynamics Research and Development Center. The experimental medium in the wind tunnel is air or nitrogen. The maximum design total pressure is 2.0 MPa, the maximum total temperature is 537 K, and the nozzle outlet diameter is 320 mm. The stable operation time of the wind tunnel is greater than 10 s, the duration of the static flow field is not less than 7 s, and the cycle of a single platform is 150~200 ms. The wind tunnel uses a Laval nozzle, and a special boundary layer suction device is set up in the upstream of the nozzle to realize the switch between the quiet flow field and the noise flow field. In quiet mode, the noise level of the Mach 6 flow field is 0.05~0.1% (adjustable), and the average Mach number is 6.03~6.15. The Mach number 8 flow field noise level is 0.07~0.1% (adjustable); that is, the sound pressure level is adjustable, the average Mach number is 7.90~7.95, and the uniform region diameter is not less than 240 mm. At the same time, the wind tunnel can operate in the conventional hypersonic Ludwig wind tunnel mode, and the noise level of the flow field is 2~3%, similar to that of the conventional hypersonic wind tunnel. The average Mach number is about 5.90, and the diameter of the uniform region is 200 mm. The main performance parameters of the wind tunnel are shown in Table 1, and the actual picture of the wind tunnel is shown in Figure 1. The actual flow conditions of the experiment are shown in Table 2, where the first row is the flow parameters of the static flow field and the second row is the flow parameters of the noise flow field.

Table 1. Main parameters of Φ300 mm hypersonic quiet wind tunnel.

Ma	Nozzle Diameter/m	P_0/MPa	T_0/K	$Re/L \times 10^6$	*t*/s	*Ma*	Flow Field Noise
6	0.32	0.1~0.45	387~422	1.28~5.99	10	6.03~6.15	0.05~0.1% air
8	0.32	0.1~0.45	478~515	0.47~1.86	10	7.90~7.95	0.07~0.1% Nitrogen

Table 2. Main parameters of incoming stream.

Ma_∞ (U_∞/c)	Re/m ($\rho U_\infty/\mu$)	U_∞ (m/s)	ρ (kg/m^3)	P_0 (MPa)	T_0 (K)	P_S (Pa)	T_S (K)	*N*
6.10	5.74×10^6	899.76	0.012	0.328	457	178.68	54.09	2%
5.90	9.38×10^6	896.11	0.017	0.410	457	286.98	57.35	0.1%

Ma_∞, Re/m, U_∞, ρ, P_0, T_0, P_S, T_S, c, and N represent the incoming Mach number, unit Reynolds number, free flow velocity, incoming density, total pressure, total temperature, static pressure, static temperature, sound velocity, and the noise level.

Figure 1. Φ300 mm hypersonic quiet wind tunnel.

2.2. Schlieren System

A schlieren display technique is used to capture the flow junction in the experiment. Compared with the traditional schlieren system, the slit and knife edge of the optical system adopts the same side structure of the off-axis projectile target. The experimental schlieren system adopts the same side structure of the off-axis parabolic mirror, and the optical path of the Schlieren system is different from the traditional Z-type optical path. The optical path diagram is shown in Figure 2. The core of the system consists of four mirrors that reflect light. The maximum resolution of the experimental camera is 2048 pixel × 2048 pixel, the target size is 20.48 mm × 20.48 mm, the pixel size is 10 μm, and the schlieren frame frequency is 5 k. According to literature [37], in the same wind tunnel test, under Mach number 8, an exposure time of 5 μs is selected. In the discharge characteristic experiment, shockwaves and hot gas masses generated by plasma actuation can be obviously observed. In this experiment, under Mach number 6, the incoming flow velocity is lower, the brightness of the field of view is guaranteed, and a lower exposure time is selected to reduce the schlieren integration effect and field of view sharpness.

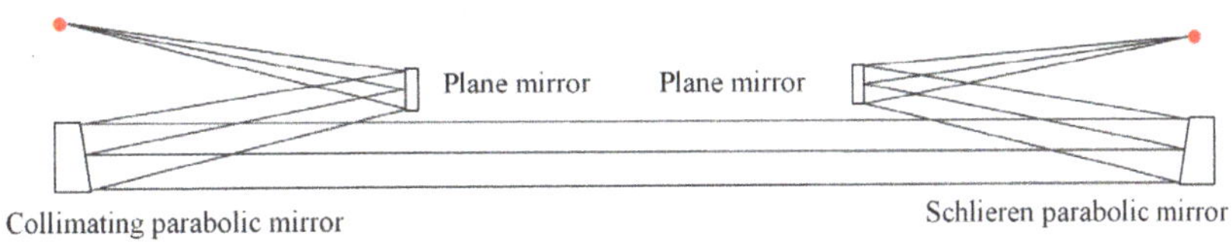

Figure 2. The diagram of schlieren light path.

2.3. Experimental Model and Actuator Setup

The experimental model was installed in the wind tunnel test section and bolted to the bottom support frame. The experimental model used in this research is shown in Figure 1, which is divided into two parts: bottom plate and double compression corner. The diagram of the assembled flat-double compression corner is shown in Figure 3a. Figure 3b shows the actuator used in the experimental research of this paper, which is formed by three arc plasma actuators in series. Each discharge can generate three plasma actuations along the flow direction to expand the actuation range of the flow direction. Figure 3c shows the schematic diagram of the double compression corner model. The angles of the two folds are, respectively, 30° and 45°, and the total length is 55 mm. In order to meet the requirements of the strength check, it is necessary to control the overall weight of the experimental model, so the spanwise width of the corner model is designed to be 40 mm, while avoiding the side wall effect.

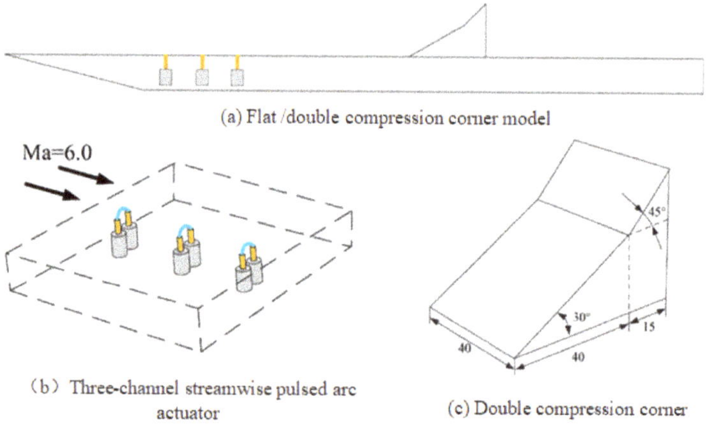

Figure 3. The diagram of the experimental model.

Before studying the control effect of a high energy drive, the discharge characteristics of the plasma arc discharge should be analyzed. Figure 4 shows the voltage and current waveform driven by high energy at a low voltage. As can be seen from the figure, the discharge time of a single pulse is not more than 20 μs, and when the voltage reaches a peak of about 8 kV, it will quickly drop to 0.8 kV. Compared with the voltage, the current rises and falls slightly behind, rapidly reaching a peak value of 110 A and then smoothly transiting to the initial value at the peak. Due to the presence of capacitors, it will decay at a relatively slow rate, and the discharge time will be extended. Finally, by integrating the current and voltage, it can be calculated that the single pulse discharge energy of the high energy arc is about 496 mJ.

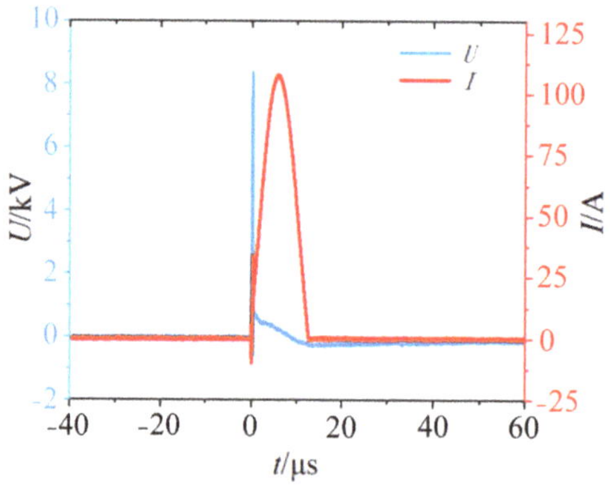

Figure 4. Volt–ampere characteristic curve.

3. The Reference Flow Field of Double Compression Corner Shock Wave/Boundary Layer Interaction at *Ma* 6.0

Firstly, schlieren measurements of a *Ma* = 6.0 noise flow field and quiet flow field are carried out. Compared with a *Ma* = 2.0 incoming flow condition [13], the corner leading edge under hypersonic conditions has a larger separation region. Figure 5a shows the instantaneous schlieren image under the noise flow field. It can be seen from the Figure that a typical SWBLI flow field structure and shock wave/shock wave interaction flow field structure are formed near the double compression corner. In Figure 5a, the flat boundary layer upstream of the corner is in a laminar flow state of linear growth, and the thickness of the boundary layer becomes obviously thicker along the flow direction. At the position of 100 mm in the flow direction, the boundary layer rises obviously, and a weak separation shock wave also appears outside the boundary layer here. From here to the point of impact between the boundary layer and the corner model is defined as the length of the separation region [9]. It is worth noting that the separation shock wave is not a clear straight line in the schlieren diagram, but there are weak compressed Mach waves in the plane region. Some research in the literature has simulated the separation region structure of an SWBLI compression corner at *Ma* 2.0 supersonic inflow by numerical simulation and found that the separation region is not strictly a two-dimensional structure but has certain three-dimensional characteristics. It is speculated that the shape of the separation region at *Ma* 6.0 is similar to it and also has certain three-dimensional characteristics. At the impact point between the boundary layer and the corner, a strong reattachment shock wave appears due to airflow compression, the separation shock wave passes through the reattachment shock wave, and there are more complex reflections behind the wave. At the leading edge of the second corner, as the turning angle of the airflow increases by 15° from the first corner to the second corner, the airflow is strongly compressed, and a second intense shock wave is formed here. When the first reattachment shock wave collides with the second shock wave, strong mutual interaction occurs. On the one hand, they intersect to form a stronger shock wave, and on the other hand, a weaker reflected shock wave is formed downward. There is a slip line between them. The slip line here is clear and obvious, indicating that the speed difference between the two regions is large.

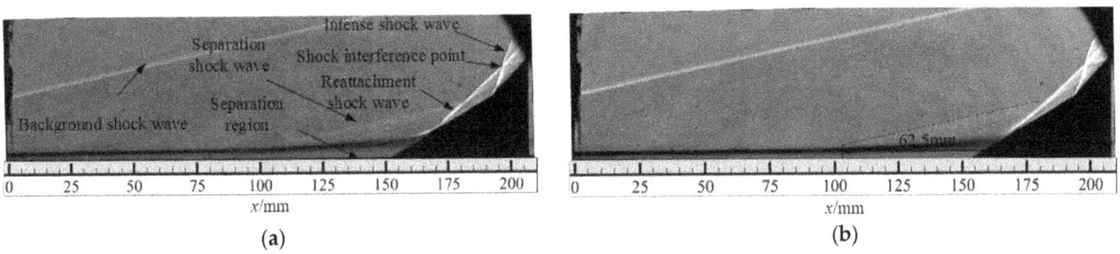

Figure 5. Schlieren image of noise flow field at *Ma* 6.0. (**a**) Instantaneous schlieren of noise flow field; (**b**) time-mean schlieren of noise flow field.

Figure 5b is the time-mean schlieren diagram of the noise flow field based on 500 instantaneous schlieren gray value averaging processing, according to which we can carry out a qualitative analysis of the above flow field structure. It can be seen from the figure that not only the boundary layer, separation shock wave, reattachment shock wave, and interaction point can be well presented but also the complex reflection process of the separation shock wave after the reattachment shock wave can be well presented, which indicates that the result of the time-mean schlieren has good reliability. Therefore, we can examine the size of the separation region in the time-mean schlieren. As shown in the Figure, the white dashed line represents the separation shock wave. According to the definition of the boundary layer separation region above, it can be concluded that in the *Ma* 6.0 noise flow field, the length of the separation region is 62.5 mm.

In the quiet flow field, the basic structure of the flow field is the same as that of the noise flow field, but the spatial position of each structure does not present a small difference. As shown in Figure 6a, in the quiet flow field stage, excluding the interaction of external factors, the separation starting point of the boundary layer at the leading edge of the corner is greatly advanced, appearing about 85 mm from the flow direction, and the size of the separation region increases significantly. Not only does the initial position of the separation shock flow advance but also the height of the normal direction increases. The angle of the reattachment shock wave decreases slightly, so the normal height of the shock wave interaction point decreases. The morphological characteristics of the intense shock wave, reflected shock wave, and the angle of the slip line produced after the intersection of shock waves hardly change, but the position and length of the wave do change. Similarly, some qualitative rules can be obtained by analyzing the time-mean schlieren diagram of the quiet flow field. As shown in Figure 6, the size of the separation region increases to 90 mm, and the size of the separation region increases by 44% compared to the noise flow field. It can be seen that under hypersonic conditions, the size of the separation region in the quiet flow field will be larger, while in the actual hypersonic flight process it is usually faced with a small noise flow field with a low Reynolds number, and under the same configuration conditions, a larger separation region will be generated. Therefore, certain flow control means are urgently needed to regulate this problem. It should be noted that the size of the separation region in the quiet flow field is larger, which is not caused by the difference in Reynolds number. The main reason is that the quiet flow field is mainly laminar flow, which has a weak ability to resist the reverse pressure gradient and is easy to separate, while the noise flow field is mainly turbulence, which has a stronger ability to resist the reverse pressure gradient than the quiet flow field, so the separation scale of the quiet flow field is larger.

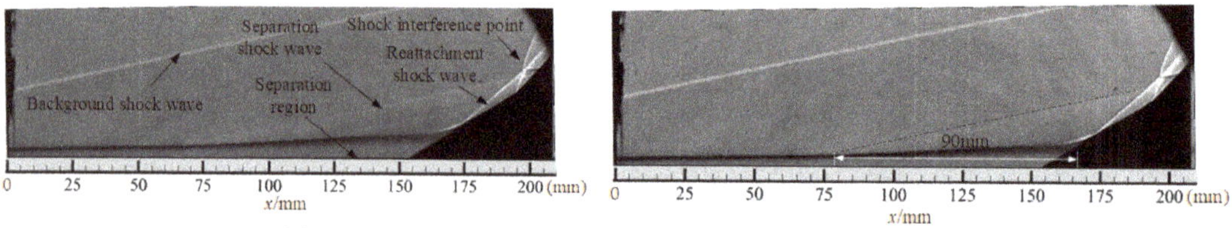

(**a**) (**b**)

Figure 6. Schlieren image of quiet flow field at *Ma* 6.0. (**a**) Instantaneous schlieren of quiet flow field; (**b**) time-mean schlieren of quiet flow field.

4. The Actuation Flow Field of Double Compression Corner Shock Wave/Boundary Layer Interaction at *Ma* 6.0

4.1. Analysis of the Control Effect

According to previous studies, even if the supersonic flow field can achieve a good control effect of high-frequency actuation, such as 5 kHz, 10 kHz, 20 kHz actuation, in the hypersonic flow field, the flow field disturbance frequency is high, far beyond the common high-frequency actuation range of plasma actuation. In addition, the extremely harsh incoming flow conditions make the plasma actuation with high frequency and low energy consumption unable to play a good control effect. Therefore, in order to regulate the strong SWBLI and shock wave/shock wave interaction in the hypersonic flow field, we naturally take into account the lower actuation frequency but use high-energy actuation with greater actuation intensity and more stable discharge. Therefore, in the experimental study of this section, a high-energy actuator with a capacitance of 1 μF and DC source voltage U_{DC} = 1 kV is used to control hypersonic noise flow field and static flow field. In order to reduce the energy consumption of arc discharge plasma actuation, and to expand the actuation range, the arc plasma actuator is arranged into two-channel streamwise; that is, two plasma discharge positions are arranged along the flow direction of the flow field. Since the frame rate of the schlieren shooting was 5000 fps, the actuation frequency was still set at 480 Hz, and the schlieren images with different phases after actuation are selected to form the schlieren sequence.

Figure 7 shows SWBLI schlieren sequence of the noise flow field controlled by high-energy actuation. As shown in Figure 7a, after the discharge channel is established, due to the low static pressure of incoming flow, the two arc actuation fuse into one actuation and inject energy into the gas on the surface of the plate to induce a large volume hot gas mass. Compared with the two smaller hot gas masses separated, this not only enlarges the range of actuation but also increases the intensity of actuation and improves the control effect of actuation.

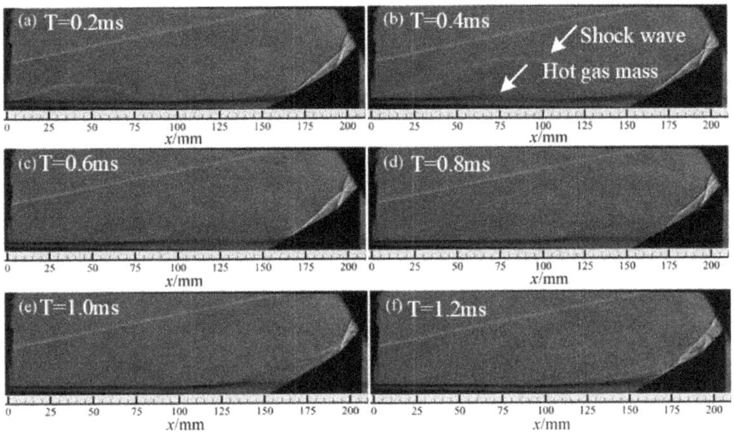

Figure 7. SWBLI schlieren image of noise flow field controlled by high-energy actuation.

In Figure 7b, the shock wave induced by actuation changes into a hemispherical hot gas mass, and the leading edge reaches the foot of the separation shock wave. The hot gas mass inside the shock wave is coupled with the laminar flow boundary layer, resulting in increased disturbance. However, the state of separation shock at this phase has not changed. As shown in Figure 7c, when the hot gas generated by the actuation flows near the separation region, n the one hand, the separation region of the leading edge of the corner is completely covered by hot gas masses. Through observation of the state of the boundary layer, it can be found that the state of the boundary layer has changed, which is due to the shock wave and hot gas mass generated under the actuation of plasma, which disturbs the boundary layer. There is a vortex structure in the hot gas mass itself [38,39], and the interaction with the boundary layer will make the boundary layer also carry a vortex structure; these vortex structures also accelerate the fusion of the boundary layer and the main stream. At the same time, the hot gas mass generated by plasma arc discharge will be accelerated along with the incoming flow. At this time, the hot gas mass will encounter the boundary layer, especially the fluid with lower speed in the separation region, and momentum exchange will inevitably occur, making the low-energy fluid accelerate. At this time, the size of the separation region is greatly reduced, so the separation shock wave disappears.

In Figure 7d, after the hot gas mass flows through the surface of the plate, it is coupled with the boundary layer at the outer edge of the separation region and hits the surface of the angle model at this phase moment. It is found in the schlieren image that the boundary layer state changes, which is due to the mass, energy, and momentum exchange between the fluid and the hot gas in the boundary layer, which makes the boundary layer chaos intensified. However, at this stage, the boundary layer is hit by the hot gas mass and the boundary layer is in a chaotic turbulent state. The hot gas mass forms a virtual compression surface connecting the plate and the angle outside the boundary layer. Different from the traditional model bulge, the virtual compression surface formed by plasma actuation is an unsteady "bulge" formed by the shock wave and the hot gas mass in the flow field, which makes the attached shock wave "truncated" by the virtual compression surface and the shock foot move upward.

At the corresponding phase moment in Figure 7e, the hot gas mass has completely covered the surface of the corner of the first stage, and the virtual compression face is compressed ahead of the incoming flow. In addition, the impact point between the boundary layer and the model surface is still within the influence range of the hot gas mass and the virtual compression surface still exists, so the intensity of the reattachment shock waves is greatly weakened and the interaction points of the shock waves are dispersed. The unsteady motion of the reflected shock wave and the slip line also deviated from the original trajectory in the above process. Because the hot gas mass did not affect the second shock wave, the intense shock wave above the interaction point still maintained a large shock wave intensity.

Figure 7f shows the structural characteristics of the flow field when the hot gas mass actuated by the plasma passes near the corner of the second stage, as shown in the figure; at this time, the hot gas mass has passed the corner of the first stage, and the boundary layer also recovers the reference state, so the separation shock wave reappears and the related structure of the seconded shock wave also reappears. A complex flow structure appears behind the attached shock wave. Due to the action of hot gas mass, the second shock wave is dispersed by the hot gas mass and it is impossible to distinguish the flow structure, such as the disturbance point and slip line, in the flow field. According to the experimental results, the plasma arc discharge produces shock waves and hot gas mass, and an impact effect is one of its characteristics. Through the shock wave, the shock wave is affected and the intensity is reduced. This also reduces the inverse pressure gradient of the boundary layer, thereby indirectly reducing the separation region, which also plays an indirect role in separation control. According to the above analysis, high-energy arc plasma actuation can play an effective role in regulating SWBLI noise flow field and shock wave interaction.

In the quiet flow field, the incoming flow has a small disturbance. As can be seen in Figure 8a,b, the hot gas mass and shock wave induced by plasma actuation are similar to those in Figure 7. However, the coupling effect with the boundary layer does not introduce a large disturbance to the boundary layer. There is no large-scale vortex structure. At the corresponding phase moment in Figure 8c,d, the hot gas mass flows near the separation region. Based on the analysis of the reference state of the quiet flow field above, we know that the size of the separation region is larger than that of the noise flow field and the range of the separation shock wave is also larger. In Figure 8e, the upstream separation shock structure reappears, and the attached shock wave almost completely disappears at this stage. Only part of the weak compression waves is attached near the corner wall, and the angle is similar to that of the first corner. The structure of the shock wave after bifurcation is more regular and there is no large oscillation after plasma actuation. At the corresponding phase moment in Figure 8f, the hot gas mass flows near the second corner, at which time the separation region reappears, the complete structure of the separation shock wave has been established, and the related structure of the attached shock wave foot also reappears, and the second shock wave is dispersed by the hot gas mass. Since the disturbance in the flow field is small, the complex flow field structure like that in the noise flow field is not formed. The Mach wave angle is smaller and closer to the surface of the corner model.

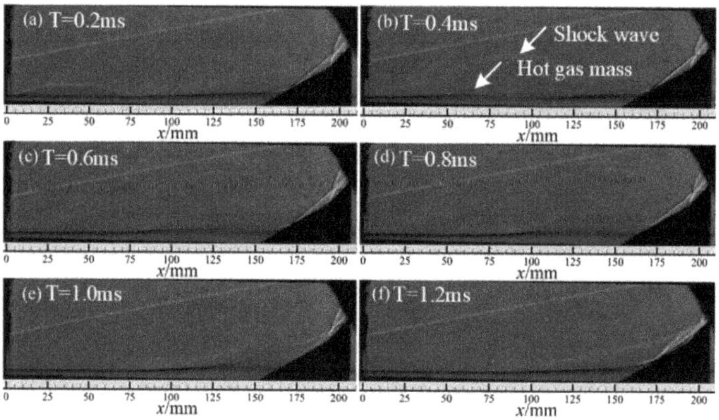

Figure 8. SWBLI Schlieren image of quiet flow field controlled by high-energy actuation.

4.2. The Influence of Actuation Energy on Control Effect

In the previous section, discharge parameters of U_{DC} = 1 kV and capacitance 1 μF were used to control the hypersonic noise flow field and static flow field, and a relatively ideal control effect was achieved. In this section, the control effect of high-energy actuation on the hypersonic double compression corner shock wave/boundary layer interaction and shock wave/shock wave interaction flow field under different actuation intensities will be explored by adjusting the DC source voltage and capacitance.

Under the condition of noise inflow, due to the large disturbance in the boundary layer, according to the experimental results of surface arc actuation to control hypersonic boundary layer transition, applying small actuation disturbance can provide a great control effect to the boundary layer, playing a "four or two" role. Therefore, firstly, the DC source voltage is set as U_{DC} = 0.5 kV and the capacitance as 1 μF. The control effect of arc actuation on the shock wave/boundary layer interaction of the hypersonic double compression angle and low energy shock wave interaction flow field was investigated. Figure 9 shows the structural evolution sequence of the flow field after actuation control. Figure 9a,c,e on the left correspond to the instantaneous schlieren images of the hot gas touching the leading edge of the corner model, hot gas reaching the boundary layer reattachment point, and the hot gas covering the shock wave region, respectively. Figure 9b,d,f on the right are the corresponding images after longitudinal gradient processing. Longitudinal gradient processing is to enhance the contrast of the longitudinal image in order to better display the boundary layer and flow field structure. In this section, the three instantaneous images with representative phase moments are selected for analysis, which will not be described below.

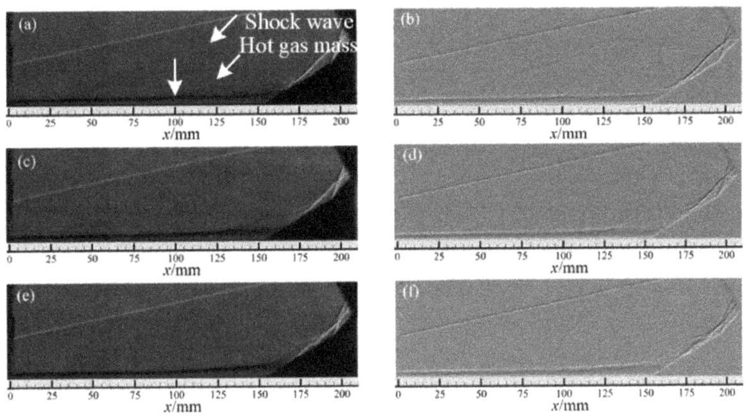

Figure 9. Structure evolution sequence of the flow field controlled by U_{DC} = 0.5 kV high-energy actuation in the noise flow field: (**a**,**c**,**e**) Original schlieren image; (**b**,**d**,**f**) schlieren image after longitudinal gradient processing.

In Figure 9a, the hot gas mass generated by actuation has a long flow coverage range. Its leading edge contacts the model at x = 165 mm, and its trailing edge is located at x = 100 mm. The influence range of the hot gas mass is about 65 mm, which is enough to cover the entire separation region of the leading edge of the corner under the condition of noise flow. Longitudinal gradient processing is used to observe the fine structure of the boundary layer and shock waves. As shown in Figure 9b, the structure of the hot gas mass becomes more chaotic after it is coupled to the boundary layer. At this time, the separation shock waves disappear, the structure of the secondary shock waves does not change significantly, and the shock interaction points and slip lines and other structures do not change. When the hot gas mass flows to the vicinity of the reattachment point, as shown in Figure 9c, it is obstructed by the corner, and the hot gas mass piles up into a bulge here. Then, the attached shock wave foot moves upwards. The density variation in the region behind the wave also increases, and the disturbance of the flow field increases. When the hot gas mass completely covers the shock wave region, as shown in Figure 9e, the secondary shock wave is dispersed, the region surrounded by the two shock waves is further reduced, and the interaction point moves down. It can be seen from Figure 9f that the reflected shock wave generated after the shock wave interaction disappears and the shock wave interaction is controlled by actuation, but the interaction point still exists and the main shock wave and the slip line after polymerization are still clear. This indicates that the actuation strength of the discharge parameter is relatively weak. Although it has a certain flow control effect on the flow field, its control effect on the shock wave structure and disturbance effect on the flow field need to be further improved.

When a higher intensity arc actuation is applied to the flow field, as shown in Figure 10, the output voltage of the DC source is set as U_{DC} = 1 kV, and the capacitance is selected as 4 μF, the flow field structure produces a more intuitive and significant control effect. In Figure 10a, when the leading edge of the hot gas mass contacts the corner model, the trailing edge is still at the position of x = 90 mm, and the flow direction influence range is at least 80 mm. Compared with the above actuation intensity, the flow direction influence range of the hot gas mass has expanded at least 23.1%. It can also be seen from the instantaneous schlieren diagram that, compared with Figure 10a, under the same phase, the longitudinal height of the hot gas mass also increased slightly, indicating that the intensity of actuation greatly increased. As can be seen from the gradient processing diagram, the disturbance to the boundary layer caused by actuation also increased with the increase in the intensity of actuation, resulting in a change in the flow structure near the reattachment point. Although the hot gas mass did not flow to the reattachment shock wave, the shock wave foot already showed the characteristics of disturbance and deformation. At the phase moment when the hot gas mass covers the reattachment point, different from Figure 9, the volume of the hot gas mass is larger after accumulation, almost covering the entire corner of the first stage. Moreover, the structure of the reattachment shock wave completely disappears, weakening into several weak compression waves, and the position of the interaction point drops to almost close to the model surface or even disappears. As can be seen from Figure 10d, the weakened compression wave kinks together with the second shock wave, and there is a slip line backwards at the interaction point. At this phase, the secondary shock wave almost completely disappears, and the strength of the shock wave structure is greatly weakened. When the hot gas mass flows to the corner surface, all the shock wave structures are weakened into weak compression waves, as shown in Figure 10e,f. Compared with small energy actuation, large energy actuation can effectively weaken the intensity of shock waves in the flow field.

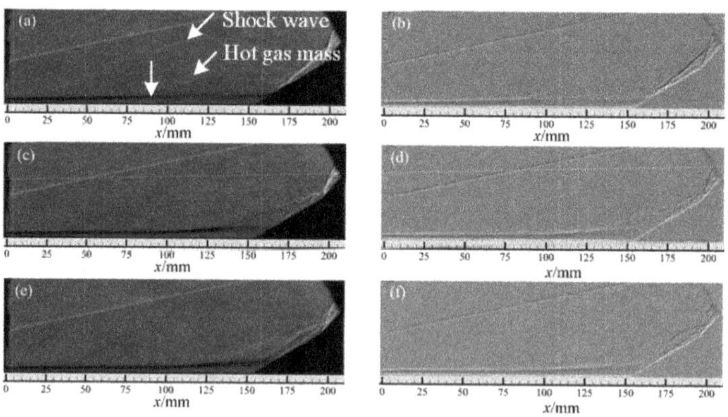

Figure 10. Structure evolution sequence of the flow field controlled by U_{DC} = 1 kV high-energy actuation in the noise flow field: (**a**,**c**,**e**) Original schlieren image; (**b**,**d**,**f**) schlieren image after longitudinal gradient processing.

Because we focused on the control effect of actuation on shock waves, only the flow field image near the corner model was captured for gradient enhancement processing, and the control effect of the two actuation intensities is further compared. Figure 11 shows the enhancement diagram of the shock wave structure evolution gradient of the noise flow field after actuation. The above three pictures show the actuated flow field when the DC power output U_{DC} = 0.5 kV and the output power is 2000 W, and the following three pictures show the actuated flow field when the output U_{DC} = 1 kV and the output power is 4000 W. In order to facilitate the observation of the shock structure and the precise observation of the evolution process, x = 160 mm on the original coordinate axis of Figures 5–10 is set as the origin of the new *X*-axis coordinate. Figure 11 shows the enhancement gradient diagram. Gradient enhancement is one of the means of image enhancement, mainly through the adjustment of image contrast; that is, the contrast intensity of light and dark. Since the schlieren image is mainly black and white toned, the gradient enhancement process is carried out. In addition, the black and white tone replacement is carried out to better reflect the flow field structure. Figure 11a,d corresponds to the phase moment of the hot gas mass touching the corner wall. It can be seen from the figure that although there is a small difference in instantaneous shock wave morphology, the flow field can still be identified as being in the same state. At this time, the hot gas mass has not had a regulating effect on the main structure of the flow field. When the hot gas mass flows to the vicinity of the reattachment point, the structure of the reattachment shock wave is still clear in Figure 11b, and the location of the interaction point of the shock wave is still clear. However, there is no structure of the reattachment shock wave in Figure 11e, and only the second shock wave and the converged shock wave have a clear structure. The structure of the other compression waves is not obvious, which also indicates that the density changes are small. The shock wave intensity decreases when the hot gas mass flows to the shock wave region; although the main flow field structure is no longer prominent in Figure 11c, the shock wave near the shock wave interaction point is still relatively clear. In Figure 11f, the overall structure of the flow field weakens and no clear interaction point appears, indicating that the actuation control effect is better.

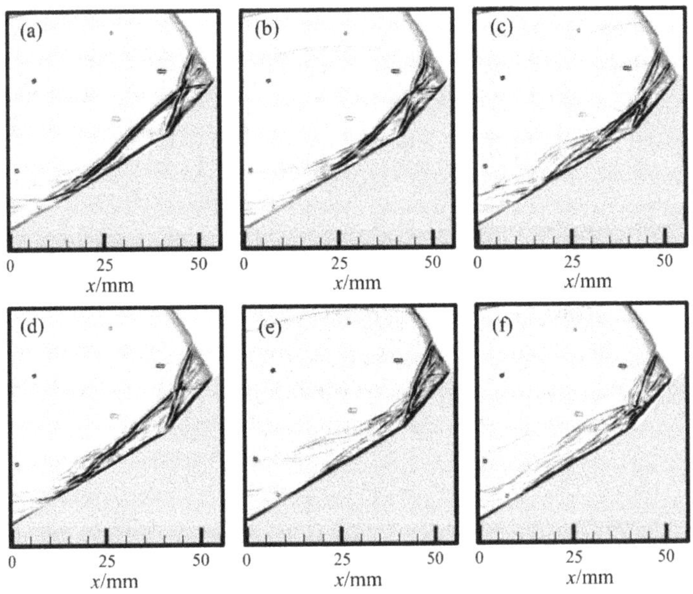

Figure 11. Enhancement gradient diagram of shock wave structure evolution gradient in noise flow field after actuation: (**a–c**) U_{DC} = 0.5 kV actuation flow field evolution; (**d–f**) U_{DC} = 1 kV actuation flow field evolution.

When the high-energy actuation is applied to the hypersonic quiet flow field, it is similar to the noise flow field. The flow field structure of three typical phase moments is extracted for longitudinal gradient processing, as shown in Figure 12. The three images on the left are the actuated flow field when U_{DC} = 0.5 kV, and the three images on the right are the actuation flow field when the output U_{DC} = 1 kV. The control effect of different actuation intensities under the condition of quiet flow is compared and analyzed.

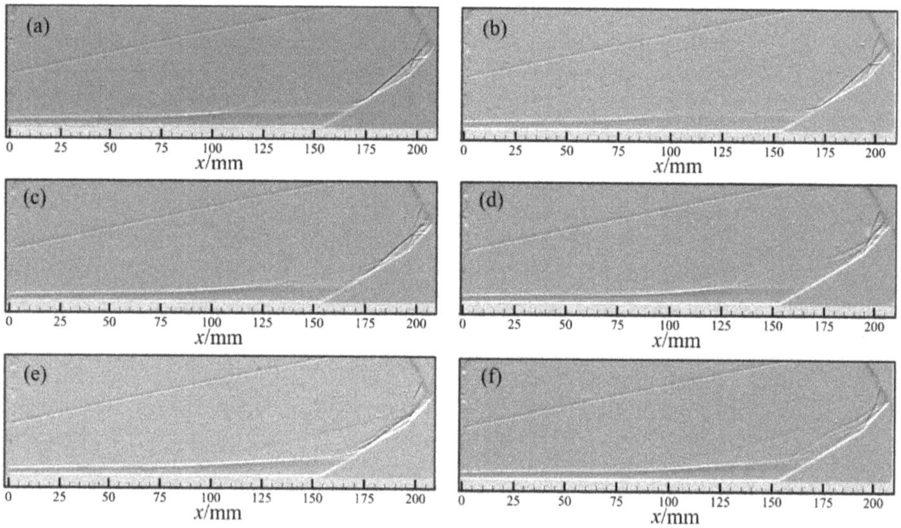

Figure 12. Longitudinal gradient processing diagram of shock wave evolution under quiet flow field: (**a–c**) U_{DC} = 0.5 kV actuation flow field evolution; (**d–f**) U_{DC} = 1 kV actuation flow field evolution.

In Figure 12a,b, the coverage range of the hot air mass is slightly different from that under the noise inflow condition. When U_{DC} = 0.5 kV, the flow range of the hot air mass is x = 107.5 mm~x = 167.5 mm, and the length is about 60 mm. The flow range of the hot gas mass increases to x = 90 mm~x = 167.5 mm, and the length is about 77.5 mm. Therefore, compared with the noise flow field, the influence range of hot gas mass in the quiet flow field is reduced under the two actuation parameters. Considering that the quiet flow field and the noise flow field are the same vehicle test, the discharge continues to the quiet flow field after the noise flow field is actuated. Continuous discharge reduces the output capacity of the power supply, which in turn reduces the influence range of the hot gas mass. However, it can be found that the influence range of the hot gas mass decreases slightly, which decreases by 7.7% and 3.1%, respectively, under the two actuation parameters, so this attenuation is ignored in this test study.

When the hot gas mass covers the reattachment point, Figure 12c shows that the control effect of the flow field under quiet actuation U_{DC} = 0.5 kV is similar to that under noise conditions. The reattachment shock wave moves upward and the interaction point drops, while the flow structure near the interaction point hardly changes. However, the control effect of U_{DC} = 1 kV actuation on the flow field is slightly different from that under noise conditions, as shown in Figure 12d. Unlike the shock wave bifurcation in Figure 9d, the reattachment shock wave here is weakened into a weak compression wave, and the control effect of actuation on the reattachment shock wave is better.

When the hot gas mass covers the shock wave region, Figure 12e shows that actuation plays a prominent role in controlling the interaction points, and the interaction points move down significantly, almost attached to the model. However, the structure of the shock waves, especially the structure of the secondary shock waves, is still clear. Figure 12f shows that higher intensity actuation has better control on the shock wave's structure, no obvious interaction points exist, and the slip line almost disappears. The detailed features will be further analyzed and studied from local gradient enhanced images.

As shown in Figure 13, the local gradient enhanced images of the flow field after actuation with two kinds of parameters are compared. The structures in Figure 13a,d are similar, with clear flow structures such as secondary shock wave, second shock wave, interaction point, main shock wave after convergence, reflected shock wave, slip line, etc. The two images can be regarded as the state of the reference flow field. There are obvious contrast differences in the images at the second phase moment. Figure 13b shows that the flow field barely changes after U_{DC} = 0.5 kV actuation is applied, and only some bending deformation occurs on the slip line, indicating that the interaction point receives high-frequency disturbance. Figure 13e shows that the secondary shock wave almost completely disappears, and the position of the interaction point moves down substantially. The control effect of the U_{DC} = 1 kV actuation is remarkable. When the hot gas mass flows to the shock wave region, the interaction point is almost attached to the model surface in Figure 13c, but the structure of the attached shock wave and the main shock wave is relatively clear. However, in Figure 13f, only one main shock wave can be observed in the whole flow field, and the other structures are greatly weakened. It shows that the actuation of this parameter can effectively control SWBLI and the shock wave/shock wave interaction in the hypersonic quiet flow field.

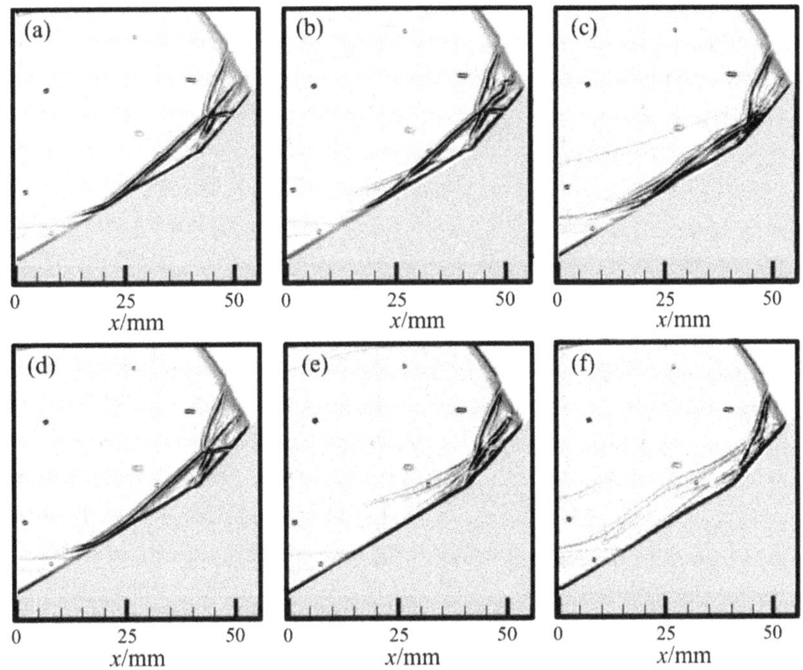

Figure 13. Enhancement gradient diagram of shock wave structure evolution gradient in quiet flow field after actuation: (**a–c**) U_{DC} = 0.5 kV actuation flow field evolution; (**d–f**) U_{DC} = 1 kV actuation flow field evolution.

5. Conclusions

In this paper, the research on dual compression corner shock/boundary layer interaction control is extended from supersonic to hypersonic, and an experimental study on dual compression corner shock/boundary layer interaction controlled by high-energy arc actuation (*Ma* = 6.0) is carried out to summarize the control law of dual compression corner shock/boundary layer interaction by arc plasma actuation under wide Mach number conditions. The specific research conclusions are as follows:

(1) The flow field structure of noise incoming flow conditions is similar to that of quiet incoming flow conditions. Compared with Ma = 2.0 incoming flow conditions, the corner leading edge under hypersonic conditions has a larger separation region, and the size of the separation region is also affected by the level of incoming flow noise. By averaging the gray value of schlieren images, it is found that the length of the separation region is 62.5 mm under the condition of noise incoming flow and 90 mm under the condition of quiet incoming flow, which may be related to the disturbance in the incoming boundary layer. In the noise flow field, the disturbance in the incoming boundary layer is large and contains more vortex structures, which promotes the energy mixing between the boundary layer and the mainstream region. The ability of the boundary layer to resist separation is enhanced.
(2) After the application of high energy arc actuation, the double compression corner shock wave/boundary layer interaction and the shock wave interaction structure under the two types of flow field can show the control effect of the shock wave disappearing and weakening under the control of hot gas masses. The hot gas mass first couples with the separation region near the leading edge of the corner, effectively promoting the momentum exchange between the boundary layer and the main flow region, and the separation shock wave weakens or even disappears. Secondly, when the hot gas mass passes through the reattachment shock region, as the reattachment region of the boundary layer is impacted by the hot gas mass, the reattachment shock wave forks and deforms, the shock wave intensity is greatly weakened, and then the interaction point of the shock wave fluctuates greatly, as well as the slip line and reflected shock wave. High-energy actuation has an effective control effect on the interaction flow field of the shock wave.
(3) The evolution characteristics of the flow structures in the noise flow field stage and the static flow field stage are similar, and the control effect of high-energy actuation on the separation shock wave and the reattachment shock wave is different. On the one hand, the separation region in noise flow field is small, and the control effect of high-energy actuation on reducing the intensity of separated shock waves is relatively good. However, in the quiet flow field, the separation region is large and the range of separated shock waves is large, and the effect of actuation is not ideal. On the other hand, the disturbance in the noise flow field is large. When the hot gas generated by actuation passes through the reattachment shock wave, the reattachment shock wave is wavy after bifurcation, while in the quiet flow field the shock wave after bifurcation is scattered and linear. Plasma actuation plays a good role in controlling the wave system.
(4) In the two types of flow fields, the higher the discharge energy is, the larger the influence range of the hot gas induced by actuation is and the better the control ability of arc plasma actuation on the hypersonic double compression corner shock wave/boundary layer interaction flow field is. When U_{DC} = 0.5 kV actuation is applied, the influence range of the hot gas mass flow direction is about 65 mm, which can weaken the shock wave intensity to a certain extent, and the shock wave interaction point oscillates; thus, the shock wave interaction can be controlled. Although it has a certain flow control effect on the flow field, its control effect on the shock wave structure and the disturbance effect of the flow field need to be further improved. When U_{DC} = 1 kV actuation is applied, the influence range of the hot gas mass flow direction expands to 80 mm, and the actuation has a significant control effect on the flow field. Even in the quiet flow field stage, the secondary shock waves almost completely disappear and the hypersonic double compression corner shock/boundary layer interaction and shock wave/shock wave interaction can be effectively controlled.

Author Contributions: Conceptualization, B.Y., H.Y. and C.Z.; methodology, B.Y. and H.Y.; software, H.Y.; validation, H.L., H.Y., N.Z. and D.Z.; formal analysis, B.Y.; investigation, H.Y.; resources, N.Z. and H.L.; data curation, D.Z.; writing—original draft preparation, B.Y., H.Y. and C.Z.; writing—review and editing, H.Y., N.Z., H.L. and D.Z.; visualization, C.Z.; supervision, H.L.; project administration, H.L.; funding acquisition, H.L. All authors have read and agreed to the published version of the manuscript.

Funding: This research was funded by the National Science and Technology Major Project (Grant No. J2019-II-0014-0035).

Data Availability Statement: The data and materials are available upon request.

Acknowledgments: Thanks to Jingqi Cao and Shaoyi Wang for their help in the experiment.

Conflicts of Interest: The authors declare no conflict of interest.

References

1. Sebastian, R.; Schreyer, A.M. Control of SWBLI in a 24deg Compression Ramp Flow with Air-jet Vortex-generator. In Proceedings of the AIAA Scitech 2023 Forum, National Harbor, MD, USA, 23–27 January 2023.
2. Kane, A.A.; Peetala, R.K.; Kulkarni, V. Investigation of pressure feedback technique to control ramp based SWBLI. *Acta Astronaut.* **2022**, *201*, 482–495. [CrossRef]
3. Yang, S.Z.; Liao, K.; Xie, W.Z. Mechanisms of Hysteresis in the Acceleration and Deceleration Processes of Hypersonic Inlets. *J. Aerosp. Eng.* **2023**, *36*, 04023007. [CrossRef]
4. Hoffman, E.N.; Kendhammer, D.M.; LaLonde, E.J. Effects of distributed roughness on shock-wave/boundary-layer interactions at Mach 7.2. In Proceedings of the AIAA Scitech 2023 Forum, National Harbor, MD, USA, 23–27 January 2023.
5. Whalen, T.J.; Schöneich, A.G.; Laurence, S.J.; Sullivan, B.T.; Bodony, D.J.; Freydin, M.; Dowell, E.H.; Buck, G.M. Hypersonic Fluid–Structure Interactions in Compression Corner Shock-Wave/Boundary-Layer Interaction. *AIAA J.* **2020**, *58*, 4090–4105. [CrossRef]
6. Verma, S.B.; Chidambaranathan, M. Transition control of Mach to regular reflection induced interaction using an array of micro ramp vane-type vortex generators. *Phys. Fluids* **2015**, *27*, 107102. [CrossRef]
7. Sullivan, B.T.; Bodony, D.J. Direct Simulation of Fluid-Structure Interaction in Compression Ramp with Embedded Compliant Panel. In Proceedings of the AIAA Aviation 2019 Forum, Dallas, TX, USA, 17–21 June 2019.
8. Sun, Z.; Gan, T.; Wu, Y. Shock-wave/boundary-layer interactions at compression ramps studied by high-speed schlieren. *AIAA J.* **2020**, *58*, 1681–1688. [CrossRef]
9. Durna, A.S.; Celik, B. Effects of double-wedge aft angle on hypersonic laminar flows. *AIAA J.* **2020**, *58*, 1–15. [CrossRef]
10. Estruch-Samper, D.; Vanstone, L.; Hillier, R.; Ganapathisubramani, B. Micro Vortex Generator Control of Axisymmetric High-Speed Laminar Boundary Layer Separation. *Shock Waves* **2015**, *25*, 521–533. [CrossRef]
11. Babinsky, H.; Li, Y.; Pitt Ford, C.W. Microramp Control of Supersonic Oblique Shock-Wave/Boundary-Layer Interactions. *AIAA J.* **2009**, *47*, 668–675. [CrossRef]
12. Tong, F.L.; Li, X.L.; Duan, Y.H. Direct numerical simulation of supersonic turbulent boundary layer subjected to a curved compression ramp. *Phys. Fluids* **2017**, *29*, 125101. [CrossRef]
13. Zhang, C.B.; Yang, H.S.; Liang, H.; Guo, S.G. Plasma-based experimental investigation of double compression ramp shock wave/boundary layer interaction control. *J. Phys. D Appl. Phys.* **2022**, *55*, 325202. [CrossRef]
14. Vatansever, D.; Celik, B. Unsteady shock interaction mechanisms of high enthalpy reacting flows over double wedges at Mach 7. *Phys. Fluids* **2021**, *33*, 056110. [CrossRef]
15. Hao, J.; Wen, C.-Y.; Wang, J. Numerical investigation of hypervelocity shock-wave/boundary-layer interactions over a double-wedge configuration. *Int. J. Heat. Mass. Transf.* **2019**, *138*, 277–292. [CrossRef]
16. Adamiak, K. Quasi-stationary modeling of the DBD plasma flow control around airfoil. *Phys. Fluids* **2020**, *32*, 085108. [CrossRef]
17. Lago, V.; Joussot, R.; Coumar, S. Plasma Flow Control in a Rarefied Mach 20 Flow over a Flat Plate. In Proceedings of the 8th European Conference for Aeronautics and Aerospace Sciences EUCASS, Kyoto, Japan, 20–25 June 2019.
18. Yates, H.B.; Matlis, E.H.; Juliano, T.J.; Tufts, M.W. Plasma-Actuated Flow Control of Hypersonic Crossflow-Induced Boundary-Layer Transition. *AIAA J.* **2020**, *58*, 2093–2108. [CrossRef]
19. Hu, H.; Meng, X.; Cai, J.; Zhou, W.; Liu, Y.; Hu, H. Optimization of Dielectric Barrier Discharge Plasma Actuators for Icing Control. *J. Aircr.* **2020**, *57*, 383–387. [CrossRef]
20. Lai, C.; Fu, H.; Hu, B.; Ling, Z.; Jiang, L. Aerodynamic Drag Reduction and Optimization of MIRA Model Based on Plasma Actuator. *Actuators* **2020**, *9*, 64. [CrossRef]
21. Yang, H.; Liang, H.; Guo, S.; Tang, M.; Zhang, C.; Wu, Y.; Li, Y. Research Progress of hypersonic boundary layer transition control experiments. *Adv. Aerodyn.* **2022**, *4*, 18. [CrossRef]
22. Bisek, N.J.; Poggie, J.; Nishihara, M. Hypersonic flow over a cylinder with a nanosecond pulse electrical discharge. *J. Thermophys. Heat Transf.* **2014**, *28*, 18–26. [CrossRef]

23. Takashima, K.; Zuzeek, Y.; Lempert, W.R. Characterization of a surface dielectric barrier discharge plasma sustained by repetitive nanosecond pulses. *Plasma Sources Sci. Technol.* **2011**, *20*, 055009. [CrossRef]
24. Leonov, S.B.; Yarantsev, D.A. Near-surface electrical discharge in supersonic airflow: Properties and flow control. *J. Propuls. Power* **2008**, *24*, 1168–1181. [CrossRef]
25. Falempin, F.; Firsov, A.A.; Yarantsev, D.A. Plasma control of shock wave configuration in off-design mode of M = 2 inlet. *Exp. Fluids* **2015**, *56*, 54. [CrossRef]
26. Leonov, S.B.; Firsov, A.A.; Yarantsev, D.A. Flow control in model supersonic inlet by electrical discharge. *Prog. Flight Phys.* **2012**, *3*, 557–568.
27. Kiyoshi, K.; Andrey, Y.S.; Richard, B.M. Control of shock wave-boundary layer interaction using nanosecond dielectric barrier discharge plasma actuators. *J. Propuls. Power* **2017**, *34*, 1–11.
28. Yang, H.; Liang, H.; Zhao, G.; Wang, B.; Zhang, S.; Kong, W. Experimental study on dynamic stall control based on AC-DBD actuation. *Plasma Sci. Technol.* **2021**, *23*, 115502. [CrossRef]
29. Li, Y.; Wu, Y. Research progress and outlook of flow control and combustion control using plasma actuation. *Sci. Sin. Technol.* **2020**, *50*, 1252–1273. [CrossRef]
30. Gan, T.; Wu, Y.; Sun, Z.; Jin, D.; Song, H.; Jia, M. Shock wave boundary layer interaction controlled by surface arc plasma actuators. *Phys. Fluids* **2018**, *30*, 055107. [CrossRef]
31. Yang, H.; Zong, H.; Liang, H.; Wu, Y.; Zhang, C.; Kong, Y.; Li, Y. Swept shock wave/boundary layer interaction control based on surface arc plasma. *Phys. Fluids* **2022**, *34*, 087119. [CrossRef]
32. Watanabe, Y.; Leonov, S.B.; Houpt, A.; Hedlund, B.E.; Elliott, S. Plasma-Assisted Control of Mach-2 Flow field over Ramp Geometry. *IOP Conf. Ser. Mater. Sci. Eng.* **2017**, *249*, 012006. [CrossRef]
33. Watanabe, Y.; Elliott, S.; Firsov, A.; Houpt, A.; Leonov, S. Rapid control of force/momentum on a model ramp by quasi-DC plasma. *J. Phys. D Appl. Phys.* **2019**, *52*, 444003. [CrossRef]
34. Tang, B.; Guo, S.; Hua, L. Experimental Study on High-Energy Surface Arc Plasma Actuation Control of Cylindrical Detached Shock Wave. *Contrib. Plasma Phys.* **2021**, *61*, e202000067. [CrossRef]
35. Luo, Y.; Li, J.; Liang, H.; Guo, S.; Tang, M.; Wang, H. Suppressing Unsteady Motion of Shock Wave by High-Frequency Plasma Synthetic Jet. *Chin. J. Aeronaut.* **2021**, *34*, 60–71. [CrossRef]
36. Tang, M.; Wu, Y.; Guo, S. Effect of the StreamwisePulsed Arc Discharge Array on Shock Wave/Boundary Layer Interaction Control. *Phys. Fluids* **2020**, *32*, 076104. [CrossRef]
37. Kong, Y.; Li, J.; Wu, Y.; Liang, H.; Guo, S.; Yang, H. Experimental study on shock-shock interaction over double wedge controlled by surface arc plasma array. *Contrib. Plasma Phys.* **2022**, *62*, e202200062. [CrossRef]
38. Tang, B.; Guo, S.; Liang, H.; Tang, M.-X. Influence of low ambient pressure on the performance of a high-energy array surface arc plasma actuator. *Chin. Phys. B* **2020**, *29*, 105204. [CrossRef]
39. Tang, M.; Wu, Y.; Zong, H.; Guo, S.; Liang, H.; Luo, Y. Experimental investigation on compression ramp shock wave/boundary layer interaction control using plasma actuator array. *Phys. Fluids* **2021**, *33*, 066101. [CrossRef]

aerospace

MDPI

Article

Experimental Study on Hypersonic Double-Wedge Induced Flow Based on Plasma Active Actuation Array

Bo Yang [1,2], Hesen Yang [3,*], Ning Zhao [1], Hua Liang [3,*], Zhi Su [3] and Dongsheng Zhang [3]

1 College of Aerospace Engineering, Nanjing University of Aeronautics and Astronautics, Nanjing 210016, China; yandr_2002@139.com (B.Y.); zhaoam@nuaa.edu.cn (N.Z.)
2 Hypervelocity Aerodynamics Institute of China Aerodynamics Research and Development Center, Mianyang 621000, China
3 National Key Lab of Aerospace Power System and Plasma Technology, Air Force Engineering University, Xi'an 710038, China; suzhi95703@163.com (Z.S.); zds13475138164@163.com (D.Z.)
* Correspondence: yanghesen96@126.com (H.Y.); lianghua82702@163.com (H.L.)

Abstract: The double-wedge configuration is a typical characteristic shape of the rudder surface of high-speed aircraft. The impact of the shock wave/boundary layer interaction and the shock wave/shock wave interaction resulting from the double wedge on aircraft aerodynamics cannot be ignored. The aerodynamic performance of the aircraft would be seriously affected. Accordingly, to reduce the wave drag, and to relieve the thermal load and pressure load, flow control is required for the shock wave/shock wave interaction and the shock wave/boundary layer interaction induced by the double-wedge configuration. In this paper, double-wedge shock wave/shock wave interaction is controlled by a high-energy surface arc discharge array and observed by high-speed schlieren flow field measurement at Mach 8. The 30-channel discharge array is set on the primary wedge plane, and actuation is generated. Hypersonic V shock wave/shock wave interaction is effectively controlled by the shock wave array induced by the high-energy surface arc discharge array, which makes the shock wave/shock wave interaction structure disappear or intermittent. The potential control mechanism is to reduce strong shock wave interaction by transforming the type of shock wave interaction. Therefore, the ability of plasma array actuation to control complex shock wave/shock wave interaction is verified, which provides a new method for hypersonic shock wave/shock wave interaction control.

Keywords: hypersonic; surface arc discharge; plasma actuation; flow control; experimental research

Citation: Yang, B.; Yang, H.; Zhao, N.; Liang, H.; Su, Z.; Zhang, D. Experimental Study on Hypersonic Double-Wedge Induced Flow Based on Plasma Active Actuation Array. *Aerospace* **2024**, *11*, 60. https://doi.org/10.3390/aerospace11010060

Academic Editor: Zhijin Wang

Received: 28 October 2023
Revised: 28 December 2023
Accepted: 4 January 2024
Published: 9 January 2024

1. Introduction

Shock wave/shock wave interaction and shock wave/boundary layer interaction are widespread problems in high-speed flows, which are key problems restricting the development of a new generation of aircraft [1,2]. The double-wedge configuration is a prominent feature of high-speed aircraft, appearing in numerous areas, such as the control rudder surface, air body–wing joint, and multi-inclined compressed inlet [3–5]. Shock wave/boundary layer interaction and shock wave/shock wave interaction occurring within these hypersonic double-wedge flows induce a range of complex flow phenomena, including boundary layer separation and reattachment, unstable shear layer, supersonic jet, nonlinear shock wave oscillation, and hysteresis phenomenon. These phenomena are major obstacles that restrict the development of a new generation of aircraft [6]. The two kinds of interaction induce extreme thermal loads and pressure loads, and the complex flow environment further leads to load fluctuations, resulting in component damage or even loss of control of the aircraft in serious cases [7,8]. Consequently, in order to reduce wave drag and relieve thermal load and pressure load, it is necessary to carry out flow control for shock wave/shock wave interaction and shock wave/boundary layer interaction induced by the double-wedge configuration [9,10].

Complex shock wave/shock wave interaction and shock wave/boundary layer interaction under the condition of high-speed incoming flow bring about severe aerodynamic thermal loads and drastic changes in flow parameters, bringing serious challenges to the aircraft body structure and thermal protection [11,12]. The current passive protection measures cannot easily ensure the safety of aircraft, but if the load can disappear or temporarily rest by changing the wave system structure, it is expected to bring new ideas for the thermal protection design of high-speed aircraft [13,14]. Therefore, a control method that can effectively change the structure of a shock wave system without changing the surface characteristic shape of aircraft is needed.

Plasma flow control is a method to control fluid flow by using plasma technology, which belongs to active flow control means. It is mainly divided into dielectric barrier discharge, synthetic jet and surface arc discharge. Especially in the hypersonic field, surface arc discharge is widely used in solving shock wave/shock wave interaction and shock/boundary layer interaction, and lots of research has been carried out by scholars in the field.

In numerous current flow control experimental studies, Wang et al. experimentally verified the control ability of a surface arc plasma actuator to compress corner-inclined shock waves under Mach 6 flow [15]. Xie et al. used plasma synthetic jet actuation to achieve effective control of type VI shock wave/shock wave interaction on a double wedge. In the three working conditions, the longest control time was only 90 μs, the energy deposited into the flow field was above 4.1 J, and the repetition frequency was only 1 Hz [16]. However, surface arc plasma actuation shows excellent control effects in supersonic shock wave control, shock wave/boundary layer interaction, and forced supersonic boundary layer transition [17–20]. Yang conducted controlling hypersonic boundary layer transition using a surface arc plasma actuation array. The influence of three different actuation frequencies (8, 34, and 55 kHz) was studied based on linear stability theory analysis, and a transition criterion under the control of plasma actuation was proposed. Finally, the corresponding control mechanism was summarized, and the transition control mechanization was refined [17]. In addition, Yang experimented on the stability of the hypersonic plate boundary layer by using a spanwise plasma actuation array. The experimental results verify the ability of extensional array plasma actuation to control the stability of the hypersonic plate boundary layer, suggesting that it has great potential in the promotion of hypersonic boundary layer transition [18]. In addition, a wind tunnel experiment was carried out under the condition of Mach number 6 to study the stability adjustment of the hypersonic blade. The results verify the ability of plasma actuation to stimulate the instability of hypersonic cone boundary layers and provide technical support for the further development of transition control methods [19]. Kong et al. conducted experiments on the control of double-wedge flow by 10 discharge actuators under hypersonic conditions and analyzed the interaction evolution process between plasma surface arc discharge and a flow field in detail [21]. Ding et al. studied the unsteady control mechanism of pulsed surface arc discharge plasma on hypersonic compression corner flow by combining numerical simulations and experiments. The interaction mechanism between arc discharge plasma and hypersonic flow was revealed. The research reveals that the local Joule heat is created by the surface arc discharge, which also triggers the separation region near the wall, leading to an increase in local displacement thickness and the formation of an unsteady virtual wedge that moves along the wall. Consequently, an oblique shock wave is produced, with the shock angle varying with time, and an unsteady shock wave/shock wave interaction including shock reflection is established between the front wedge shock wave. The hot gas mass generated by the discharge demonstrates a significant capacity to control the oblique back wedge shock wave [22]. Although the conventional surface arc discharge electrode plays a certain role in controlling the hypersonic flow field, there is still a more urgent need for plasma discharge with more paths and larger affected areas, in order to obtain better control effects.

In order to further improve the ability of surface arc actuation to control complex shock wave/shock wave interaction and shock wave/boundary layer interaction induced by a double-wedge configuration in a wide velocity domain, a spanwise and flow-direction combination array actuator is designed in this paper to broaden the disturbance range and achieve effective control for typical shock wave interaction induced by a double-wedge configuration. The ability of array plasma actuation to control complex shock wave/shock wave interaction is verified.

2. Experimental System

The experiment is conducted in Φ0.5 m conventional hypersonic wind tunnel in the Hypervelocity Aerodynamics Institute of China Aerodynamics Research and Development Center. The whole experiment system mainly includes hypersonic wind tunnel, double-wedge model, plasma actuation system, high-speed schlieren system, synchronous control system, etc.

2.1. Hypersonic Wind Tunnel and Test System

The experiment was carried out in a hypersonic wind tunnel with a diameter of 0.5 m at the Hypervelocity Aerodynamics Institute of China Aerodynamics Research and Development Center. The wind tunnel is a conventional hypersonic wind tunnel with a simulated Mach number of 5–10 and a diameter of 0.5 mm (Mach number 5–8) and 0.6 mm (Mach number 9, 10), respectively. The total pressure range of the wind tunnel is 0.1–3 MPa, the total temperature range is 301–1073 K, and the unit Reynolds number is 1.5×10^5~6.0×10^7/m. The wind tunnel test section is shown in Figure 1. There are relatively few studies on the double-wedge flow control under hypersonic conditions. The maximum Mach number of wind tunnel operation is Mach 8, and studies on the double-wedge problem under higher Mach number are more valuable; there are fewer studies on the double-wedge problem under Mach 8 conditions. Therefore, Mach 8 is chosen as the experimental condition, and the experimental model is designed under this condition. In addition, for other flow conditions, considering the high-altitude aircraft environment and plasma discharge environment, a total pressure of 3 MPa and a total temperature of 410 K are selected. After determining the total pressure, total temperature, and Mach number, other test conditions can be determined. The experimental flow parameters are shown in Table 1.

Figure 1. Hypersonic wind tunnel and test section.

Table 1. Main parameters of incoming stream.

Ma_∞ (U_∞/c)	Re/m ($\rho U_\infty/\mu$)	U_∞ (m/s)	ρ (kg/m³)	P_0 (MPa)	T_0 (K)	P_S (Pa)	T_S (K)
8.0	6.16×10^6	874.292	0.036	3	410	307.287	29.710

Ma_∞, Re/m, U_∞, ρ, P_0, T_0, P_S, T_S, c represent incoming Mach number, unit Reynolds number, free flow velocity, incoming density, total pressure, total temperature, static pressure, static temperature, and sound velocity.

2.2. Double-Wedge Model and Actuator

The model consists of a first-stage wedge, a second-stage wedge, a support, and a plasma actuator. Among them, because the primary wedge is at the front of the model, it will be subjected to strong aerodynamic heating in the hypersonic flow field, so alumina ceramic material is used to prevent its tip from deformation or damage at high temperature. In the flow field, double wedge will generate separation shock waves, and there are shock waves on its surface. However, compared with the tip of the first-stage wedge, the surface of the second-stage wedge is not easily deformed. The second-stage wedge is made of PEEK (poly-ether-ether-ketone) material. The actuator is embedded in the support, and in order to prevent the ablation caused by discharge, the whole alumina ceramic is processed. The model support is made of plexiglass material.

The size of the model is shown in Figure 2. The streamwise direction length is 240 mm, the normal height is 177 mm, and the angle of the first-stage wedge is 30°; the angle of the second-stage wedge is 60°. The actuator array is located 15 mm upstream of the second-stage wedge, the electrode gap is 0.5 mm, and the flow distance between the two adjacent actuators is 9 mm. The arrangement of two columns × 15 channels of flow direction is adopted, and there are 30 channels of actuators in total. For the driver design, in addition to the needle electrode at both ends, the middle electrode is U-shaped, and the ceramic cover plate is used for packaging.

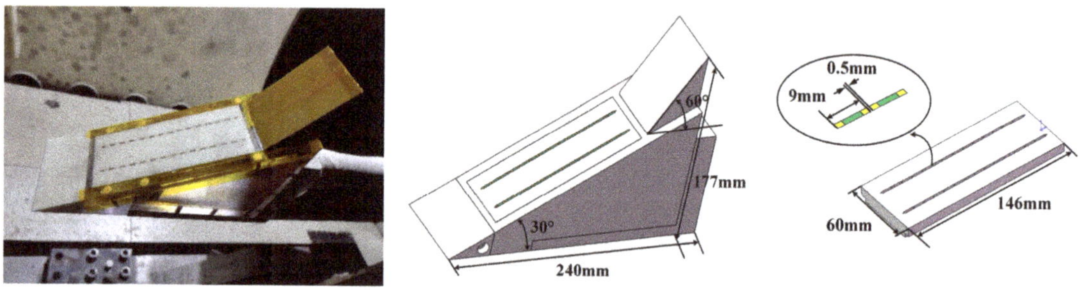

(a) (b) (c)

Figure 2. Double wedge model. (**a**) Installation of the model in the wind tunnel; (**b**) model parameter. (**c**) The details of the actuator.

2.3. The Power Supply System and Schlieren System

The power supply system consists of a nanosecond pulse power supply and a low-power DC power supply. The input voltage of both power supplies is 220 V and the input frequency is 50 Hz. The nanosecond pulse power supply is manufactured by China Xi'an Lingfeng Yuan Electronic Technology Co. The peak voltage range of the nanosecond pulse power supply is 0–20 kV, and the discharge pulse frequency is 0–20 kHz. The low-power DC power supply is manufactured by Jiangsu, China of Zhengjie power factory. DC power output ranges from 0 to 1.2 kV. The electrical parameter-measuring equipment is composed of an oscilloscope, voltage probe, and current ring. The discharge loop is connected to a micronormal capacitor, a diode, and a current-limiting resistor. Two power sources are connected in the circuit, in which the nanosecond pulse power supply can provide instantaneous pulse high voltage, quickly break down the discharge gap between the positive and negative electrodes of the driver, and establish a discharge channel. The DC power supply charges the capacitor through the resistance, storing the energy in the capacitor. When the discharge gap is broken down, the capacitor quickly releases the energy through the discharge channel, producing a high-energy plasma actuation. Under the operation of this circuit, the peak output current can reach an order of 100 A, and the deposited energy can reach the order of 1 J.

Instantaneous snapshot by high-speed schlieren is the most commonly used test method in high-speed flow experiments. It does not interfere with the flow field, maintains the authenticity and accuracy of the original flow field, and can capture the change in density gradient in the flow field, thus revealing the detailed structure in the flow field. In order to observe the flow field structure and wave system of shock wave interaction, a high-speed schlieren system is used in this experiment. The schlieren light path is set in a Z-shape light path. The structure of the flow field is captured by a high-speed schlieren system. The schlieren light path is set in a Z-shape light path. The high-speed camera uses a Phantom V2512 high-speed CCD camera (Phantom, Vision Research, Wayne, NJ, USA) to record images at a sampling frequency of 75,000 fps and an exposure time of 9 μs. For the optical setting, through the early debugging, the coupling effect of exposure time, aperture, and light source intensity is the best, which better reflects the experimental results.

2.4. Research Methods

A grayscale image is a two-dimensional data structure that represents an image through the intensity value of each pixel in the range of grayscale values, generally used to depict information on black-and-white or grayscale intensity but not color. In this experiment, we utilize the mean gray-level processing method to process the spatial gray-level image. This entails calculating the average intensity value of each pixel in the image time series. Based on the time-resolved schlieren space gray image, this paper also analyzes the decomposition through root mean square (RMS) and Snapshot proper orthogonal decomposition (SPOD). Although schlieren display technology is mainly used for qualitative flow field diagnosis, some post-processing methods based on schlieren snapshot sequences have also been developed rapidly in recent years, which are used to extract semi-quantitative data from a large number of qualitative schlieren data and then initially reveal some quantitative flow field results [23]. In this paper, in addition to analyzing the instantaneous schlieren results, statistical processing is performed on the obtained schlieren image sequence to obtain its average schlieren intensity field (I_{mean}) and RMS schlieren intensity field (I_{rms}), which are specifically defined as follows:

$$I_{mean} = \sum_{k=1}^{N} I_k / N \tag{1}$$

$$I_{rms} = \sqrt{\sum_{k=1}^{N} (I_k - I_{mean})^2 / N} \tag{2}$$

where I_k is the gray value matrix of pixels in the kth schlieren snapshot; N is the total sample number of the schlieren snapshot sequence.

The average gray value is the schlieren diagram of the steady flow field. The flow pulsation state can be obtained by calculating the root mean square of the time series of the gray level of spatial pixels.

Proper orthogonal decomposition (POD), as a principal component analysis method, decomposes the flow field into different modes according to the contribution rate of the flow field, which can filter out the secondary structure and noise in the flow field and obtain the main flow structure. POD method is widely used in experimental and simulation data to study turbulence characteristics. Data obtained by PIV technology and LES simulation method can be processed by POD method to obtain spatial flow topology of velocity and vortex structure [24]. Berry et al. [25] used POD method to process time series of time-resolved schlieren images and studied the characteristic structure of supersonic rectangular jet. Chaganti et al. [26] processed a time series of color schlieren images and studied the non-steadiness of shock wave/boundary layer interaction.

However, fast orthonormal decomposition (SPOD) mainly analyzes the correlation in multiple pictures distributed along the time axis for points in the same position [27]. The modes obtained by SPOD decomposition have the following characteristics: the lower the mode order, the higher the contribution rate to the flow field; the modes with high contribution rate represent the dominant structure of the flow field. The first mode has the highest contribution rate and usually reflects the steady-state information of the flow field. Steady-state characteristics have been studied in gray mean or root-mean-square analysis. The advantage of SPOD analysis method is to extract unsteady characteristics of flow field [26].

In the process of experiment, in order to ensure that statistical results are obtained under the same number of samples, 300 instantaneous schlieren snapshots after actuation are uniformly selected as statistical sample values (300 instantaneous schlieren snapshots are also selected for the base flow field).

3. Study on Base Interaction Flow Field

In order to clarify the structure of the shock wave in the flow field and determine the type of shock wave/shock wave interaction, the schlieren image is analyzed. As shown in Figure 3, the instantaneous schlieren snapshot of the base flow field at Mach 8.0 is given, where the flow direction and normal coordinate scales are dimensionless, processed by the model length and height, respectively. According to the schlieren display results, the base flow field of the double-wedge model is the normal intersection of the same side shock wave at the two Mach numbers. The oblique shock wave AC generated by the first-stage wedge and the oblique shock wave BC generated by the second-stage wedge intersect at point C, and the transmitted shock wave CD is generated. The shock angle of CD is smaller than BC but larger than AC. Subsequently, in order to balance the pressure between regions ④ and ③ behind shock wave CD and shock wave BC, the wave structure CF is induced. Notice that CE here is not a simple slip line but a supersonic jet. At the same time, it can be clearly observed that there are two multi-wave points, C and C′, in the interfering flow field. The results show that Edney's typification of the shock/shock interaction phenomenon is also applicable to hypersonic ranges. Therefore, according to a comprehensive judgment, the benchmark interfering flow field shown in Figure 3 is the type V shock wave/shock wave interaction.

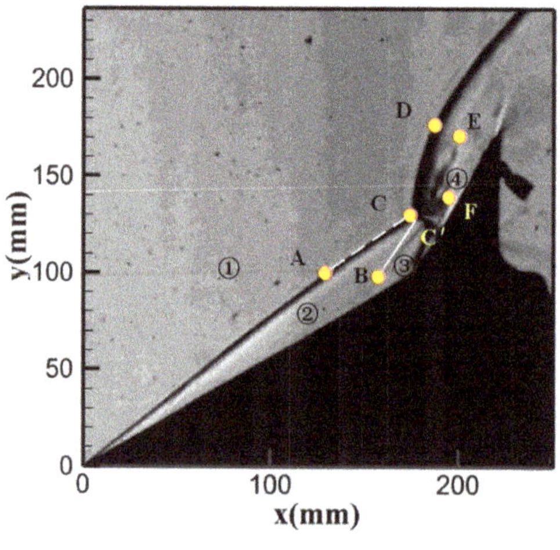

Figure 3. Base flow field of double-wedge model at Mach 8.

4. The Control Effect of Array Plasma Actuation

The interaction flow field induced by a double wedge under the actuation of 30-channel discharge is shown in Figure 4. Since the discharge arrangement of the actuator is two rows and 15 channels, 15 shock wave and hot gas cluster arrays can be clearly seen from the figure, indicating that the discharge is successful at Ma 8. It can also be seen from the figure that with the generation of plasma actuation, the flow field structure changes, the shape of the first-stage wedge precursor of the experimental model bends and deforms, and the type V shock wave/shock wave interaction on the surface of the second-stage wedge is significantly weakened, which indicates that the arc plasma actuation achieves effective control of the flow field wave system.

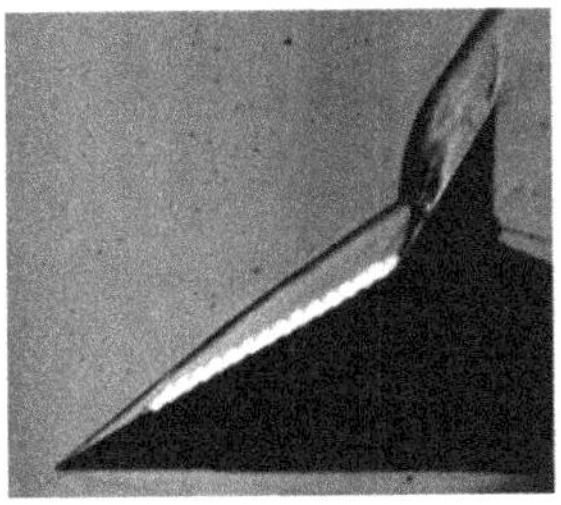

(**a**) t = 6.66 μs

(**b**) t = 13.32 μs

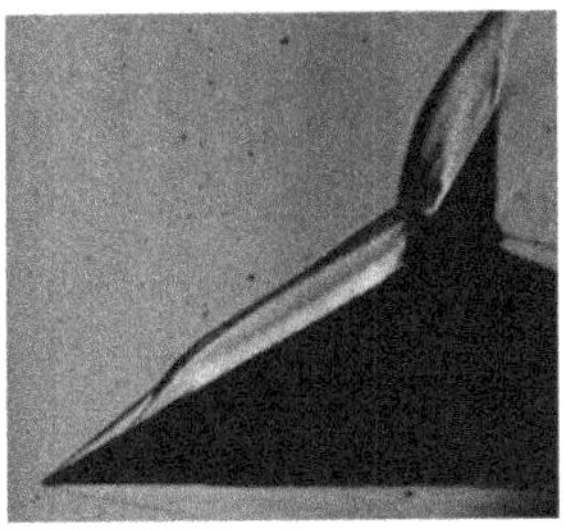

(**c**) t = 26.64 μs

(**d**) t = 39.96 μs

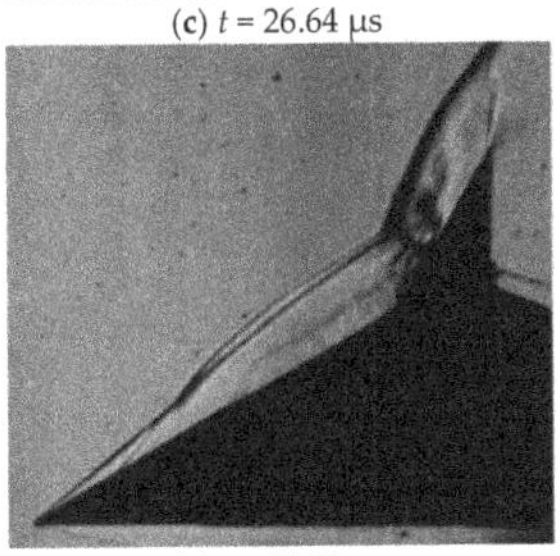

(**e**) t = 53.28 μs

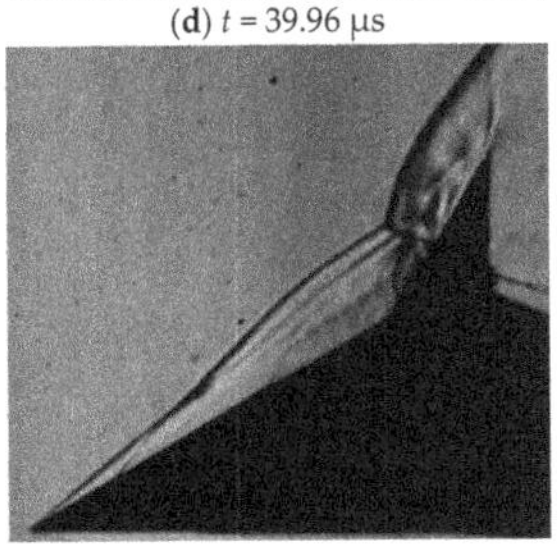

(**f**) t = 66.60 μs

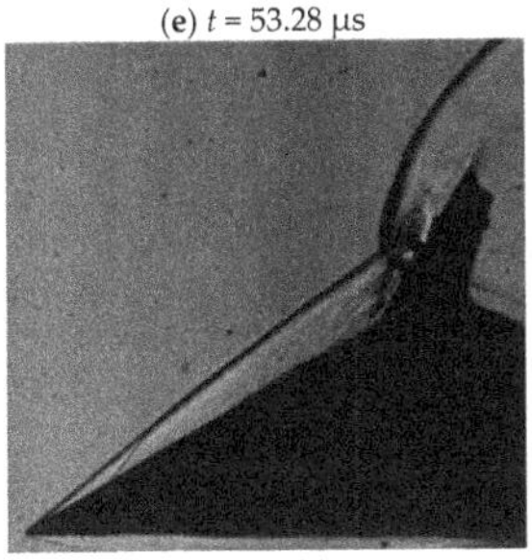

(**g**) t = 93.24 μs

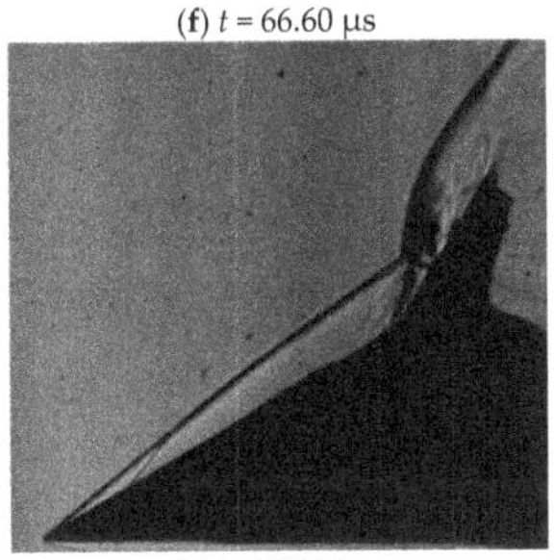

(**h**) t = 106.56 μs

Figure 4. Thirty-channel discharge control of shock wave/boundary layer interaction induced by double wedge.

Through detailed comparative analysis, the control evolution process of a discharge pulse on a double-wedge flow field can be divided into two stages, namely, the disappearance of the shock wave interaction structure and the reconstruction of the shock wave interaction structure.

As shown in Figure 4b, at the initial discharge moment (t = 0 μs), 30 arc arrays with spanwise overlap are observed on the surface of the model. After 6.66 μs, the arc array develops into a dazzling white light due to energy deposition in a short time. Subsequently, as shown in Figure 4d, the actuation induced by the plasma bulge makes the impact shock wave AC appear arched, while the second-stage pre-wedge oblique shock wave BC disappears. In the next two moments, with the expansion of the plasma bulge, the shock wave and interaction structure in front of the wedge disappear, leaving only a bow shock wave. As shown in Figure 4f, the plasma bulge changes into a plasma wedge, and the bow shock wave in the flow field also changes into an oblique shock wave.

At the time of t = 53.28 μs, as shown in Figure 4e, the second-stage pre-wedge oblique shock wave BC begins to reconstruct. Then, as shown in Figure 4f, the oblique shock wave BC appears completely but its shape is distorted, and the interaction structure does not appear. At the time of t = 93.24 μs, the transmitted shock wave appears, and the interaction structure begins to rebuild. The flow field returns to the state before actuation until t = 106.56 μs; then, the control of array plasma actuation flow field ends.

According to the analysis of static discharge results, at Mach 8, the shock wave array successively forms a plasma bulge and plasma wedge, resulting in shock wave deformation, while the hot gas mass directly causes the disappearance of the shock wave interaction structure in the process of downstream expansion, and the array plasma actuation shows the control ability of double-wedge flow.

4.1. Results of Gray Average and RMS

Firstly, the overall control effect of thirty surface arc actuations on hypersonic interaction flow field induced by a double-wedge is measured by gray average results and RMS results, as shown in Figures 5 and 6.

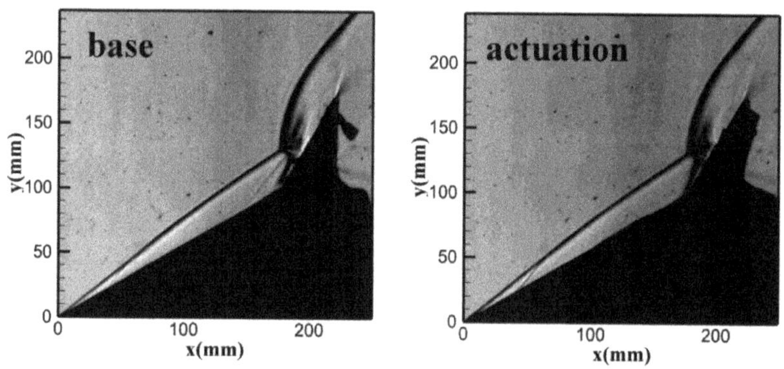

Figure 5. The control effect of gray average.

In the gray average results, it is obvious that the intensity of the oblique shock wave BC generated by the second-stage wedge decreases obviously under the shock wave induced by the array actuation, while the interaction point of the type V shock wave induced by the double wedge does not change significantly. In the control effect of RMS, the pulsation region of the shock wave induced by actuation can be observed. Most importantly, the pulsation level of a type V shock wave induced by the double wedge is enhanced near the shock disturbance point. Therefore, the introduction of array plasma actuation has a positive control effect on the shock disturbance point.

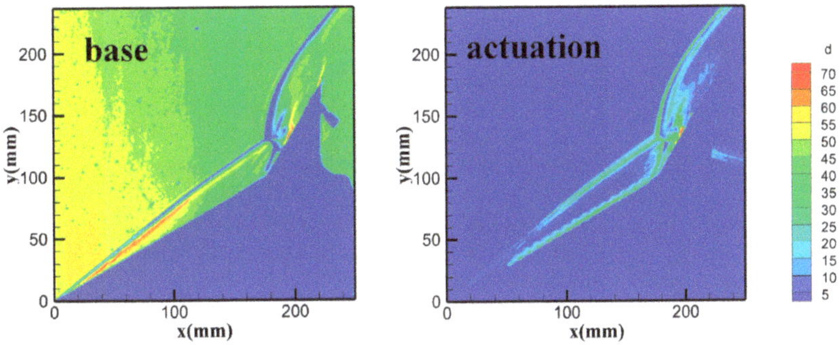

Figure 6. The control effect of RMS.

4.2. Results of SPOD Analysis

SPOD decomposes the flow field into a series of modes to extract the coherent structure in the flow field. The time series of 300 images of the base flow field is decomposed by SPOD, and each mode is sorted according to the energy contribution rate from the largest to the smallest. The lower the mode order, the greater the energy contribution rate to the original flow field. Thus, 300, 500, and 1000 photos are selected for processing, and the results are found to be basically the same, so the processing results of 300 photos are displayed.

As shown in Figure 7, the energy proportion of each unsteady mode analyzed by SPOD is shown. It can be seen that the energy proportion of the first several unsteady modes is very high, and the unsteady mode energy accumulation value of the first ten modes reaches more than 99.94%. Therefore, the first ten modes are selected to analyze the modal results of the base interaction flow field and the actuation control flow field. The results are shown in Figure 8.

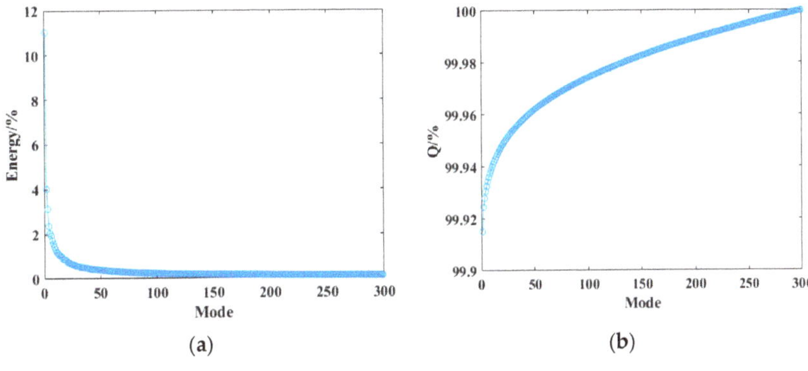

Figure 7. The energy proportion of each mode analyzed by SPOD. (**a**) Unsteady modal energy distribution; (**b**) unsteady mode energy accumulation value.

From the SPOD results, it can be seen that in the benchmark SPOD results, the obvious second wedge oblique shock wave and shock disturbance point are observed both in the steady-state mode (mode 1) and other unsteady modes. Since MOD1 is a steady-state mode, it can be found that the color of the red region in the actuation flow field is lighter than that of the base flow field, indicating that the shock wave intensity is weakened under the condition of plasma actuation, and the plasma arc discharge plays a role in regulating the shock wave. In the unsteady mode results, mode 2, mode 3, mode 4, and mode 5 are the actuation modes. The first shock wave structure or the second shock wave structure can be obviously observed in both the base flow field and the actuation flow field, and the energy of the shock wave structure is still dominant. In addition, the shock wave driven by the array plasma can also be clearly observed from MOD2–5. It shows that the energy stimulated by the MOD2–5 plasma also occupies a certain dominant position in the actuation flow field. The latter unsteady mode is the shock wave/shock wave interaction control mode, and it can be observed that the display degree, which decreases after the shock wave interaction point, is affected. Therefore, on the basis of the SPOD analysis results, the display of the double-wedge oblique shock wave controlled by plasma is reduced after processing, the control effect of the array plasma actuation on the shock wave/shock wave interaction is verified, and the intensity of the double-wedge oblique shock wave is weakened.

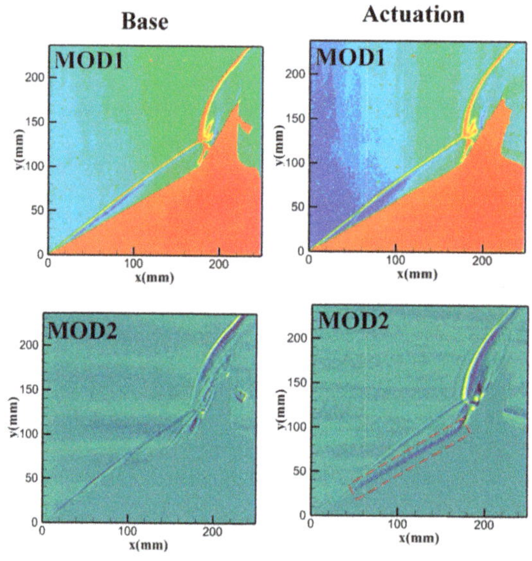

(a)

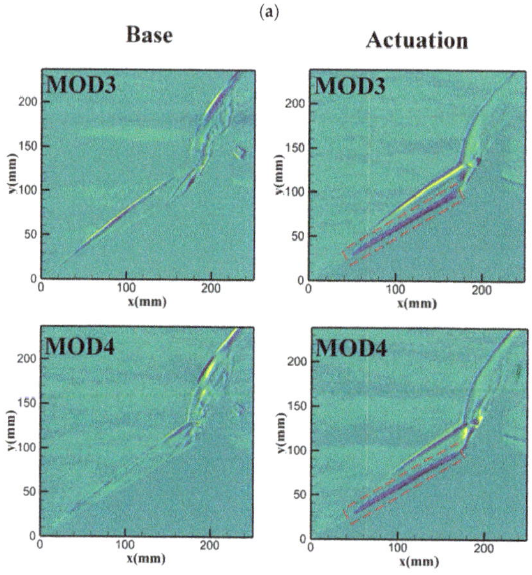

(b)

Figure 8. *Cont.*

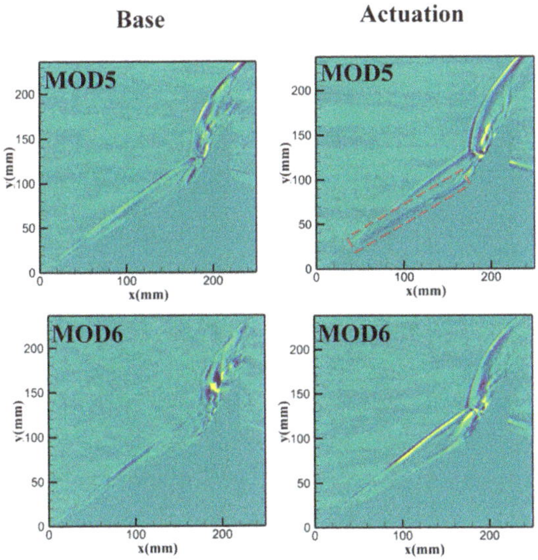

(c)

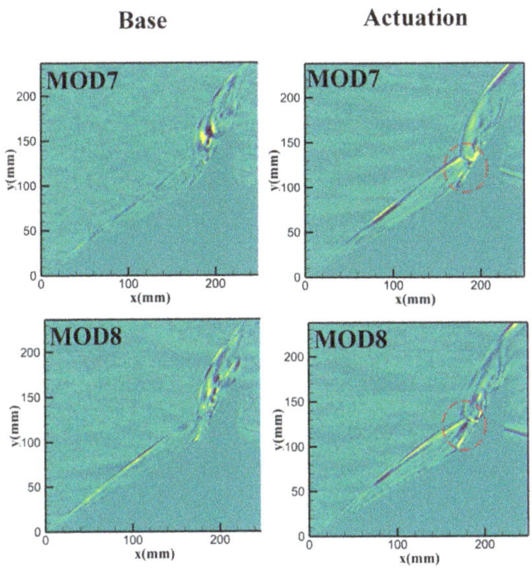

(d)

Figure 8. *Cont.*

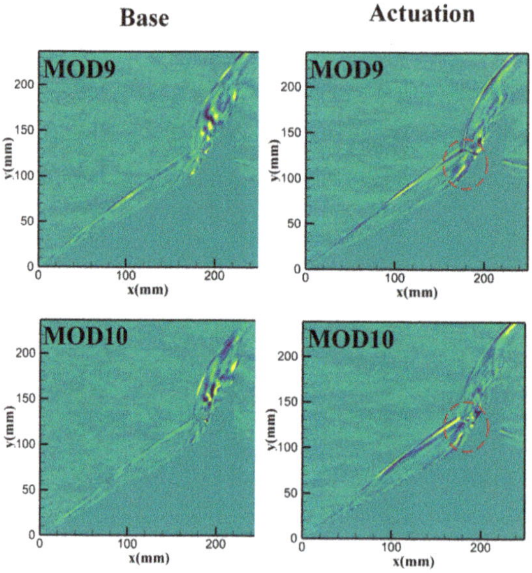

(e)

Figure 8. The results of each mode analyzed by SPOD. (**a**) MOD1, 2; (**b**) MOD3, 4; (**c**) MOD5, 6; (**d**) MOD7, 8; (**e**) MOD9, 10.

Another control effect that needs attention is the ④ region defined in Figure 3, where each SPOD mode shows that the supersonic jet emitted by point C has been effectively controlled and has a tendency to develop into a complete slip line. Therefore, it can be inferred that, by the control of the shock wave/shock wave interaction structure, the final control mechanism formed by the array surface arc actuation may be to convert the type V shock wave/shock wave interaction type, with more intense interaction and more complex flow field into the type VI shock wave/shock wave interaction type during the change in the shock mechanism.

5. Conclusions

In this paper, a high-speed schlieren flow field test method is used to study the Ma8 hypersonic double-wedge shock wave interaction controlled by a high-energy surface arc discharge array. As a new type of super multipath discharge, 30 channels are used in the double-wedge mode. A 30-channel discharge array is set on the first-stage wedge surface, and actuation is applied to the interaction flow field. The shock wave array and the hot gas array induced by the high-energy surface arc discharge array develop into a virtual bulge and wedge in the hypersonic double wedge. Under the action of the virtual wedge, the complex double-wedge shock disturbance structure evolves into a bow shock wave or an oblique shock wave. The results show that the ultra-multi-array surface arc actuation used in this paper can effectively control hypersonic Type V shock wave/shock wave interaction, making the shock wave/shock wave interaction structure disappear or be intermittent, and the new surface arc discharge plays a better role in regulating the shock wave. The potential control mechanism is to mitigate the strong interaction caused by shock wave/shock wave interaction by transforming the type of shock wave/shock wave interaction. This verifies the ability of array plasma actuation to control complex shock wave/shock wave interaction and provides a new type of plasma surface arc discharge and a new method for hypersonic shock wave/shock wave interaction control.

Author Contributions: Conceptualization, B.Y. and H.Y.; methodology, B.Y. and H.Y.; software, H.Y.; validation, H.L., H.Y. and D.Z.; formal analysis, B.Y.; investigation, H.Y.; resources, N.Z.; data curation,

D.Z.; writing—original draft preparation, B.Y.; writing—review and editing, H.Y.; visualization, Z.S.; supervision, H.L.; project administration, H.L.; funding acquisition, H.L. All authors have read and agreed to the published version of the manuscript.

Funding: This research was funded by the National Science and Technology Major Project (Grant No. J2019-II-0014-0035).

Data Availability Statement: The data and materials are available upon request.

Acknowledgments: Thanks to Cao Jingqi and Wang Shaoyi for their help in the experiment.

Conflicts of Interest: The authors declare no conflicts of interest.

References

1. Boyd, I.D.; Chen, G.; Candler, G.V. Predicting failure of the continuum fluid equations in transitional hypersonic flows. *Phys. Fluids* **1995**, *7*, 210–219. [CrossRef]
2. Lee, C.; Chen, S. Recent progress in the study of transition in the hypersonic boundary layer. *Natl. Sci. Rev.* **2019**, *6*, 155–170. [CrossRef] [PubMed]
3. Castrogiovanni, A. Review of "The Scramjet Engine, Processes and Characteristics". *AIAA J.* **2010**, *48*, 2173–2174. [CrossRef]
4. Devaraj, M.K.K.; Jutur, P.; Rao, S.M.V.; Jagadeesh, G.; Anavardham, G.T.K. Experimental investigation of unstart dynamics driven by subsonic spillage in a hypersonic scramjet intake at Mach 6. *Phys. Fluids* **2020**, *32*, 026103. [CrossRef]
5. Kimmel, R.L. Aspects of hypersonic boundary layer transition control. In Proceedings of the 41st Aerospace Sciences Meeting and Exhibit, Reno, NV, USA, 6–9 January 2003.
6. Raj, N.O.P.; Venkatasubbaiah, K. A new approach for the design of hypersonic scramjet inlets. *Phys. Fluids* **2012**, *24*, 086103. [CrossRef]
7. Borg, M.P.; Schneider, S.P. Effect of Freestream Noise on Roughness-Induced Transition for the X-51A Forebody. *J. Spacecr. Rockets* **2008**, *45*, 1106–1116. [CrossRef]
8. Everhart, J.L.; Alter, S.J.; Merski, N.R. Pressure gradient effects on hypersonic cavity flow heating. In Proceedings of the 44th AIAA Aerospace Sciences Meeting and Exhibit, Reno, NV, USA, 6–9 January 2006.
9. Brès, G.A.; Inkman, M.; Colonius, T.; Fedorov, A.V. Second-mode attenuation and cancellation by porous coatings in a high-speed boundary layer. *J. Fluid. Mech.* **2013**, *726*, 312–337. [CrossRef]
10. Fujii, K. Experiment of the Two-Dimensional Roughness Effect on Hypersonic Boundary-Layer Transition. *J. Spacecr. Rocket.* **2006**, *43*, 731–738. [CrossRef]
11. Zhang, S.; Li, X.; Zuo, J.; Qin, J.; Cheng, K.; Feng, Y.; Bao, W. Research progress on active thermal protection for hypersonic vehicles. *Prog. Aerosp. Sci.* **2020**, *119*, 100646. [CrossRef]
12. Zhu, Y.; Peng, W.; Xu, R.; Jiang, P. Review on active thermal protection and its heat transfer for airbreathing hypersonic vehicles. *Chin. J. Aeronaut.* **2018**, *31*, 1929–1953. [CrossRef]
13. Le, V.T.; Ha, N.S.; Goo, N.S. Advanced sandwich structures for thermal protection systems in hypersonic vehicles: A review. *Compos. B Eng.* **2021**, *226*, 109301. [CrossRef]
14. Terentieva, V.S.; Astapov, A.N. Conceptual Protection Model for Especially Heat-Proof Materials in Hypersonic Oxidizing Gas Flows. *Russ. J. Non-Ferr. Met.* **2019**, *59*, 709–718. [CrossRef]
15. Wang, H.; Hu, W.; Xie, F.; Li, J.; Jia, Y.; Yang, Y. Control effects of a high-frequency pulsed discharge on a hypersonic separated flow. *Phys. Fluids* **2022**, *34*, 066102. [CrossRef]
16. Xie, W.; Luo, Z.; Zhou, Y.; Gao, T.; Wu, Y.; Wang, Q. Experimental study on shock wave control in high-enthalpy hypersonic flow by using SparkJet actuator. *Acta Astronaut.* **2021**, *188*, 416–425. [CrossRef]
17. Yang, H.; Liang, H.; Zhang, C. Plate boundary layer transition regulation based on plasma actuation array at Mach 6. *Phys. Fluids* **2023**, *35*, 064104.
18. Yang, H.; Liang, H.; Zhang, C.; Wu, Y.; Li, Z.; Zong, H.; Su, Z.; Yang, B.; Kong, Y.; Zhang, D.; et al. An experimental study on the stability of hypersonic plate boundary layer regulated by a plasma actuation array. *Phys. Fluids* **2023**, *35*, 026112. [CrossRef]
19. Yang, H.; Liang, H.; Zhang, C.; Wu, Y.; Zong, H.; Su, Z.; Kong, Y.; Zhang, D.; Li, Y. Investigation of hypersonic cone boundary layer stability regulation with plasma actuation. *Phys. Fluids* **2023**, *35*, 024112. [CrossRef]
20. Yang, H.; Zong, H.; Liang, H.; Wu, Y.; Zhang, C.; Kong, Y.; Li, Y. Swept shock wave/boundary layer interaction control based on surface arc plasma. *Phys. Fluids* **2022**, *34*, 087119. [CrossRef]
21. Kong, Y.; Li, J.; Wu, Y.; Liang, H.; Guo, S.; Yang, H. Experimental study on shock-shock interaction over double wedge controlled by surface arc plasma array. Contrib. *Plasma Phys.* **2022**, *62*, e202200062.
22. Ding, B.; Chen, Z.; Jiao, Z.; Wang, J.; Li, Z.; Bai, G. Unsteady control mechanisms of hypersonic compression corner using pulsed surface arc discharg. *Acta Aeronaut. Astronaut. Sin.* **2023**, *44*, 127744. (In Chinese)
23. Zhang, C.; Yang, H.; Liang, H.; Guo, S. Plasma-based experimental investigation of double compression ramp shock wave/boundary layer interaction control. *J. Phys. D Appl. Phys.* **2022**, *55*, 325202. [CrossRef]
24. Von Terzi, D.; Sandberg, R.; Sandberg, R.; Fasel, H. Identification of large coherent structures in supersonic axisymmetric wakes. *Comput. Fluids* **2009**, *38*, 1638–1650. [CrossRef]

25. Berry, M.; Magstadt, A.; Glauser, M. Application of POD on time-resolved schlieren in supersonic multi-stream rectangular jets. *Phys. Fluids* **2017**, *29*, 020706. [CrossRef]
26. Chaganti, N.; Kurup, A.; Olcmen, S. Study of unsteadiness of shock wave boundary layer interaction using Rainbow Schlieren Deflectometry and Proper Orthogonal Decomposition. In Proceedings of the AIAA Aerospace Sciences Meeting Including the New Horizons Forum & Aerospace Exposition, Grapevine, TX, USA, 7–10 January 2013.
27. Wang, H.; Min, F.; Xie, Z.; Li, J.; Dai, J.; Yang, Y. Quantitative study of the control of hypersonic aerodynamics using millisecond pulsed discharges. *Phys. Fluids* **2022**, *34*, 021701. [CrossRef]

aerospace

Article

Noise Prediction and Plasma-Based Control of Cavity Flows at a High Mach Number

Hongming Cai [1,*], Zhuoran Zhang [1], Ziqi Li [2] and Hongda Li [3]

1 Academy of Astronautics, Nanjing University of Aeronautics and Astronautics, Nanjing 210016, China; zhangzhuoran@nuaa.edu.cn
2 Aircraft Design Institute, Jiangxi Hongdu Aviation Industry Co., Ltd., Nanchang 330095, China; songnifeixiang650@163.com
3 School of Aeronuatica Engineering, Nanjing Vocational University of Industry Technology, Nanjing 210023, China; 2021101181@niit.edu.cn
* Correspondence: caihongming@nuaa.edu.cn

Abstract: Cavity flows are a prevalent phenomenon in aerospace engineering, known for their intricate structures and substantial pressure fluctuations arising from interactions among vortices. The primary objective of this research is to predict noise levels in high-speed cavity flows at Mach 4 for a rectangular cavity characterized by an aspect ratio of L/D = 7. Moreover, this study delves into the influence of the plasma actuator on noise control within the cavity flow regime. To comprehensively analyze acoustic characteristics and explore effective noise reduction strategies, a computational fluid dynamics technique with the combination of a delayed detached eddy simulation (DDES) and plasma phenomenological model is established. Remarkably, the calculated overall sound pressure level (OASPL) and plasma-induced velocity closely align with the experimental data, validating the reliability of the proposed approach. The results show that the dielectric barrier discharge (DBD) plasma actuator changes the movement range of a dominating vortex in the cavity to affect the OASPL at the point with the maximum noise level. The control of excitation voltage can reduce the cavity noise by 2.27 dB at most, while control of the excitation frequency can only reduce the cavity noise by 0.336 dB at most. Additionally, the increase in excitation frequency may result in high-frequency sound pressure, but the influence is weakened with the increase in the excitation frequency. The findings highlight the potential of the plasma actuator in reducing high-Mach-number cavity noise.

Keywords: cavity flows; noise prediction; flow control; delayed detached eddy

Citation: Cai, H.; Zhang, Z.; Li, Z.; Li, H. Noise Prediction and Plasma-Based Control of Cavity Flows at a High Mach Number. *Aerospace* **2023**, *10*, 922. https://doi.org/10.3390/aerospace10110922

Academic Editor: Konstantinos Kontis

Received: 6 September 2023
Revised: 25 October 2023
Accepted: 25 October 2023
Published: 29 October 2023

1. Introduction

Cavity structures find extensive application in advanced aircraft and propulsion systems, such as internal weapon bays and landing gear bays of aircraft [1]. On one hand, cavity structures play a proactive role in enhancing stealth performance and reducing the aerodynamic drag during aircraft cruising. On the other hand, cavity structures also introduce aerodynamic noise-related challenges [2–4], which could potentially lead to structural fatigue or electronic equipment damage within weapon bays, posing a significant impact on flight safety. Research indicates that the generation of cavity noise is closely linked to unsteady flow patterns and complex wave systems within the cavity. Under specific conditions, cavity flows are prone to undergo flow–acoustic coupling phenomena, resulting in distinct periodic pressure oscillations and producing high-amplitude pure-tone noise [5].

In recent years, there has been in-depth experimental research into the spatiotemporal evolution process of complex wave systems in supersonic cavities, leading to the formulation of corresponding mechanistic models [6,7]. Ryan F. Schmit et al. [8] investigated the flow physics phenomena within cavities using wind tunnel experiments. Taro

Handa et al. [9] explored the relationship between shear layer motion, pressure wave generation, and cavity trailing edge pressure oscillations through wind tunnel experiments. N. T. Clemens et al. [10] conducted wind tunnel experiments to explore the physical foundation of oscillation periods in high-Mach-number, turbulent, open cavity flows.

In the simulation regime, Rizzetta et al. [11] employed large eddy simulation techniques to investigate the noise feedback mechanism in closed cavities. Kyoung Sik Chang et al. [12] employed Reynolds-averaged Navier–Stokes/large eddy simulation (RANS/LES) methods to study the acoustic characteristics of deep cavities. Shia-Hui Peng et al. [13] conducted cavity flow analysis using DES and a hybrid RANS/LES approach, achieving favorable results. Hamed et al. [14] employed an SST-based DES method to compute pressure fluctuations in a Ma = 1.19 cavity and compared the results with the experimental data. Yin Tang et al. [15] utilized IDDES to study flow control methods for acoustic characteristics of open cavities at Ma = 3.51. Xuhong Jin et al. [16] employed direct simulation Monte Carlo methods to investigate the impact of inlet boundary layer thickness on cavity flow characteristics. Larsson J et al. [17] directly solved the compressible Navier–Stokes equations to study laminar flow and near-field acoustics in a two-dimensional open cavity at Ma = 0.15. S. J. Lawson et al. [18] conducted DES-based research on the geometries of the M219 experimental cavity and the 1303 uninhabited combat air vehicle cavity geometry.

Although a series of experimental and numerical studies have been conducted on the unsteady flow characteristics of cavities, most of the research has focused on cavities with Ma < 2. As aircraft move towards higher speeds and longer distances, the engineering applications of high-Mach-number (Ma > 2) cavities have become more prominent. Research on high-Mach-number (Ma > 2) cavity flows is relatively limited and further studies are warranted.

Noise is the byproduct of high-Mach-number cavity flows, and the advancement in understanding the mechanisms behind cavity noise generation has spurred the development of noise control techniques. In recent years, a variety of supersonic cavity noise control methods have been devised. Zhang et al. [19] proposed a control method that modifies the shape of the cavity leading-edge surface to increase the leading-edge height. Zhuang [20] employed the approach of microjets at the leading edge to modify the incoming Mach number, thereby reducing aerodynamic noise within the cavity. Vakili and Gauthier [21] utilized leading-edge jets to alter the cavity's inlet boundary layer thickness and consequently reduce cavity noise.

Luo et al. [22] conducted numerical simulations of the M219 cavity flow at Ma = 1.5 using different serrated leading-edge sawtooth spoilers. Alam et al. [23] achieved pressure oscillation control within the cavity flow by utilizing a sub-cavity at the leading edge. Schmit et al. [24] experimentally studied supersonic cavity flow using leading-edge vortex generators. Danilov et al. [25] achieved noise reduction by perturbing the cavity shear layer using vortex generators. Thangamani et al. [26] investigated cavity noise control through modifications at the trailing edge. Vikramaditya et al. [27] explored the influence of different rear-wall inclination angles on the cavity's internal noise characteristics.

Despite these studies, there remains a notable research gap in the field of noise control for high-Mach-number cavity flows, necessitating further investigation and exploration. As such, there is an evident need to bridge this gap in knowledge and advance our understanding of noise generation and control mechanisms within high-Mach-number cavity flows. This study endeavors to contribute to filling this void by comprehensively exploring the complexities of noise generation and developing effective noise control strategies within the context of high-Mach-number cavity flows.

This paper is organized as follows: Section 2 establishes the numerical methods used for analysis, encompassing a comprehensive literature review and methodology overview. Validation of these methods follows in Section 3 through rigorous experimental and numerical verification. Section 4 delves into the heart of the study, presenting detailed analyses of noise characteristics, flow structures, and the intricate interplay within high-Mach-number cavity flows. Parameters such as excitation voltage and frequency are systematically investigated to elucidate their impact on noise generation and propagation. Finally, Section 5 synthesizes our findings into insightful conclusions, discussing both the fundamental understanding gained and the practical implications for noise control in high-Mach-number cavity flows. This structured approach not only advances our understanding of noise mechanisms but also provides a roadmap for addressing cavity noise challenges in high-Mach-number environments.

2. Numerical Simulation Methods

2.1. Flow Simulation Method

The numerical investigation of the intricate characteristics of high-Mach-number cavity flows requires an appropriate computational method. In this section, we established a DDES-based method to capture the complex flow structure and acoustic characteristics. The detached eddy simulation (DES) [28] technique emerges as a compelling approach to simulate cavity flows due to its inherent capability to bridge the gap between the Reynolds-averaged Navier–Stokes (RANS) [29] and large eddy simulation (LES) [30] methods. This coupling algorithm offers a refined way of resolving the intricate flow structures present in cavity flows, ranging from the near-wall boundary layer to the regions experiencing significant flow separation.

Extending the foundational framework of DES, the DDES method introduces a novel refinement—the integration of a delay function. This function serves as a sophisticated criterion for transitioning between RANS and LES modes based on the local flow characteristics. By incorporating this delay function, the DDES method enhances the accuracy and fidelity of the simulation by seamlessly blending the strengths of both RANS and LES, while effectively mitigating their respective limitations.

Building upon the DES framework, the DDES [31] method introduces a delay function to modify the criteria for transitioning between RANS and LES modes accordingly:

$$\widetilde{d} = d - f_d \max(0, d - C_{des} L_g) \tag{1}$$

where the delay function f_d is defined as

$$f_d = 1 - \tanh\left[(8r_d)^3\right] \tag{2}$$

$$r_d = \frac{\upsilon_T + \upsilon}{\sqrt{U_{i,j} U_{i,j}} \kappa^2 d^2} \tag{3}$$

In the equation, C_{des} is the adaptive parameter, L_g is the grid scale, υ_T is the viscosity of vortex motion, υ is the viscosity coefficient of molecular motion, $U_{i,j}$ is the velocity gradient, κ is the Karman length, and d is the distance from the object surface. r_d represents the ratio of turbulent scale at a given location to the distance from that location to the wall. This parameter ensures that LES is not used within the boundary layer, and the DDES model transitions to the LES mode in regions of pronounced separation. In areas farther from the wall where $r_d << 1$ and $f_d = 1$, LES is utilized, while RANS is employed in other regions where $f_d = 0$.

2.2. Plasma Actuator Model

This section introduces an employed phenomenological model [32] and showcases its significance in capturing the effects of plasma actuation on cavity flows. Through this approach, we aim to elucidate the intricate mechanisms of plasma-induced flow control. Plasma actuation stands as a promising avenue for actively manipulating the flow characteristics within cavities. The pivotal concept underpinning this model revolves around the direct integration of experimental and theoretical findings into the computational framework. Specifically, the range and distribution of electric field forces, gleaned from meticulous experiments and theoretical analyses, are seamlessly incorporated as source terms within the computational equations.

The integration of plasma-induced effects directly into the computational framework introduces a dynamic and interactive component to the simulation. This empowers us to effectively emulate the intricate interplay between plasma actuation and the surrounding flow field. By intertwining the electric field forces with the governing equations, we gain a holistic perspective on how plasma actuation influences the flow's behavior, leading to alterations in vortical structures, pressure distributions, and, ultimately, acoustic signatures.

A simplification of the plasma interaction region, represented as a right-angled triangle OAB with width b and height a, is depicted in Figure 1. Within this region, the electric field forces exhibit a linear distribution, attaining maximum strength E_0 at the origin O and breakdown electric field strengths E_b along the OA, OB, and AB edges. When the electric field strength E_b falls below the truncated electric field strength, the electric field forces are insufficient to penetrate the gas, thereby inhibiting plasma formation, and, thus, their influence can be disregarded. Throughout the triangular region, the electric field forces align parallel to the AB edge.

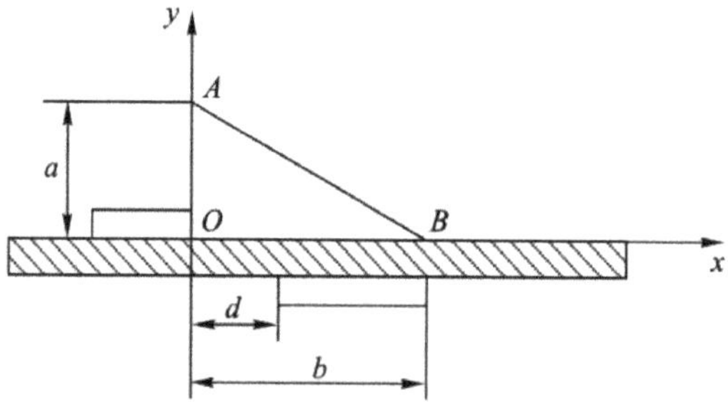

Figure 1. Schematic of plasma actuator and volume-force acting region.

The phenomenological model efficiently encapsulates the influence of the plasma excitation region on the flow field by incorporating the electric field forces. The resulting simulation outcomes will be critically assessed against the experimental data to validate the efficacy of the proposed plasma actuator model.

The electric field strength can be described by the following equations:

$$|E| = E_0 - k_1 x - k_2 y \tag{4}$$

$$k_1 = \frac{E_0 - E_b}{b}, k_2 = \frac{E_0 - E_b}{a} \tag{5}$$

where E_0 is the electric field strength at point O, representing the maximum electric field strength; the electric field strength at the AB boundary (denoted as E_b) corresponds to the breakdown electric field strength. If the electric field strength falls below this value, the effect of the electric field forces is neglected.

The electric field force is oriented parallel to the hypotenuse AB of the triangle. This leads to the magnitudes of the electric field strengths in the x and y directions, which are defined by:

$$E_x = \frac{Ek_2}{\sqrt{k_1^2 + k_2^2}}, E_y = \frac{Ek_1}{\sqrt{k_1^2 + k_2^2}} \tag{6}$$

According to the phenomenological simulation model proposed in reference [33], the volume forces in the x and y directions within the plasma interaction region are defined by:

$$\begin{cases} F_{tavex} = v\alpha\rho_c e_c E_x \Delta t\delta \\ F_{tavey} = v\alpha\rho_c e_c E_y \Delta t\delta \end{cases} \tag{7}$$

where v represents frequency. α denotes the effective coefficient for elastic collisions. ρ_c stands for charge density. e_c signifies the elementary charge constant. Δt signifies the

discharge time. δ represents the Dirac delta function, utilized to delineate the extent of the electric field force. When $E \geq E_b$, its effect is accounted for as $\delta = 1$; otherwise, it is treated as 0.

3. Validation of the Established Method

3.1. Validation of DDES Approach

To assess the reliability and accuracy of the DDES approach, we embark on a rigorous validation exercise utilizing a standard high-Mach-number cavity case with the experimental data. This benchmark case serves as a touchstone, allowing us to ascertain the ability of the DDES method to replicate known acoustic characteristics. In the subsequent section, we will present and discuss the outcomes of our DDES simulations on the standard cavity case.

The validation case utilized the standard cavity benchmark from reference [34]. The cavity model has a length (L) of 0.4572 m, and a width (W) and depth (D) of 0.1016 m each, resulting in an aspect ratio of 4.5:1. Computational parameters in this study match the experimental conditions to ensure the feasibility and accuracy of the numerical simulation. Specifically, the following parameters are used: incoming flow Mach number of 3.51, zero angle of attack, incoming static pressure of 2495.56 Pa, incoming static temperature of 93.08 K, and a Reynolds number based on the cavity length of 9.9×10^9. A structured grid is employed, with a finer mesh within the cavity to capture small-scale vortex structures, and a coarser mesh in the far-field region to expedite computations. The total number of grid cells is approximately 13 million (as shown in Figure 2). The SST-based DDES method is employed for unsteady turbulent calculations. The Roe flux-difference splitting (Roe-FDS) scheme is employed for flux splitting, with a second-order upwind scheme used for convective fluxes and a second-order central difference scheme for viscous diffusion terms. A second-order implicit dual-time-stepping scheme is adopted for time integration. The physical time step is set to 2×10^{-5} s, and a total of 2000 physical time steps are computed (corresponding to a total physical time of 0.04 s). The monitoring point corresponds to the pressure measurement location K18, as defined in reference [34]. The K18 point is located on the rear wall of the cavity, $x/L = 1$, $y/L = 0$, $z/L = -0.1706$, as shown in Figure 3.

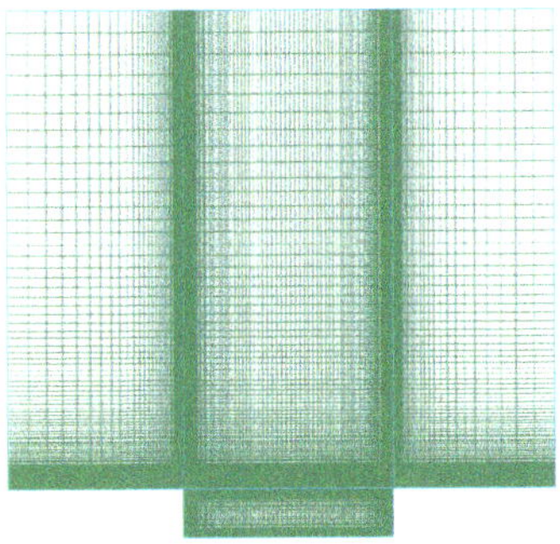

Figure 2. Computational grid of the cavity model.

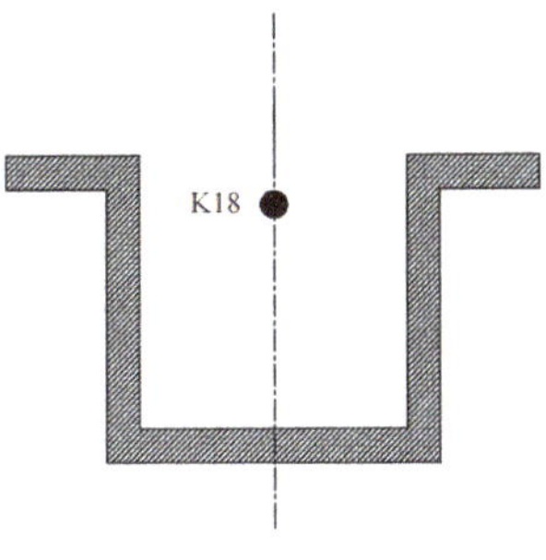

Figure 3. $x/L = 1$. Cross-sectional view.

Figure 4 illustrates the comparison between experimental and simulation results at monitoring point 18, while Table 1 provides the frequencies and sound pressure level amplitudes corresponding to dominant modes at monitoring point K18. The computed overall sound pressure level (OASPL) is 155.05 dB, whereas the experimental counterpart stands at 151.94 dB, yielding a discrepancy of 3.08 dB. The first-order mode shows a frequency error of 21.9% with a relative error in SPL of 4.08%. The second-order mode has a frequency error of 10.3% and a relative SPL error of 10.1%. The third-order mode indicates an error of 3.1% in frequency and a 3.2% relative error in SPL. The fourth-order mode displays a discrepancy of 3.97% in frequency and a 1.07% relative error in SPL.

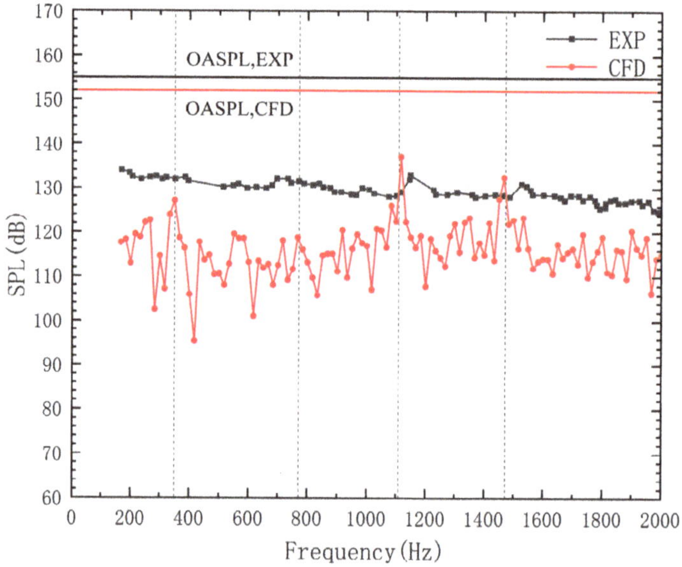

Figure 4. Comparison of simulated and experimental cavity sound pressure levels at the K18 point.

Table 1. Noise data comparison at monitoring point K18.

	Modal	1_{st}	2_{nd}	3_{rd}	4_{th}
Experiment	Frequency (Hz)	287	695	1152	1527
	Sound Pressure Level (dB)	132.5	132.1	132.8	131
CFD	Frequency (Hz)	349.9	766.5	1116	1466.4
	Sound Pressure Level (dB)	127.1	118.8	137.1	132.4

As shown in Figure 5, the horizontal axis represents frequency, while the vertical axis represents power spectral density (PSD). PSD reflects the power distribution of noise components at different frequencies. From the figure, it is evident that the contributions of first-, second-, third-, and fourth-order modes to OASPL can be clearly observed. The impact of the third- and fourth-order modes on OASPL is significantly greater than that of the first- and second-order modes. Therefore, we believe that our simulation method is reliable overall.

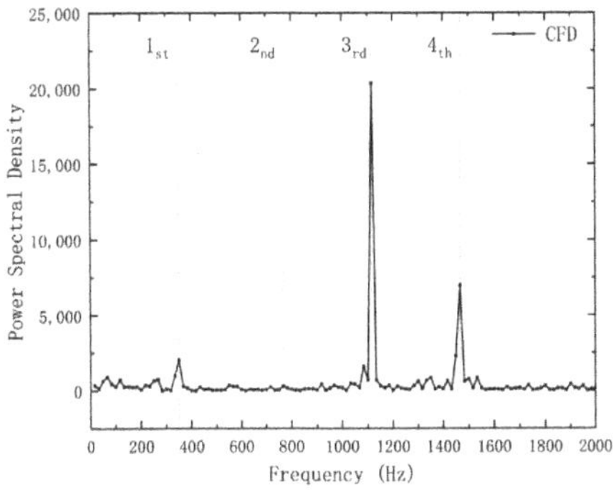

Figure 5. Power spectral density at point K18.

3.2. *Validation of the Plasma Actuator Model*

To validate the plasma phenomenological model, we turn to a comprehensive case study and subject the computational results to rigorous comparison with the experimental data in reference [32] so we can ascertain the method's ability to replicate the observed phenomena and trends. Through this validation process, we gain confidence in the predictive capabilities of the plasma actuator model, paving the way for its application in unraveling the complex cavity flows. In the subsequent section, we will delve into the outcomes of our plasma-related simulations, shedding light on the impact of plasma actuation on cavity flow characteristics.

Based on reference [32], the upper electrode is positioned on a two-dimensional flat plate, as depicted in Figure 6. The total length of the plate is 20.5 mm, and the incoming flow velocity is $U_\infty = 5$ m/s, aligned parallel to the plate surface along the positive X-axis direction. Sections ST1, ST2, ST3, and ST4 are located at distances of 4.75 mm, 12 mm, 13.9 mm, and 17.3 mm from the leading-edge wall, respectively, as illustrated in Figure 6.

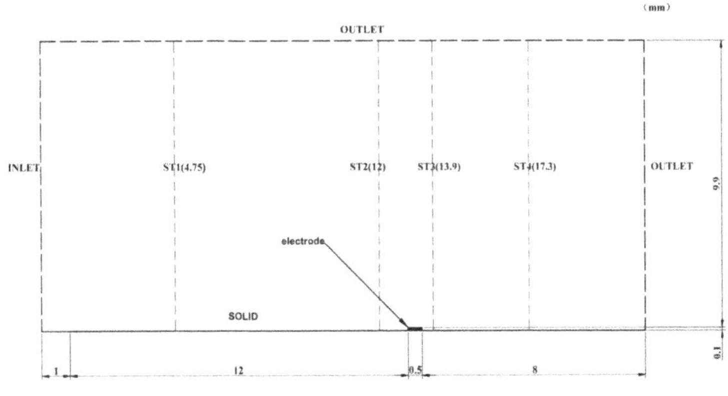

Figure 6. Computational domain of the two-dimensional model.

The validation of the computational results is shown in Figure 7, where the horizontal axis represents the magnitude of the X-direction velocity, and the vertical axis represents the height of the vertical wall. From Figure 7, it can be observed that for cross-sections ST1, ST2, and ST3, the maximum error occurs at a distance of 0.63mm from the wall. The maximum error for each of these sections is less than 9%, and the errors at other points are all below 6%. For cross-section ST4, the maximum error occurs at a distance of 0.857mm from the wall, with an error of 8.17%. The errors at other points are also below 6%. These results indicate a good agreement between our computational findings and the reference [32], demonstrating an accurate simulation of the plasma's excitation effects on the flow field.

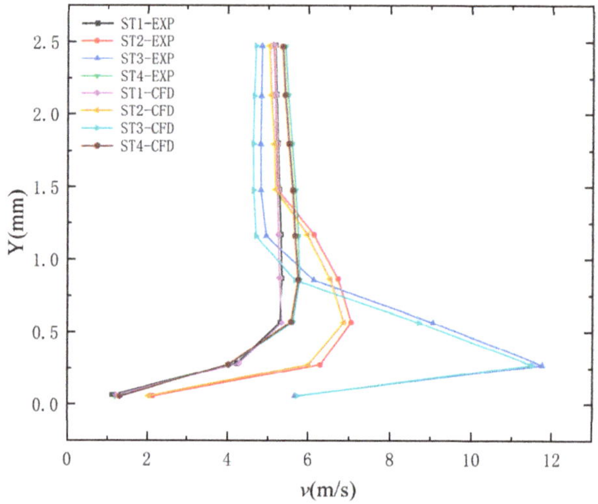

Figure 7. Comparison of the X-component velocities at different sections between the literature data and the validation computation.

In summary, the plasma phenomenological model emerges as a powerful tool in our arsenal, enabling us to delve deep into the effects of the plasma actuator on the cavity flows. The integration of experimental findings into the computational framework marks a pivotal advancement, allowing us to unravel the intricate flow control mechanisms with a high degree of accuracy and fidelity.

4. The High-Speed Cavity Model and Independence Analysis

4.1. The Studied High-Speed Cavity Model

The two-dimensional cavity designed in this study (as shown in Figure 8) has a length of 4.2 m and a depth of 0.6 m, with an aspect ratio (L/D) of seven, falling within the category of typical long-depth ratio configurations for open cavity flows. The chosen operating condition corresponds to Ma = 4 and atmospheric parameters at an altitude of 25 km, aiming to simulate the aerodynamic noise characteristics of a high-Mach-number cavity. Along the upstream direction, a total of 10 pressure monitoring points, labeled as K20-K29, are uniformly distributed along the bottom surface of the cavity within the region of $x/L = 0.05$ to 0.95. These monitoring points are strategically placed to measure the variations in internal cavity noise.

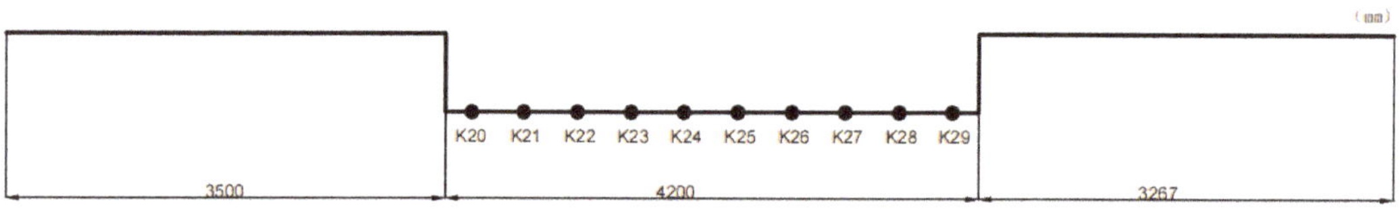

Figure 8. Two-dimensional cavity model and location of monitoring points.

4.2. Grid Independence Analysis

Grid independence analysis is crucial for ensuring the accuracy and reliability of numerical simulations. In our study of cavity flows and plasma-induced flow control, this analysis is essential to validate the robustness of our simulation framework. The main purpose of grid independence analysis is to confirm that our results are not heavily influenced by the choice of grid resolution. By systematically varying the grid and observing the convergence of flow features, we can ensure that our findings are consistent and accurate. This analysis is especially important for capturing the intricate interactions in cavity flows, especially vortices. Moreover, grid independence analysis helps optimize computational resources. Striking a balance between grid resolution and efficiency is key to obtaining meaningful results without excessive computational costs. By identifying the minimum grid refinement needed for convergence, we can ensure efficient simulations.

The computational conditions include an incoming flow velocity of Ma = 4, a static pressure of $P\infty$ = 2509.98 Pa, and a static temperature of T = 93.08 K. The chosen time step for simulations is 5×10^{-5} s. To investigate grid independence, the cavity grid model is divided into four sets with varying grid densities labeled as Models A, B, C, D, and E, corresponding to grid quantities of 69,000, 126,000, 163,000, 210,000, and 300,000, respectively.

Figure 9 illustrates a comparison of the total sound pressure level at ten monitoring points distributed along the cavity bottom surface for the four mesh models. Among these, Models C, D, and E exhibit relatively close results, while Models A and B fail to accurately predict the increase in the total sound pressure level at x/L = 0.25 and the decrease at x/L = 0.85. As the grid quantity reaches approximately 163,000, the trend in internal cavity noise stabilizes, and the total sound pressure level error becomes sufficiently small. Considering computational cost and resources, Model C is selected for subsequent investigations.

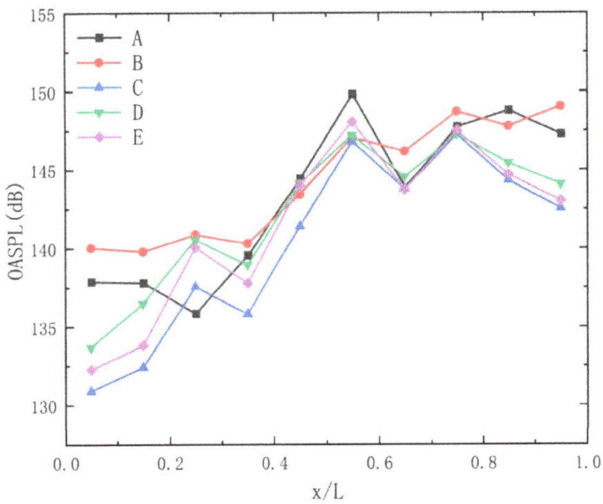

Figure 9. Sound pressure level data at the bottom of the cavity for different meshes.

4.3. Time Step Independence Analysis

In the realm of numerical simulations, investigating the independence of results with respect to the chosen time step is paramount. Time step independence analysis, akin to grid independence analysis, plays a crucial role in affirming the reliability and accuracy of our computational findings. The essence of time step independence analysis lies in comprehending the impact of temporal discretization on simulated flow behavior. A smaller time step might provide greater temporal resolution, capturing intricate flow features with precision. However, excessively small time steps can escalate computational costs substantially. Conversely, a larger time step may offer computational efficiency but could potentially lead to the loss of critical flow details.

By systematically varying the time step size and observing the convergence of relevant flow parameters, we can ascertain the appropriate temporal discretization for our simulations. This ensures that our numerical results are not unduly influenced by the choice of time step, strengthening the credibility of our investigation into cavity flows and plasma-induced flow control. Time step independence analysis also facilitates a balance between accuracy and computational efficiency. Achieving convergence with a reasonable time step optimizes the utilization of computational resources while providing accurate insights into the temporal evolution of flow phenomena.

In the upcoming section, we will elaborate on our approach to time step independence analysis for the selected Model C considering five distinct time step values.

Figure 10 displays the computational results obtained using the five different time steps: 2×10^{-4} s, 1×10^{-4} s, 8×10^{-5} s, 5×10^{-5} s, and 2.5×10^{-5} s. As depicted in the figure, when the time step exceeds 5×10^{-5} s, the total sound pressure levels at each monitoring point tend to be higher. Conversely, when the time step is less than 5×10^{-5} s, the variations in sound pressure levels at the monitoring points stabilize in terms of both trend and magnitude. Therefore, the time step of 5×10^{-5} s, which offers higher computational efficiency, is employed for the numerical simulations.

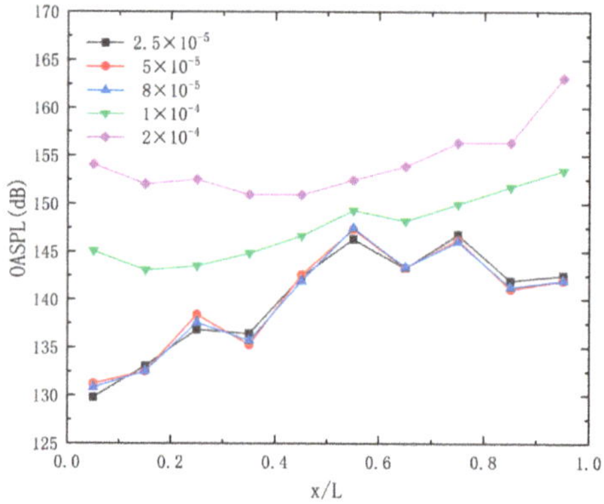

Figure 10. Sound pressure level data at the bottom of the cavity for different time steps.

5. Noise Prediction and Plasma-Based Control of Cavity Flows

5.1. Calculation Setups

In this section, the DBD plasma actuator is employed with parameters identical to those specified in reference [32]. The actuation region has dimensions a = 1.5 mm and b = 3 mm, operating at a voltage frequency of 3000 Hz, with a charge density of 1×10^{11} cm^3, and the elementary electron charge e = 1.602×10^{-19}. The discharge duration is 67 μs, and the breakdown electric field strength at the electric field boundary is set at 30 kV/cm. The distance between the electrodes is 0.25 mm, and the initial excitation voltage is 4 kV. The plasma actuation effect is incorporated into the Navier–Stokes equations as a volumetric force source term.

The incoming flow conditions remain consistent with those described in Section 2, utilizing the same numerical simulation methodology. The mesh quantity is based on Model C (as shown in Figure 11) as a baseline of 163,000 cells. Mesh refinement is applied within the plasma actuation region (as shown in Figure 12), resulting in a final mesh quantity of approximately 200,000 cells. The time step is set to 5×10^{-5} s, totaling 20,000 steps. The initial 6000 steps are discarded to eliminate initial oscillations.

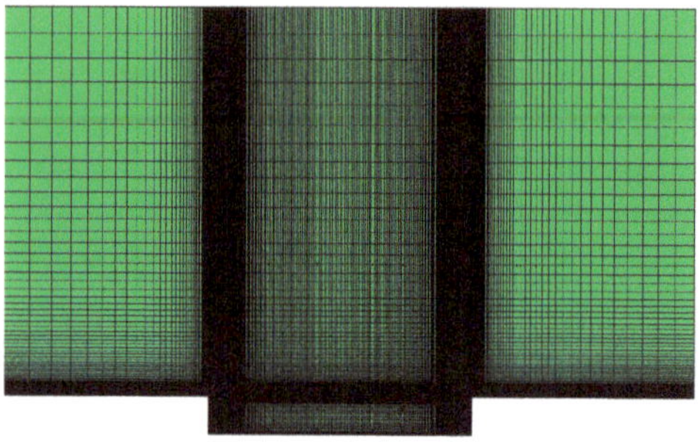

Figure 11. Cavity mesh configuration.

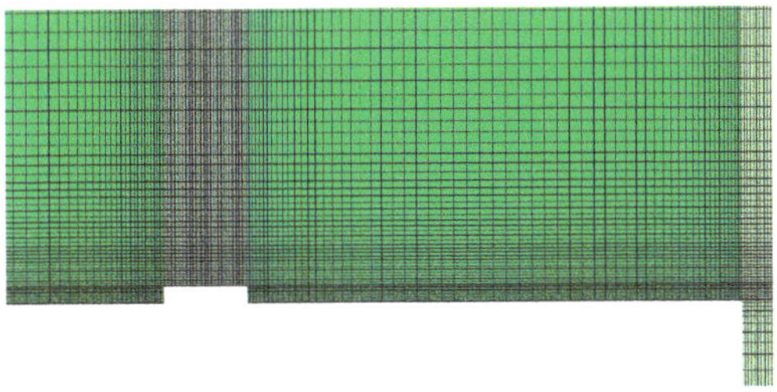

Figure 12. Plasma actuator mesh configuration.

In accordance with the diagram presented in Figure 13, the boundary conditions for the numerical simulation are configured as follows: boundary a, b, and c represent the pressure far-field conditions, enabling the flow to exit the computational domain smoothly without introducing any disturbances. Boundary d corresponds to the wall condition, characterized by a no-slip constraint, which ensures that the flow does not penetrate or slip through the solid boundary. The height of the computational domain extends 10 times the cavity depth based on the dimensions of the design model.

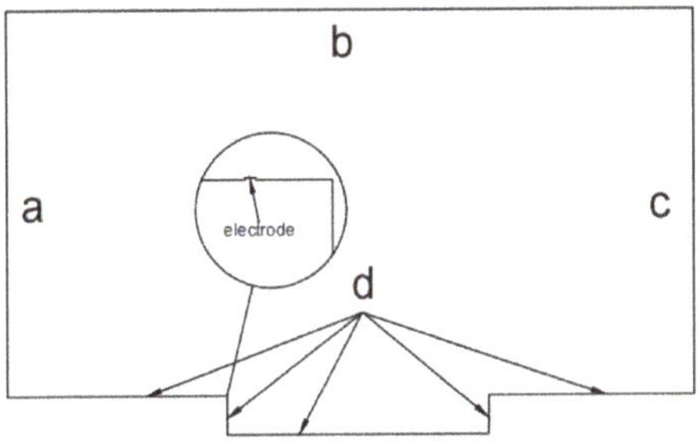

Figure 13. Boundary conditions of the plasma-excited cavity model.

5.2. *Aerodynamic and Acoustic Characteristics of the Uncontrolled Cavity*

Figure 14 illustrates the OASPL distribution at ten positions along the cavity bottom, ranging from K20 to K29. Notably, there are three distinct peaks in the OASPL, and the highest OASPL occurs at point K25, corresponding to $x/L = 0.55$.

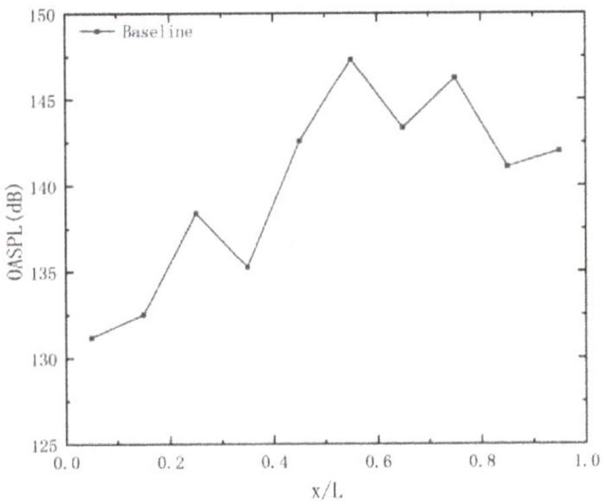

Figure 14. OASPL at various monitoring points along the bottom of the uncontrolled cavity.

In Figure 15, we can observe that the flow pattern in the examined cavity exhibits typical characteristics of open cavity flow. The airflow separates at the cavity's leading edge, part of it entering the cavity while the rest forms a shear layer above. This shear layer extends over the central part of the cavity and then reattaches at the trailing edge. Importantly, there are multiple peaks in the global sound pressure level, both at the front (x/L = 0.25) and rear (x/L = 0.55, x/L = 0.75) sections of the cavity.

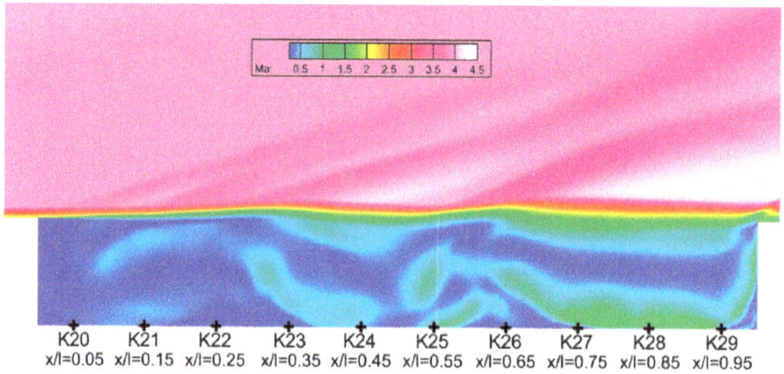

Figure 15. Mach number contour map of the uncontrolled cavity flow.

Figure 16 shows vorticity maps at different phases within a single cycle using the Q-criterion, as indicated in Figure 16a, which illustrates the vorticity direction. In Figure 16a, a sequence of alternating clockwise and counterclockwise vortices can be observed in the incoming flow direction within the cavity. These vortices alternate with smaller vortices of opposite directions, and the entire flow process involves both the breakdown of larger vortices and the merging of smaller ones.

At a specific moment, denoted as T (representing a single flow cycle's duration), several vortex interactions occur. Firstly, a counterclockwise small vortex E detaches from a clockwise large vortex B, moving along the negative X direction of the cavity bottom. Eventually, it merges with the counterclockwise large vortex A at its upper section. Simultaneously, vortex C and vortex F combine as well. Vortex F, which is a counterclockwise small vortex resulting from the detachment of a clockwise large vortex D near the cavity's rear wall along the negative X direction of the bottom, moves to the upper part of the counterclockwise large vortex C. This results in the complete fusion of vortex C and F. Following the collision of the shear layer with the rear cavity wall, a large vortex D forms. As vortex F periodically merges with vortex C, shed from the large vortex D, the dissipative and fusion processes allow vortex C to remain stable within the cavity flow. However, its position in the X direction undergoes periodic variations.

A noticeable observation is the relative stability of vortices at the front of the cavity, where vortex fusion is minimal within a single cycle. In contrast, the middle and rear portions of the cavity are more susceptible to vortex breakdown and fusion phenomena. Interestingly, the locations of vortex breakdown and fusion roughly align with regions where extreme values of cavity OASPL noise occur. It can be inferred that vortex breakdown and fusion play a significant role in the generation of aerodynamic noise.

5.3. *Influence of Excitation Voltage*

In order to investigate the impact of excitation voltage on noise control effectiveness, numerical simulations are conducted with varying excitation voltages at the same excitation frequency. The specific excitation voltage settings are presented in Table 2.

Table 2. Excitation voltage configuration.

Case	1	2	3	4	5	6	7
Excitation Frequency (kHz)				3			
Excitation Voltage (kV)	4	6	8	10	12	14	16

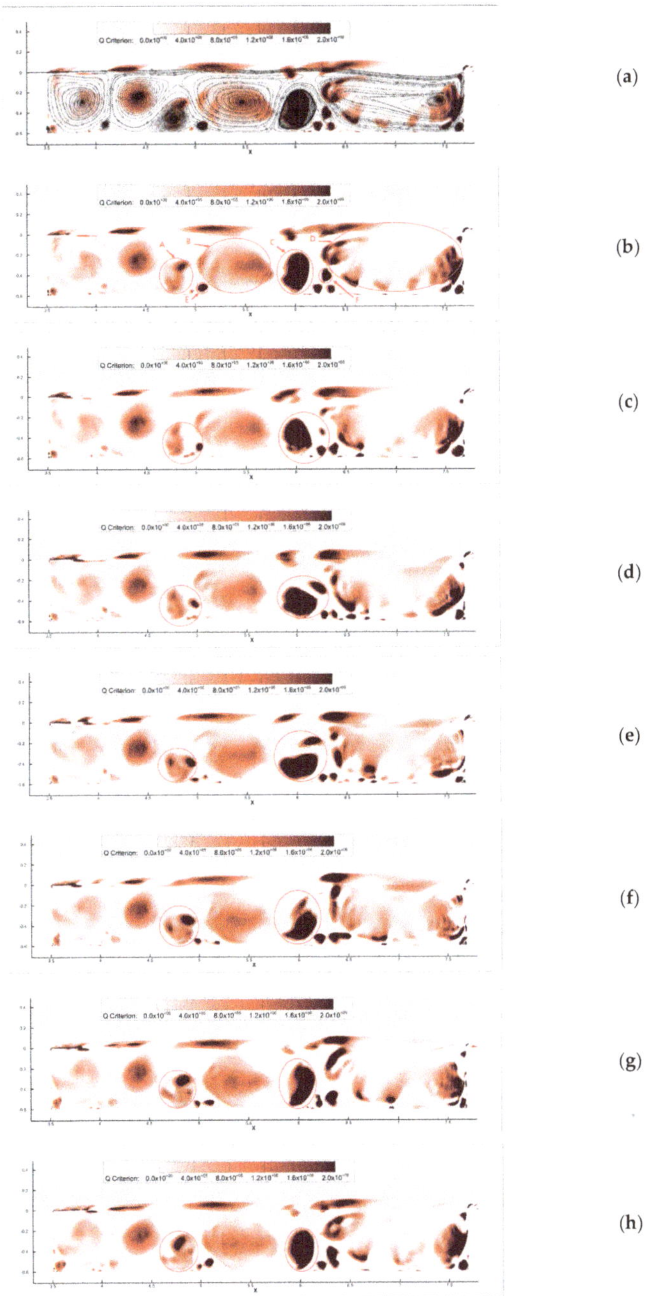

Figure 16. Vorticity contour plot of the uncontrolled cavity flow: (**a**) $t = \frac{1}{7}$T, (vortex rotation direction); (**b**) $t = \frac{1}{7}$T; (**c**) $t = \frac{2}{7}$T; (**d**) $t = \frac{3}{7}$T; (**e**) $t = \frac{4}{7}$T; (**f**) $t = \frac{5}{7}$T; (**g**) $t = \frac{6}{7}$T; (**h**) t = T.

Figure 17 shows the changes in the total sound pressure level at various monitoring points along the bottom of the cavity under different excitation voltages at a constant excitation frequency of 3 kHz. With the exception of point K28 (x/L = 0.85), the optimal excitation voltage is found to be 12 kV. At point K20 (x/L = 0.05), the maximum reduction can be 3.073 dB. Notably, different excitation voltages do not effectively reduce the global sound pressure level at point K28 (x/L = 0.85).

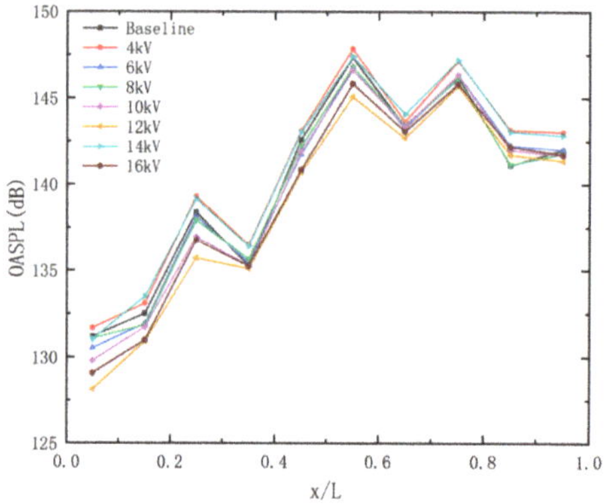

Figure 17. OASPL values at different monitoring points along the bottom surface of the cavity under different excitation voltages.

In Figure 18, we observe the variation in OASPL at point K25 under different excitation voltages while maintaining a consistent excitation frequency of 3 kHz. When the excitation voltage is set at 4 kV, a noticeable increase in high-frequency noise levels is observed, resulting in an overall rise in OASPL. As the excitation voltage gradually increases from 6 kV to 12 kV, there is a slight reduction in SPL amplitude in both the low-frequency range (≤100 Hz) and the high-frequency range (≥1000 Hz), despite the SPL value of the dominant mode at 315 Hz remaining relatively stable. These combined effects contribute to the decrease in OASPL at point K25 from 147.329 dB under uncontrolled conditions to 145.057 dB under 12 kV excitation voltage.

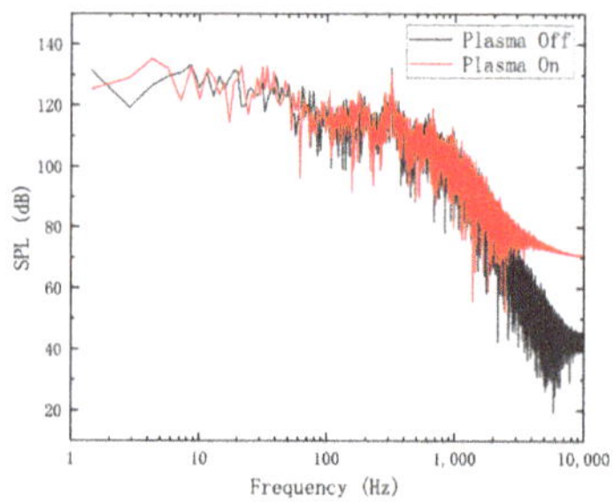

(**a**) V = 4 kV

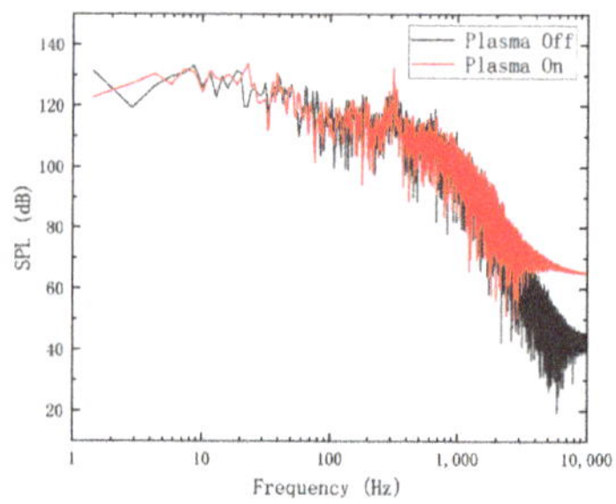

(**b**) V = 6 kV

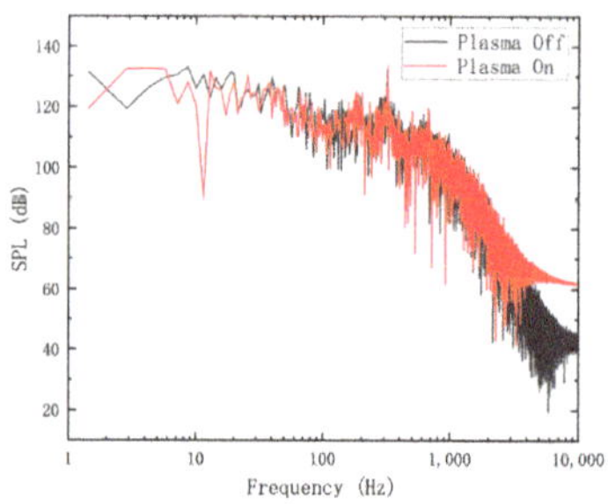

(**c**) V = 8 kV

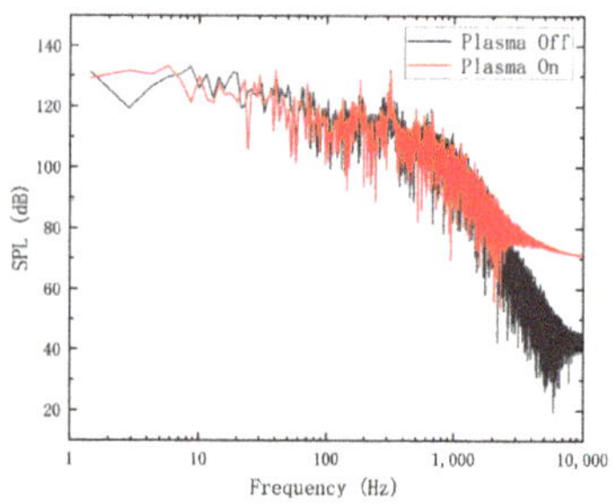

(**d**) V = 10 kV

Figure 18. *Cont.*

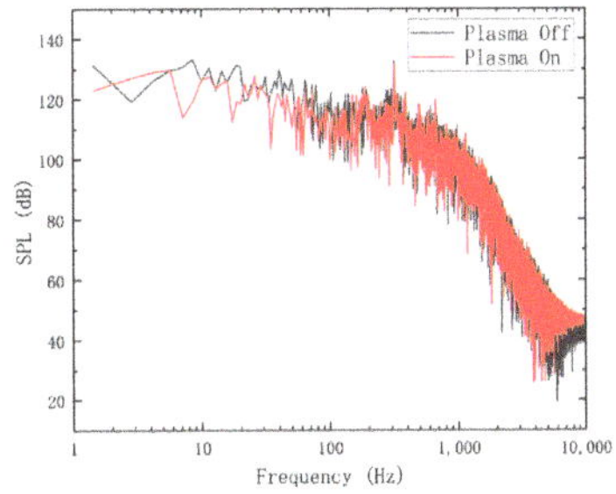

(**e**) V = 12 kV

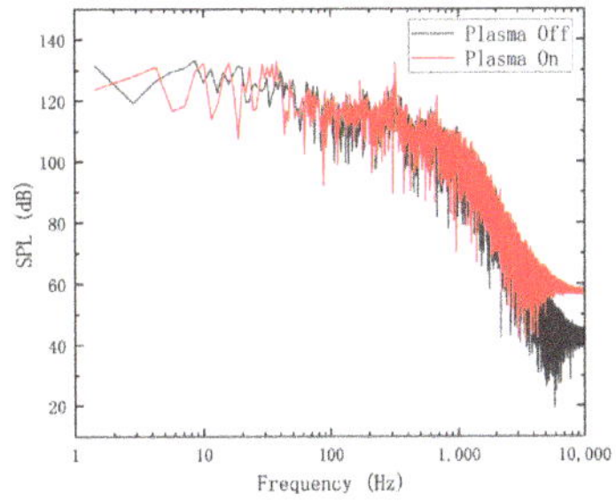

(**f**) 14 kV

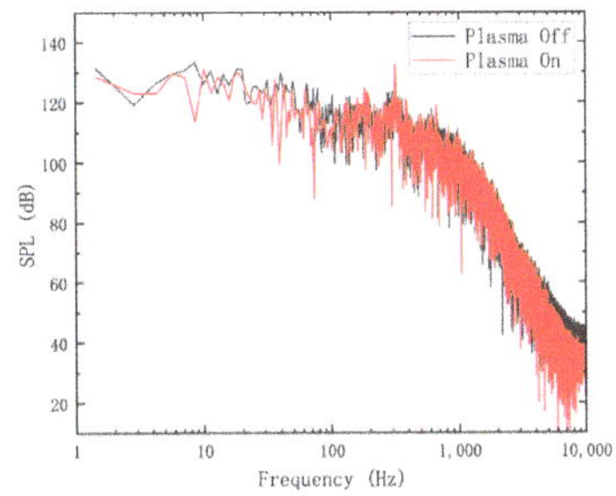

(**g**) V = 16 kV

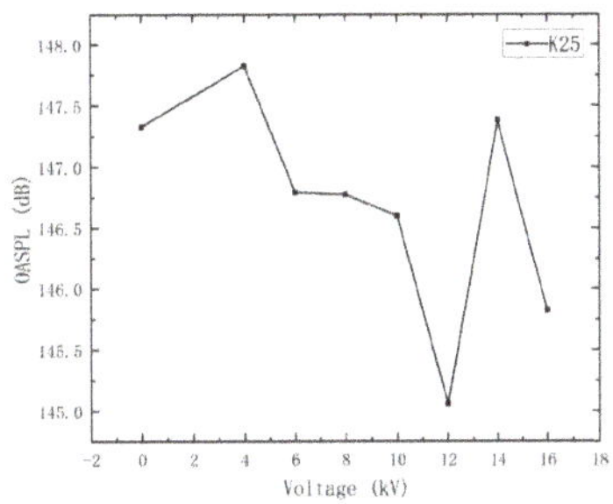

(**h**)

Figure 18. Noise characteristics at point K25 under different excitation voltages: (**a**) f = 3 kHz, V = 4 kV; (**b**) f = 3 kHz, V = 6 kV; (**c**) f = 3 kHz, V = 8 kV; (**d**) f = 3 kHz, V = 10 kV; (**e**) f = 3 kHz, V = 12kV; (**f**) f = 3 kHz, V = 14 kV; (**g**) f = 3 kHz, V = 16 kV; (**h**) OASPL of different voltages of excitation.

However, at 14 kV excitation voltage, there is a notable increase in SPL amplitude in the high-frequency range, resulting in an increase in OASPL to 147.38 dB. When the excitation voltage is further increased to 16 kV, the frequency distribution characteristics resemble those observed at a 12 kV excitation voltage, leading to a decrease in OASPL to 145.819 dB.

To investigate whether plasma excitation impacts the internal noise of the cavity by altering the boundary layer thickness, we conducted measurements of the boundary layer thickness at the leading edge of the cavity lip. These measurements were taken at excitation voltages of 4 kV, 6 kV, 8 kV, 10 kV, 12 kV, 14 kV, and 16 kV, all at a frequency of 3 kHz. The results consistently indicated that the boundary layer thickness remained approximately 50mm under different excitation voltages. This suggests that plasma voltage excitation does not significantly affect noise levels at various points within the cavity by modifying the boundary layer thickness.

In Figure 19, the variation of the relative limit position of the dominant vortex throughout the entire flow duration (t = 0.3–1 s) is depicted for different excitation voltages, as indicated by the red circles. The figures (a,c,e,g,i,k) in Figure 19 show the left limit position of the dominant vortex under different excitation voltages. Meanwhile, the figures (b,d,f,h,j,l) in Figure 19 show the left limit position of the dominant vortex under different excitation voltages. The occurrence time of the dominant vortex's limit position can be influenced by different excitation voltages and variations in timing. It can be observed that the flow pattern within the cavity remains consistent and the vortex system structure undergoes negligible changes under different voltage levels. The position of the vortices within the cavity remains relatively stable over one cycle. However, due to the continuous interaction with the shear layer impinging on the rear wall, the positions of the vortex experience some displacement over multiple flow cycles. This displacement leads to an alteration in the extent of vortex shedding and collapse, resulting in an expansion or contraction of the radiating region, as indicated by the shifting of the red circles. It is evident that the stability of the dominant vortex position, as demonstrated by the red circles, influences OASPL of the K25 point.

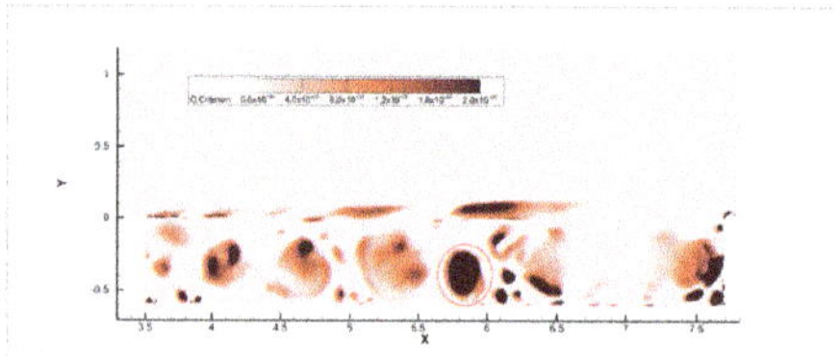

(**a**) V = 0 kV

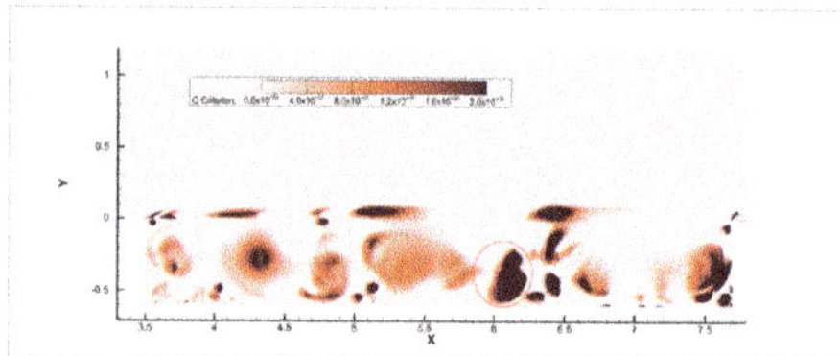

(**b**) V = 0 kV

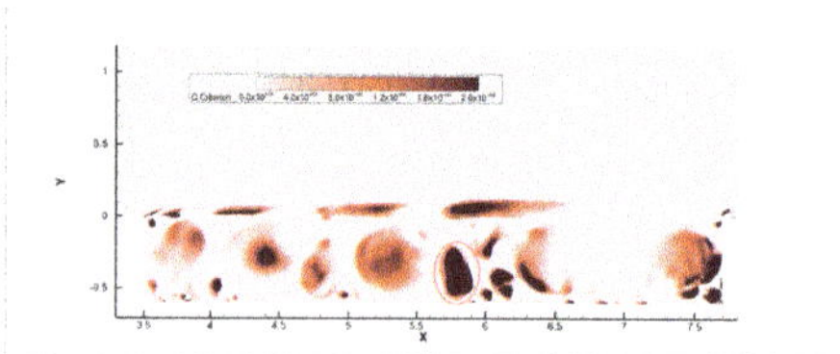

(**c**) *V* = 4 kV

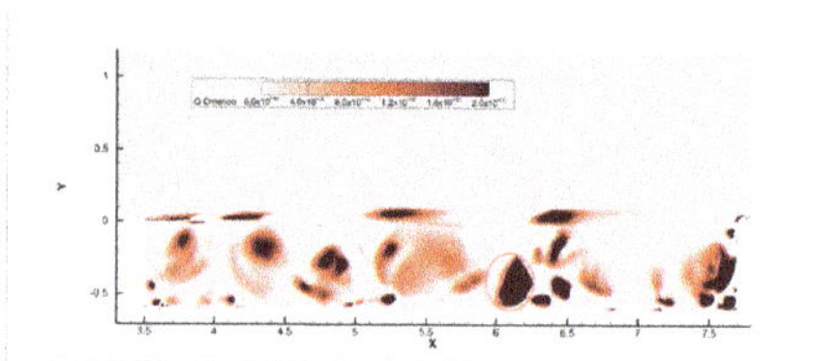

(**d**) *V* = 4 kV

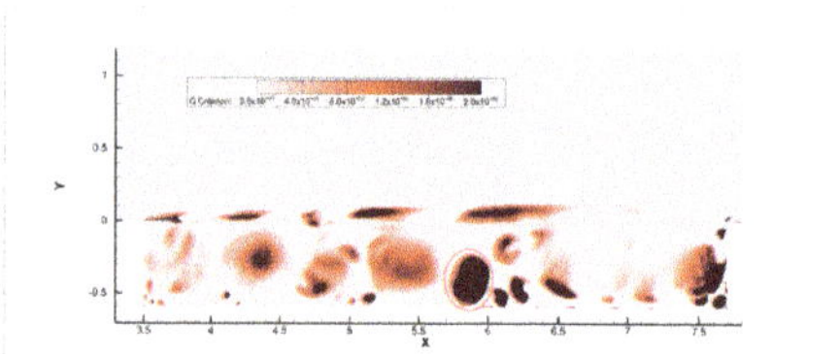

(**e**) V = 6 kV

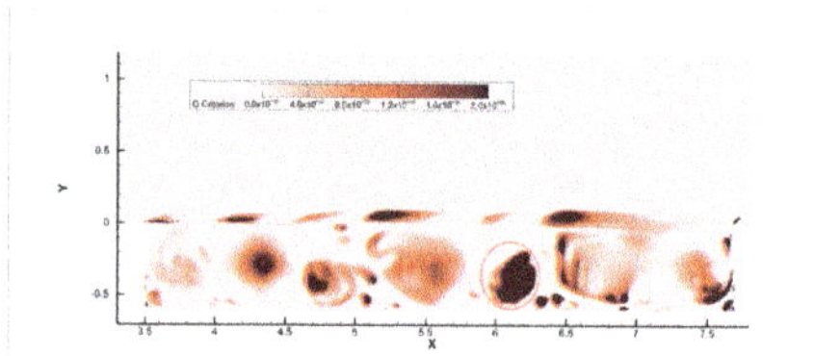

(**f**) V = 6 kV

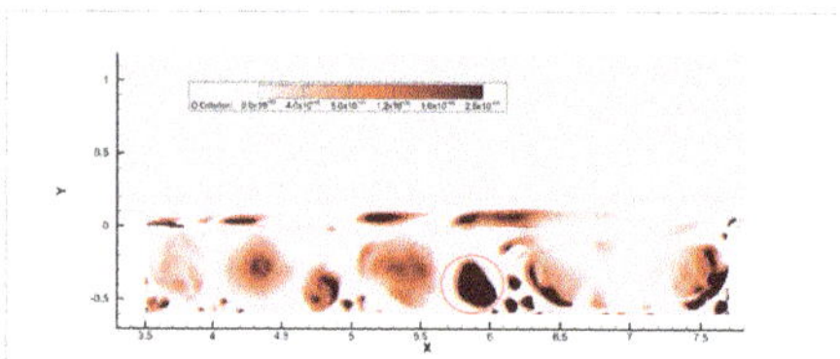

(**g**) V = 12 kV

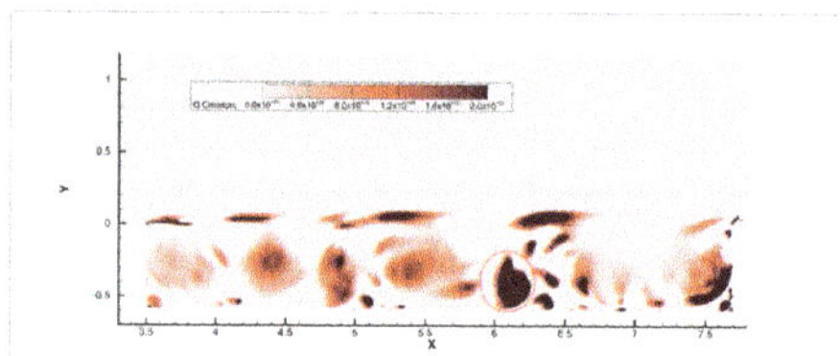

(**h**) V = 12 kV

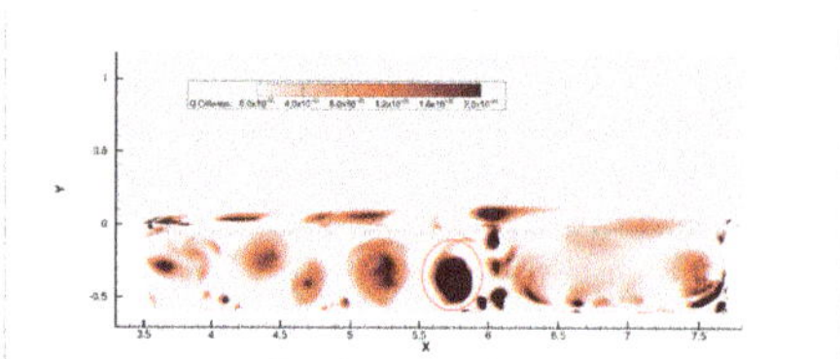

(**i**) V = 14 kV

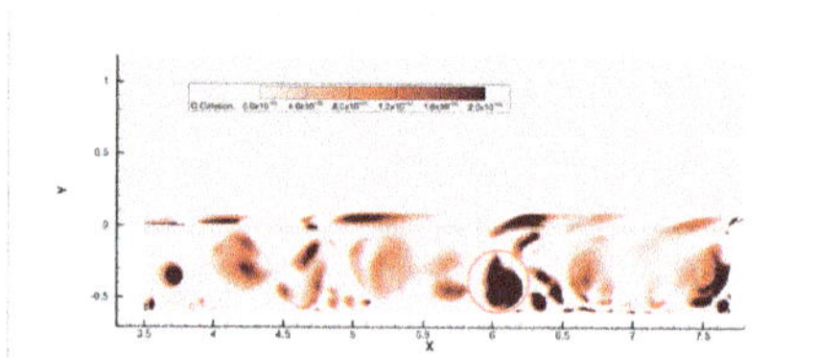

(**j**) V = 14 kV

Figure 19. *Cont.*

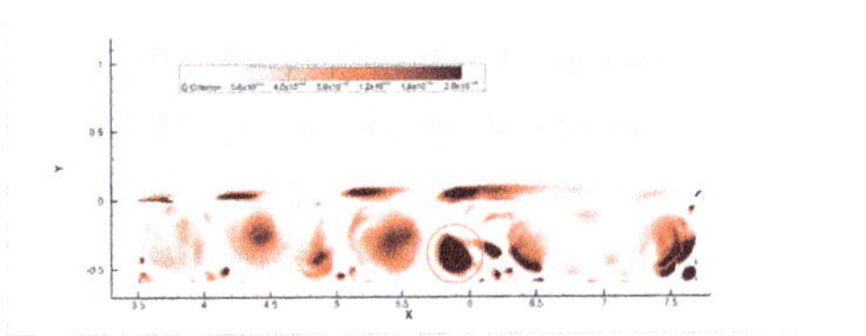

(**k**) V = 16 kV

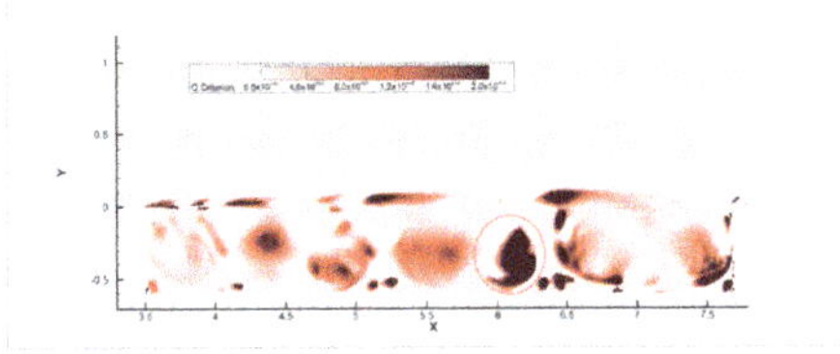

(**l**) V = 16 kV

Figure 19. Contour plot of cavity vorticity under different excitation voltages: (**a**) $V = 0$ kV, left limit position; (**b**) $V = 0$ kV, left limit position; (**c**) $V = 4$ kV, right limit position; (**d**) $V = 4$ kV, right limit position; (**e**) $V = 6$ kV, left limit position; (**f**) $V = 6$ kV, right limit position; (**g**) $V = 12$ kV, left limit position; (**h**) $V = 12$ kV, right limit position; (**i**) $V = 14$ kV, left limit position; (**j**) $V = 14$ kV, right limit position; (**k**) $V = 16$ kV, left limit position; (**l**) $V = 16$ kV, right limit position.

Table 3 presents the corresponding data between the OASPL values at the K25 point and the X-direction displacement range of the characteristic vortex for various excitation voltages. It can be observed that there exists a positive correlation between the OASPL values and the X-direction displacement range of the vortex under different excitation voltages. As the relative displacement becomes smaller, the vortex structure exhibits higher stability, leading to reduced high-frequency noise caused by vortex shedding and merging, ultimately resulting in a lower total sound pressure level.

Table 3. Dominant vortex X-direction displacement range and corresponding OASPL data under different excitation voltages.

Excitation Voltage (kV)	**0**	**4**	**6**	**12**	**14**	**16**
OASPL (dB)	147.329	147.827	146.791	145.057	147.38	145.819
X-direction Displacement Range (mm)	304.8	347.6	276.5	239.6	338.2	267.8

5.4. Effect of Excitation Frequency

To investigate the influence of excitation frequency on noise control effectiveness, numerical simulations were conducted with different excitation frequencies under the same excitation voltage, as shown in Table 4.

Table 4. Excitation frequency conditions.

Case	**8**	**9**	**10**	**11**	**12**
Excitation Voltage (kV)			4		
Excitation Frequency(kHz)	3	6	9	12	15

Figure 20 illustrates the impact of different excitation frequencies on the OASPL at points K20 to K29. It can be observed that the aerodynamic noise level at the K25 point does not exhibit a significant decrease under different plasma excitation frequencies. In some cases, such as f = 3 kHz and f = 12 kHz, the OASPL even increased by approximately 0.5 dB. Notably, as depicted in Figure 21, with increasing plasma excitation frequency, its effect on the mid- to low-frequency range (below 1000 Hz) is relatively small. The dominant mode around 315 Hz is minimally affected, and in the high-frequency range, plasma excitation may even amplify the SPL values, leading to an increase in OASPL. When the excitation frequency is 15 kHz, the SPL amplitude decreases in the mid-frequency range (100~200 Hz) and slightly increases in the high-frequency range, causing the K25 point's OASPL to decrease from 147.329 dB to 146.407 dB. In comparison to the influence of excitation voltage, the effect of excitation frequency is less significant. Similarly, variations in excitation frequency do not notably change the frequency and SPL amplitude of the dominant mode but primarily affect the SPL amplitude at high frequencies.

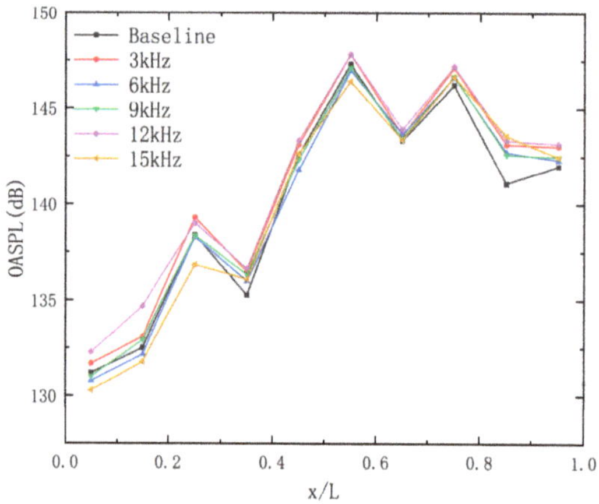

Figure 20. OASPL distribution at different bottom surface monitoring points for various excitation frequencies.

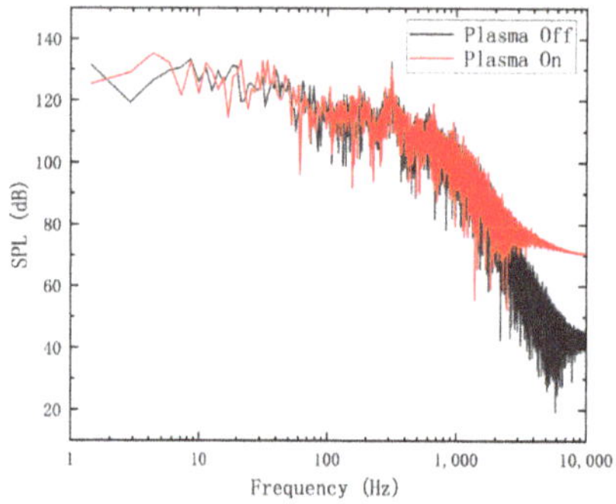

(**a**) $f = 3$ kHz

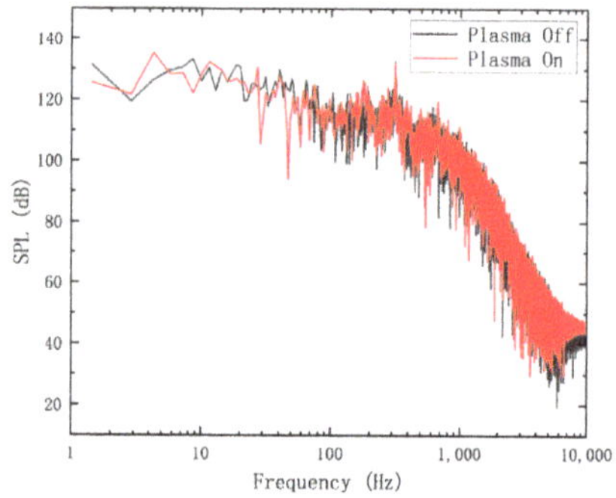

(**b**) $f = 6$ kHz

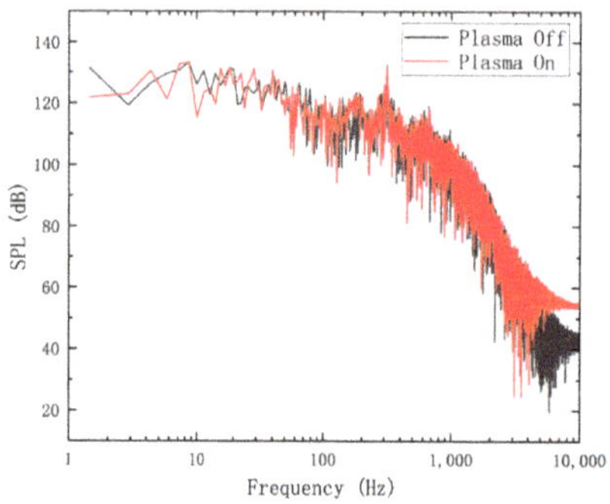

(**c**) $f = 9$ kHz

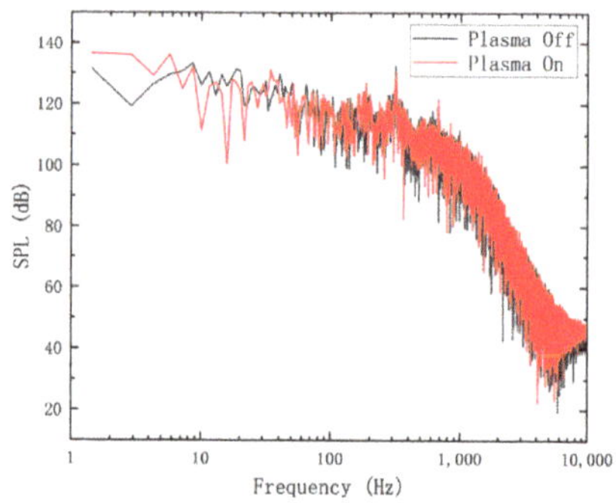

(**d**) $f = 12$ kHz

Figure 21. *Cont.*

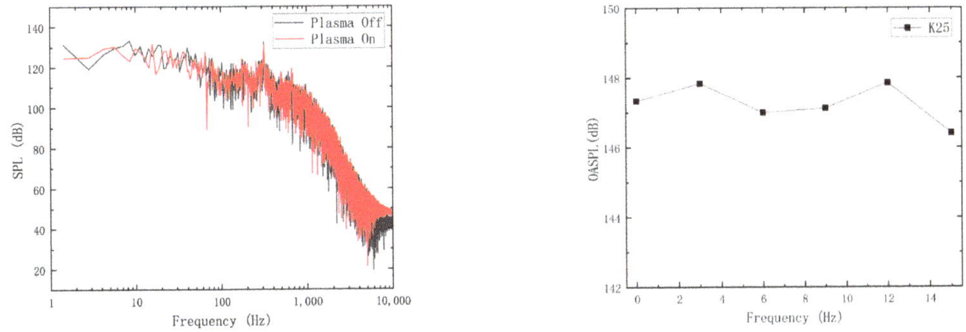

(**e**) f = 15 kHz (**f**)

Figure 21. Noise variation characteristics at monitoring point K25 for different excitation frequencies: (**a**) f = 3 kHz, V = 4 kV; (**b**) f = 6 kHz, V = 4 kV; (**c**) f = 9 kHz, V = 4 kV; (**d**) f = 12 kHz, V = 4 kV; (**e**) f = 15 kHz, V = 4 kV; (**f**) OASPL of different frequency of excitation.

We measured the boundary layer thickness at the lip of the cavity leading edge corresponding to excitation voltages of 3 kHz, 6 kHz, 9 kHz, 12 kHz, and 15 kHz, all at an excitation voltage of 4 kV. The results indicate that the boundary layer thickness remains consistent at approximately 50 mm across different excitation frequencies. It can be observed that the effect of the plasma excitation frequency on the boundary layer thickness is similar to the situation with excitation voltage; it does not produce a significant influence. This suggests that plasma excitation does not primarily impact the cavity noise level by altering the boundary layer thickness.

Building upon the findings from the previous section, we proceed to observe the behavior characteristics of the dominant vortex near point K25. As illustrated in Figure 22, the left and right extreme positions of the dominant vortex are presented throughout the entire flow duration (t = 0.3–1 s) for different excitation frequencies. Figures (a,c,e) in Figure 22 illustrate the left limit position of the dominant vortex under different excitation frequencies. Similarly, Figures (b,d,f) in Figure 22 depict the left limit position of the dominant vortex under different excitation frequencies. The occurrence time of the dominant vortex's limit position can be influenced by different excitation frequencies and variations in timing. It is evident that the flow pattern within the cavity remains consistent, and the vortex system structure undergoes minimal changes across different voltage levels. The positions of the vortices within the cavity exhibit relative stability over a single cycle, aligning with the observed trend in plasma excitation voltage.

Table 5 provides a comparison of the OASPL values at point K25 for different excitation frequencies and the corresponding X-direction displacement range of the dominant vortex. It is evident that the X displacement of the dominant vortex is positively correlated with their OASPL magnitude, similar to the observed trend with plasma excitation voltage. A smaller relative displacement indicates a more stable vortex structure, resulting in reduced high-frequency noise generated by vortex shedding and merging.

Table 5. X-direction displacement range of dominant vortex and corresponding OASPL data at different excitation frequencies.

Excitation Frequency (kHz)	0	3	12	15
OASPL (dB)	147.329	147.827	147.843	146.407
X-direction Displacement Range (mm)	304.8	347.6	374.9	262.5

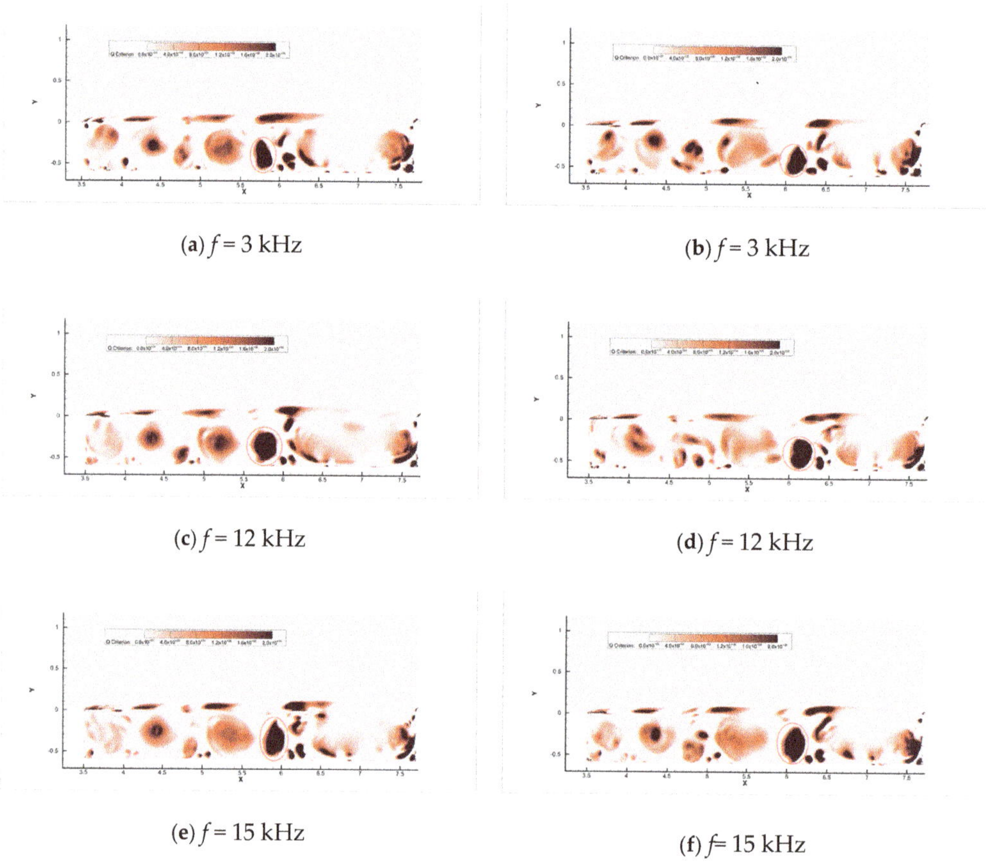

Figure 22. Contour plot of cavity vorticity under different excitation frequencies: (**a**) f = 3 kHz, left limit position; (**b**) f = 3 kHz, right limit position; (**c**) f = 12 kHz, left limit position; (**d**) f = 12 kHz, right limit position; (**e**) f = 15 kHz, left limit position; (**f**) f = 15 kHz, right limit position.

6. Conclusions

In this study, the primary objective is to investigate the acoustic characteristics and noise reduction strategies of a rectangular cavity under high-Mach-number conditions. Through comprehensive simulations conducted at Mach 4, employing a cavity with dimensions of 4.2×0.6 m^2, the influence of plasma-based control on noise reduction was meticulously studied. The following key conclusions can be drawn from the conducted research:

1. This paper studies the suppression of aerodynamic noise in high-speed cavities using a combined DDES and DBD method for the first time. Comparing with experimental data, the calculation error of the OASPL in high-speed cavities is within 2%, and the calculation error of the X-direction velocity of the plasma actuator model is within 9%.
2. The cavity with L/D = 7 exhibits distinct open flow characteristics at Ma=4 and an altitude of 25 km. Multiple OASPL peaks are observed in the front, middle, and rear regions of the cavity. The maximum OASPL reaches 147.329 dB, occurring at x/L = 0.55. The locations of vortex breakdown and fusion correspond to the regions and roughly align with regions where extreme values of cavity noise occur.
3. Relative to the excitation frequency, the excitation voltage of the plasma actuator has a more pronounced effect on noise suppression. Appropriate excitation voltage can reduce the OASPL by up to 2.27 dB by suppressing low-frequency noise. The

excitation voltage can reduce the sound pressure level amplitude of the dominant mode, thereby decreasing the OASPL of the high-speed cavity.

4. The effect of the excitation frequency of the plasma actuator on noise suppression is weaker, yet an optimal frequency exists. Variations in the excitation frequency have a less noticeable impact on the frequency and sound pressure level amplitude of the dominant mode, primarily affecting high-frequency sound pressure levels, with a maximum reduction of 0.336 dB in the OASPL.
5. Plasma actuators can alter the lateral movement range of the dominant vortex within the high-speed cavity. As the lateral displacement of the dominant vortex decreases, the OASPL of the cavity also decreases.

Author Contributions: Conceptualization, H.C., Z.Z. and Z.L.; methodology, H.C. and Z.Z.; software, H.C. and Z.Z.; validation, H.C., Z.Z. and H.L.; formal analysis, H.C. and Z.Z.; investigation, H.C., Z.Z. and Z.L.; data curation, H.C. and Z.Z.; writing—original draft preparation, H.C., Z.Z. and Z.L.; writing—review and editing, H.C., Z.Z. and H.L.; All authors have read and agreed to the published version of the manuscript.

Funding: This work was supported by Aeronautical Science Foundation of China (Grant No. 2020Z006052001), Rotor Aerodynamics Key Laboratory Fund Project (RAL20190101), and the Fundamental Research Funds for the Central Universities (Grant No.1011-YAH21001).

Data Availability Statement: Not applicable.

Conflicts of Interest: The authors declare that there is no conflict of interest.

References

1. Rowley, C.W.; Williams, D.R. Dynamics and Control of High-Reynolds-Number Flow over Open Cavities. *Annu. Rev. Fluid Mech.* **2006**, *38*, 251–276. [CrossRef]
2. Heller, H.; Delfs, J. Cavity Pressure Oscillations: Cavity Pressure Oscillations: The Generating Mechanism Visualized. *Lett. Ed. J. Sound Vib.* **1996**, *196*, 248–252. [CrossRef]
3. Lee, B.H.K. Effect of Captive Stores on Internal Weapons Bay Floor Pressure Distributions. *J. Aircr.* **2010**, *47*, 732–736. [CrossRef]
4. Lawson, S.J.; Barakos, G.N. Review of Numerical Simulations for High-Speed, Turbulent Cavity Flows. *Prog. Aerosp. Sci.* **2011**, *47*, 186–216. [CrossRef]
5. Knotts, B.D.; Selamet, A. Suppression of Flow–Acoustic Coupling in Sidebranch Ducts by Interface Modification. *J. Sound Vib.* **2003**, *265*, 1025–1045. [CrossRef]
6. Zhuang, N. Experimental Investigation of Supersonic Cavity Flows and their Control. Ph.D. Thesis, The Florida State University, Tallahassee, FL, USA, 2007. Available online: https://www.proquest.com/dissertations-theses/experimental-investigation-supersonic-cavity/docview/304872228/se-2 (accessed on 24 October 2023).
7. Heller, H.; Bliss, D.B. *Aerodynamically Induced Pressure Oscillations in Cavities: Physical Mechanisms and Suppression Concepts*; Wright-Patterson Air Force Base: Dayton, OH, USA, 1975.
8. Schmit, R.; Semmelmayer, F.; Grove, J.; Haverkamp, M. Fourier Analysis of High Speed Shadowgraph Images Around a Mach 1.5 Cavity Flow Field. In Proceedings of the 29th AIAA Applied Aerodynamics Conference, Honolulu, HI, USA, 27–30 June 2011; American Institute of Aeronautics and Astronautics: Washington, DC, USA, 1996. [CrossRef]
9. Handa, T.; Miyachi, H.; Kakuno, H.; Ozaki, T. Generation and Propagation of Pressure Waves in Supersonic Deep-Cavity Flows. *Exp. Fluids* **2012**, *53*, 1855–1866. [CrossRef]
10. Unalmis, O.H.; Clemens, N.T.; Dolling, D.S. Cavity Oscillation Mechanisms in High-Speed Flows. *AIAA J.* **2004**, *42*, 2035–2041. [CrossRef]
11. Rizzetta, D.P. Numerical simulation of supersonic flow over a three-dimensional cavity. *AIAA J.* **1988**, *26*, 799–807. [CrossRef]
12. Chang, K.-S.; Park, S.O. Hybrid RANS/LES Simulation of Deep Cavity Flow. In Proceedings of the 42nd AIAA Aerospace Sciences Meeting and Exhibit, Reno, NV, USA, 5–8 January 2004; American Institute of Aeronautics and Astronautics: Washington, DC, USA, 1996. [CrossRef]
13. Davidson, L.; Peng, S.H. Hybrid LES–RANS Modelling: A One–equation SGS Model Combined with a $k-\omega$ Model for Predicting Recirculating Flows. *Int. J. Numer. Methods Fluids* **2003**, *43*, 1003–1018. [CrossRef]
14. Hamed, A.; Basu, D.; Das, K. Detached Eddy Simulations of Supersonic Flow over Cavity. In Proceedings of the 41st Aerospace Sciences Meeting and Exhibit, Reno, NV, USA, 6–9 January 2003; American Institute of Aeronautics and Astronautics: Washington, DC, USA, 1996. [CrossRef]
15. Tang, Y.; Luo, L.; Zhang, P.; Ma, M. Effect of Rear Wall Inclination on Cavity Acoustic Characteristics at High Mach Numbers. *J. Phys. Conf. Ser.* **2022**, *2252*, 012007. [CrossRef]

16. Jin, X.; Huang, F.; Miao, W.; Cheng, X.; Wang, B. Effects of the boundary-layer thickness at the cavity entrance on rarefied hypersonic flows over a rectangular cavity. *Phys. Fluids* **2021**, *33*, 036116. [CrossRef]
17. Larsson, J.; Davidson, L.; Olsson, M.; Eriksson, L.-E. Aeroacoustic Investigation of an Open Cavity at Low Mach Number. *AIAA J.* **2004**, *42*, 2462–2473. [CrossRef]
18. Lawson, S.J.; Barakos, G.N. Computational Fluid Dynamics Analyses of Flow over Weapons-Bay Geometries. *J. Aircr.* **2010**, *47*, 1605–1623. [CrossRef]
19. Zhang, X.; Chen, X.; Rona, A.; Edwards, J. Attenuation of cavity flow oscillation through leading edge flow control. *J. Sound Vib.* **1999**, *221*, 23–47. [CrossRef]
20. Zhuang, N.; Alvi, F.S.; Alkislar, M.B.; Shih, C. Supersonic Cavity Flows and Their Control. *AIAA J.* **2006**, *44*, 2118–2128. [CrossRef]
21. Vakili, A.D.; Gauthier, C. Control of cavity flow by upstream mass-injection. *J. Aircr.* **1994**, *31*, 169–174. [CrossRef]
22. Luo, K.; Zhe, W.; Xiao, Z.; Fu, S. Improved delayed detached-eddy simulations of sawtooth spoiler control before supersonic cavity. *Int. J. Heat Fluid Flow* **2017**, *63*, 172–189. [CrossRef]
23. Alam, M.; Matsuob, S.; Teramotob, K.; Setoguchib, T.; Kim, H.-D. A new method of controlling cavity-induced pressure oscillations using sub-cavity. *J. Mech. Sci. Technol.* **2007**, *21*, 1398–1407. [CrossRef]
24. Schmit, R.; Semmelmayer, F.; Haverkamp, M.; Grove, J.; Ahmed, A. Examining Passive Flow Control Devices with High Speed Shadowgraph Images around a Mach1.5 Cavity Flow Field. In Proceedings of the 6th AIAA Flow Control Conference, New Orleans, LA, USA, 25–28 June 2012; American Institute of Aeronautics and Astronautics: Washington, DC, USA, 1996. [CrossRef]
25. Danilov, P.; Quackenbush, T. Flow Driven Oscillating Vortex Generators for Control of Cavity Resonance. In Proceedings of the 49th AIAA Aerospace Sciences Meeting Including the New Horizons Forum and Aerospace Exposition, Orlando, FL, USA, 4–7 January 2011; American Institute of Aeronautics and Astronautics: Washington, DC, USA, 1996. [CrossRef]
26. Thangamani, V.; Saddington, A.; Knowles, K. An Investigation of Passive Control Methods for a Large Scale Cavity Model in High Subsonic Flow. In Proceedings of the 19th AIAA/CEAS Aeroacoustics Conference, Berlin, Germany, 24 May 2013; American Institute of Aeronautics and Astronautics: Washington, DC, USA, 1996. [CrossRef]
27. Vikramaditya, N.S.; Kurian, J. Pressure Oscillations from Cavities with Ramp. *AIAA J.* **2009**, *47*, 2974–2984. [CrossRef]
28. Spalart, P.R. Comments on the Feasibility of LES for Wings, and on a Hybrid RANS/LES Approach. In Proceedings of the First AFOSR International Conference on DNS/LES, Ruston, Louisiana, LA, USA, 4–8 August 1997.
29. Baldwin, B.; Lomax, H.T.-l. Approximation and Algebraic Model for Separated Turbulentflows. In Proceedings of the 16th Aerospace Sciences Meeting, Huntsville, AL, USA, 16–18 January 1978; American Institute of Aeronautics and Astronautics: Washington, DC, USA, 1996. [CrossRef]
30. Larcheveque, L.; Sagaut, P.; Lê, T.-H. Large-Eddy Simulations of Flows in Weapon Bays. In Proceedings of the 41st Aerospace Sciences Meeting and Exhibit, Reno, NV, USA, 6–9 January 2003; American Institute of Aeronautics and Astronautics: Washington, DC, USA, 1996. [CrossRef]
31. Rodriquez, G.; Carlos, V.; Marcel, I. Numerical Studies of High-Speed Cavity Flows Using LES, DDES and IDDES. In Proceedings of the 51st AIAA Aerospace Sciences Meeting Including the New Horizons Forum and Aerospace Exposition, Orlando, LA, USA, 1 January 2013.
32. Shyy, W.; Jayaraman, B.; Andersson, A. Modeling of glow discharge-induced fluid dynamics. *J. Appl. Phys.* **2002**, *92*, 6434–6443. [CrossRef]
33. Sinha, N.; Dash, S.; Chidambaram, N.; Findlay, D. A Perspective on the Simulation of Cavity Aeroacoustics. In Proceedings of the 36th AIAA Aerospace Sciences Meeting and Exhibit, Reno, NV, USA, 12–15 January 1998; American Institute of Aeronautics and Astronautics: Reno, NV, USA, 2012. [CrossRef]
34. Bauer, R.C.; Dix, R.E. *Engineering Model of Unsteady Flow in a Cavity*; Arnold Engineering Development Center, US Air Force: Arnold AFB, TN, USA, 1991.

aerospace

MDPI

Article

Evolution of Shock Waves during Muzzle Jet Impinging Moving Bodies under Different Constrained Boundaries

Zijie Li * and Hao Wang

School of Energy and Power Engineering, Nanjing University of Science and Technology, Nanjing 210016, China; wanghao1960@126.com
* Correspondence: lizijie@njust.edu.cn

Abstract: A recently developed launching device called the gun–track launch system is affected by its constrained track, such that the form of the muzzle jet changes from the state of free development in the entire space to a constrained state, where this lends unique characteristics of development to its flow field. In this study, the authors establish the corresponding model for numerical simulations based on the dynamic mesh method. We also considered a model of simulation of the muzzle jet with an "infinitely" constrained track to analyze its performance under real launch conditions to explore the mechanism of development and the disturbance-induced propagation of the shock wave when the muzzle jet impinges on moving bodies. The results showed that the muzzle jet exhibited a circumferential asymmetric shape that tilted toward the area above the muzzle and generated transverse air flow that led to the generation of a vortex on it. Because the muzzle was close to the ground, the jet was reflected by it to enhance the development and evolution of the shock waves and vortices and to aggravate the rate of distortion and asymmetry of the jet. The wave reflected from the ground was emitted once again when it encountered the infinitely constrained track. No local low-pressure area or a prominent vortex was observed after multiple reflections. Because the track in the test model was short, the waves reflected by the ground were not blocked, and vortices were formed in the area above the ground. Significant differences in the changes in pressure were also observed at key points in the domain. The results of a comparative analysis showed that the infinitely constrained track increased the Mach number of the moving body from 1.4 to 1.6. The work provides a theoretical basis and the requisite technical support for applications of the gun–track launch system.

Keywords: shock wave/vortex; muzzle jet; constrained boundary; reflection; dynamic mesh

Citation: Li, Z.; Wang, H. Evolution of Shock Waves during Muzzle Jet Impinging Moving Bodies under Different Constrained Boundaries. *Aerospace* **2023**, *10*, 908. https://doi.org/10.3390/aerospace10110908

Academic Editor: Raffaello Mariani

Received: 30 August 2023
Revised: 11 October 2023
Accepted: 17 October 2023
Published: 25 October 2023

1. Introduction

A new type of test device called the gun–track launch system has been developed for the non-destructive recovery of high-value warheads by taking full advantage of the characteristics of the testing capabilities of the gun and the rocket track and aiming to satisfy the requirements of a warhead with a muzzle with high kinetic energy [1,2]. This launch system connects the track outside the muzzle, where this constitutes a unique form of motion for the warhead. That is, its state changes from free, high-speed motion in the entire space to constrained, high-speed motion [3,4]. The muzzle jet also exhibits three-dimensional (3D) circumferential asymmetry at this time, which induces vortices of different forms [5,6]. Against this backdrop, the authors of this study raise the problem of the transient development and evolution of the shock wave when the muzzle jet impinges on a constrained moving body. This is important for investigating the complex phenomenon of flow in the muzzle jet with constrained boundaries, which is in turn critical for conducting basic scientific research and solving key technical problems [7,8].

Researchers have used a variety of test platforms for launching projectiles to investigate the performance of the muzzle based on numerical simulations and visual experiments [9,10]. The characteristics of the transient evolution of shock waves due to the impact of the muzzle jet on a moving object have long been a subject of interest in the area [11,12]. Florio et al. [13] numerically simulated the influence of the area and layout of the exhaust port in the muzzle device on the flow field of the muzzle and considered a configuration with a single opening of different sizes (located near the muzzle). The results showed that increasing the area of the opening reduced the radial range of the muzzle jet, and the radial flow of the lateral opening enhanced the expansion of the gas and its radial diffusion into the environment. Moreover, a large, single-auxiliary opening was found to be more conducive to reducing the pressure of the muzzle.

Schmidt et al. [14] measured the effects of characteristic parameters of the brake of the muzzle on a 20 mm small-caliber gun by using time-accumulated shadow photography. They obtained clear images of the muzzle jet and analyzed its characteristics and mechanism of flow. They reported a strong coupling between the axial flow fields and a weak coupling between the transverse flow fields.

Guo [15,16] used the direct shadow method to carry out visualization experiments on the muzzle jet of a 7.62 mm small-caliber gun and obtained a large number of clear, high-resolution time-series shadow photographs that displayed the characteristics of the typical muzzle jet, including the shock wave, weak compression wave, contact discontinuity, and boundary of the jet, under various conditions. They also discussed the dynamic development of the flow field under different conditions and provided direct experimental comparisons as well as a reference for numerical calculations and related weapons research.

Zhang [17,18] used the shadow method to conduct experimental and numerical simulations on the characteristics of the development of the shock wave of an impinging jet on a moving projectile with a muzzle device. The processes of mutual collision, formation of the lateral jet, and its eventual coupling with the main flow field of the muzzle device were systematically described. The authors observed that a typical, circumferential, symmetrical, multi-level shock wave and a wave system with overlapping discontinuities were formed in the flow field of the muzzle device without a cavity. However, the diverting effect of the open-ended device caused part of the jet to discharge from the side hole in the flow field of the open-ended muzzle device, and its rate of initial axial expansion was greater than that of initial transverse expansion.

Du [19] analyzed the flow field of a 200 mm caliber projectile that was dynamically launched by an electromagnetic track launcher by using the multi-block, structured, overlapping mesh method. The results showed that the dynamic launch of the electromagnetic orbit launcher involved the flow field of a complex shock wave system, with the combined action of a pre-projectile shock wave, a moving spherical shock wave, and a coronal shock wave. The pressure distribution at the midpoint of the head of the projectile was "symmetric," the distribution of the drag coefficient was correlated with the pressure distribution at this point, and the "weak symmetry" of the shock wave at the muzzle first decreased and then increased.

Zhou [20] used a high-pressure gas propulsion device to launch 130 mm caliber projectiles and analyzed the muzzle jet. The results of simulations showed that the jet pushed the projectile forward, and the air flow dispersed and expanded into the space on both sides of the bottom of the projectile to generate vortices. As the projectile accelerated, a Mach disk appeared at its bottom, and the vortices on both sides diffused outward. The Mach disk subsequently moved and shrank, and this led to the appearance of prominent reflected shock waves and slip surfaces.

In sum, researchers have extensively investigated the characteristics of the flow field of a muzzle jet impacting a freely moving body, but few studies have examined the development and mechanism of evolution of the shock and vortex of a muzzle jet impacting a constrained moving body. The innovation of this article is that the authors designed a small-scale test platform with a constrained track to tackle the problem of the evolution of the shock wave of a supersonic jet as it impacts a high-speed moving body. Meanwhile, we established a model with an infinitely long track to simulate the real engineering environment, with the aim of better understanding the complex phenomenon of flow caused by the special, constrained boundary of the muzzle jet of the gun-track launch system. We compare and analyze the characteristics of the transient evolution of the shock wave by establishing models of the muzzle jet with short and "infinitely" constrained tracks.

The remainder of this paper is structured as follows: Section 2 details the relevant physical model, model of the mesh, boundary conditions, and methods to obtain solutions. Section 3 briefly introduces the small-scale test of the muzzle jet impacting a moving body, including the device and the scheme applied. Section 4 compares and analyzes the results of the experiments and simulations under different constraints on the structure of the muzzle jet, the characteristics of the evolution of the shock wave, the generation and evolution of the vortex, transient changes in pressure at key points, and key parameters of the moving body. Section 5 summarizes the conclusions of this study.

2. Physical and Mesh Models

2.1. Physical Model

We established two models to realistically analyze the launch environment. Case 1 corresponded to the structure of the small-scale test, while Case 2 involved an infinitely constrained track.

Figure 1a–c show the simulated physical structure of the models considered here. It mainly included a constrained track, a tube, a moving body, and a supporting plate. The diameter of the muzzle was 65 mm, and the constrained track was simplified to a I-shaped, ignoring small grooves. The launcher was cylindrical, 600 mm long, and fitted the structure of the track. We have appropriately simplified the moving body to a cylinder. It was 500 mm above the ground and was 6.0 m long along the direction of motion and 2.5 m long along the other directions.

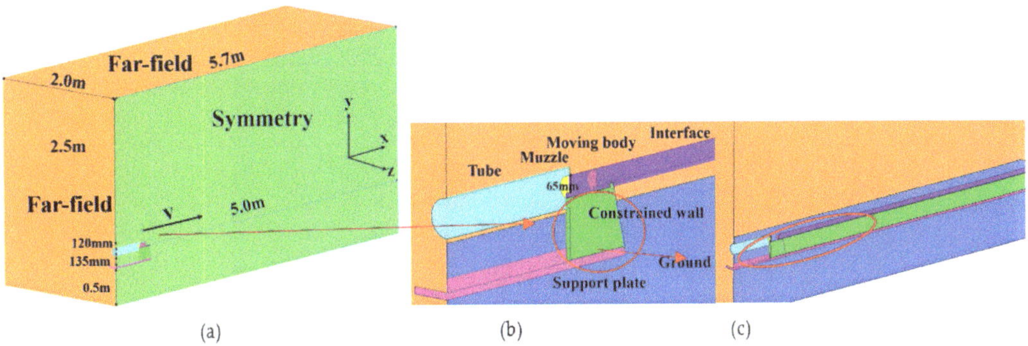

Figure 1. Simulated physical structure of the model: (**a**) Physical structure and size of the computational domain. (**b**) Local physical structure of Case 1. (**c**) Local physical structure of Case 2.

2.2. Dynamic Mesh Method

The dynamic mesh method was used to handle the changes in the mesh caused by the motion of objects during the calculation. The dynamic mesh algorithm was used to determine the requisite adjustments to nodes in the mesh. Any of three algorithms can be used for this purpose: the layering method, the smoothing method, and the remeshing method. As the object only exhibited translational motion in one direction, we determined that the layering method could best simulate its motion.

It is necessary to set the segmentation and merging factors of the mesh in the layering method. Due to the movement of the body, the mesh behind it moves forward along the X-axis, and this causes the mesh layer near the bottom of the mobile body to stretch, with the length of the edge of the mesh of hs. The mesh layer near the top of the moving body is compressed at the same time, with an edge of length hc. Assuming that the ideal length of the mesh edge is hi, the mesh is divided into two parts when the size of the newly generated mesh satisfies the condition h > (1+Cs) hi. When its size satisfies h < Cshi, the meshes adjacent to it are merged into one mesh. In this study, we set hi = 2.0 mm, Cs = 0.4, and Cl = 0.2.

2.3. Mesh Model

To reduce the cost of the calculations as well as the time needed to perform them, we simulated only half of the model. Figure 2a–c shows a diagram of the mesh model. The entire computational domain was divided into 9.8 million meshes. It is difficult to generate a single, high-quality mesh in many cases when simulating a complex flow field. The mesh needs to be segmented and spliced to facilitate processing. The computational domain was divided into domains with static and dynamic meshes, and their interface was used for numerical exchange to this end.

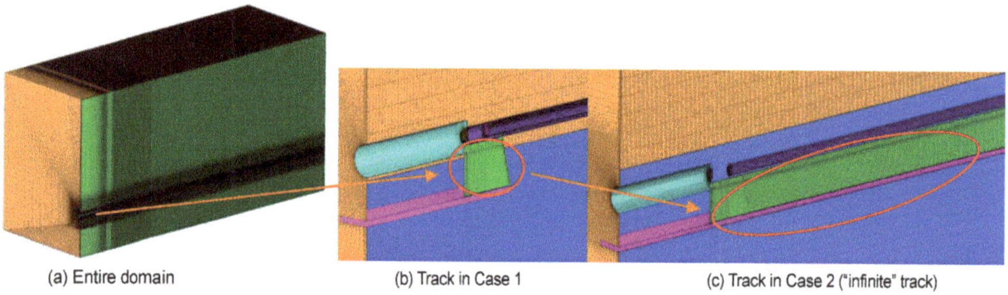

Figure 2. Schematic diagram of the mesh model.

2.4. Boundary Conditions and Solution Methods

We simulated the model of a supersonic jet impinging on a high-speed moving body by using the commercial software Fluent. The domains of calculation were set as the boundary conditions at the outlet of pressure, while the tube and track were assumed to be non-sliding insulation walls. The moving body was assumed to adhere to the boundary condition of the moving wall and was controlled by a user-defined function (UDF) program with six degrees of freedom (six-DOF). The initial velocity was set to 500 m/s, according to the test data. The boundary condition for the muzzle was the boundary at the inlet of pressure, which was 45 MPa according to the test. According to the after-effect model, the pressure of the muzzle changed with time as $p = 45.0 \times 10^6 e^{-24.886}t$. The environmental pressure was 101,325 Pa, and the temperature was 300 K.

We processed the motion in the given block-based division of the mesh based on the finite volume method in combination with the structural dynamic method. The Navier–Stokes (N–S) equation solver based on density correction and the K-ε realizable turbulence model were used. The inviscid convective flux was split by using the Roe-FDS scheme, and the implicit scheme was used with the time marching method to accelerate the convergence of the numerical calculations. The equation of flow control was discretized by using the second-order upwind scheme, and the material was assumed to be an ideal gas.

2.5. Mesh Independence Verification

To verify the independence of the mesh, we designed three sets of meshes, namely 2.2 million, 6.0 million, and 9.8 million. They were respectively defined as Coarse mesh, Fine mesh, and Dense mesh. Regarding the variation in Ma of the moving body in Case 1, Figure 3 demonstrates the results of numerical simulations of the different mesh models considered. From the figure, the results obtained by the Fine mesh and Dense mesh were very close, with an error of less than 1%. On the contrary, there were significant differences in the results obtained by Coarse mesh model. Therefore, it can be concluded that when using Fine mesh model, the simulation results were independent of the size of the mesh. Therefore, we used Fine mesh model in our calculations.

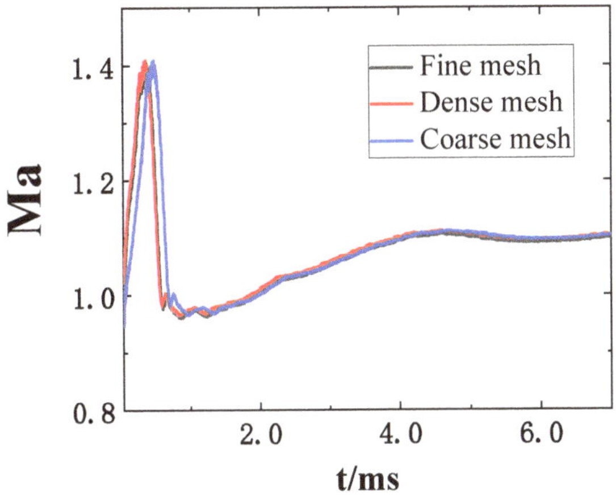

Figure 3. Comparison of Ma of the moving body to verify mesh independence.

3. Analysis of Test Results

3.1. Test Equipment

We developed a small-scale test of the muzzle jet as it impacted a high-speed moving body in the presence of a constrained track. The test device shown in Figure 4 was a 65 mm caliber gun with a constrained track installed at its muzzle to construct a special constrained boundary.

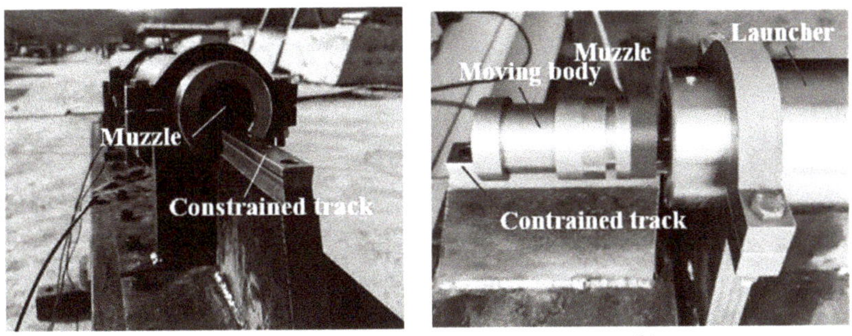

Figure 4. Photographs of the test mode: (**a**) Photograph of the constrained boundary. (**b**) Photograph of the moving body.

3.2. Test Plan

Figure 5 illustrates the test plan. The test platform was the main generator of the jet, and the moving body was pushed out of the muzzle at a speed of about 500 m/s. The muzzle was externally connected to a constrained track, and the body moved along it after having been discharged. A piezoelectric dynamic pressure sensor, manufactured by the Kistler company, was used to measure the overpressure signals. All pressure signals were collected using the DEWE2-A13 transient data recorder that was manufactured by DEWETRON Industrial Measurement System (Grambach, Austria). The formation and development of the overpressure field were photographed by using a high-speed photon camera at a frame rate of 5000 fps.

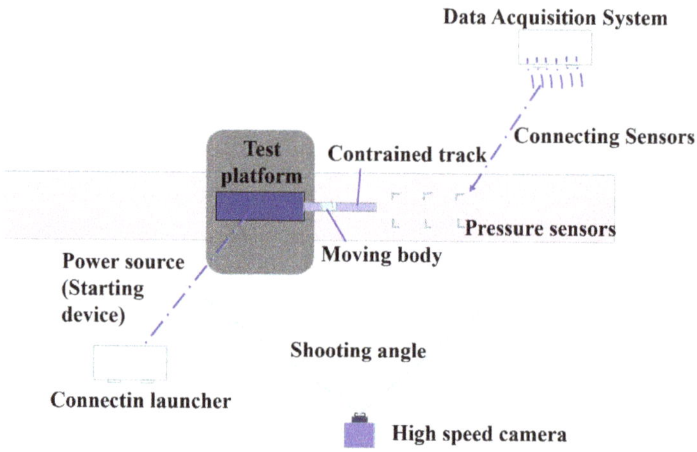

(a)

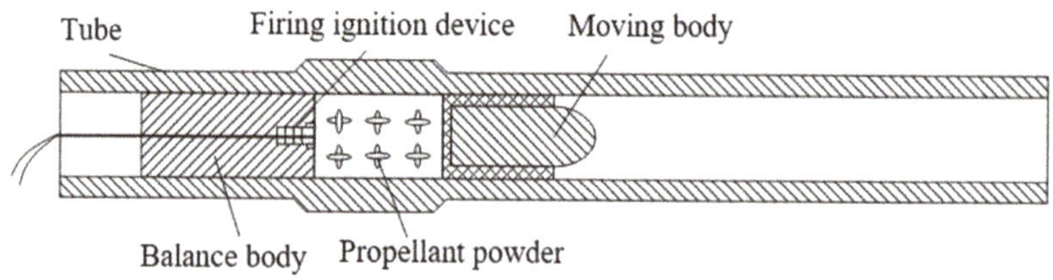

(b)

Figure 5. Schematic diagram of the test. (**a**) Schematic diagram of the test plan, (**b**) Schematic diagram of the internal loading of the launch device.

3.3. Test Results

During the experiment, we triggered the ignition system to ignite the gunpowder inside the chamber, generating high-temperature and high-pressure gas that propelled the moving body forward at high speed. Meanwhile, we applied the high-speed camera to capture the development posture of the muzzle jet and the movement posture of the moving body to determine the operating speed of the moving body. At the same time, the pressure sensor is triggered to monitor the pressure at key points. Figure 6 shows the test procedure as captured by the high-speed camera. The moving object had just appeared from the muzzle in Figure 6a, marking the beginning of the simulation, and the muzzle jet had just begun to form at this time. Figure 6b shows that the moving body flew out of the muzzle, and high-pressure and high-temperature gas and smoke were ejected from it and impacted the moving body. With the movement of the high-speed projectile and the development of the jet, the muzzle jet came into contact with the ground and generated a reflected wave front in the opposite direction. The new wave front continued to develop upward and is clearly displayed on the background plate. It significantly interfered with the development of the muzzle jet. As the latter developed, the reflected wave formed a local high-pressure area under the muzzle. The numerical simulations provided below account for this phenomenon in detail. Figure 6c shows that the moving object gradually flew out of the muzzle over time, and the muzzle jet finally disappeared as the gunpowder was exhausted.

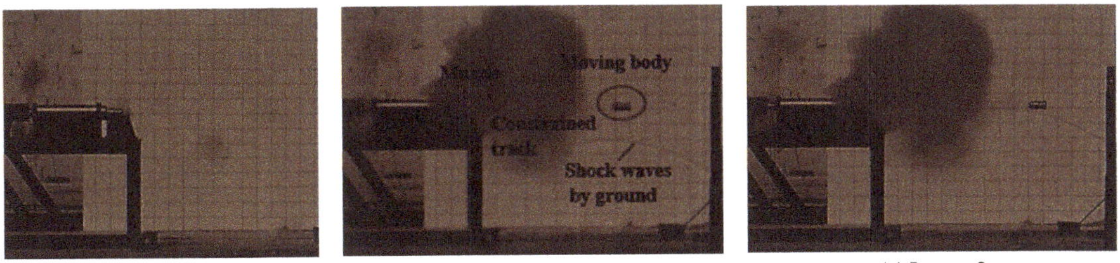

(**a**) Image 1 (**b**) Image 2 (**c**) Image 3

Figure 6. Representative visualized results of the test.

Changes in the pressure at key points were also monitored during the test. Figure 7 compares the curves of pressure versus time between the test and the simulation at the points (1500, 30, 10) mm, and it is clear that they were consistent. Therefore, the method of simulation used in this paper could accurately simulate the characteristics of the muzzle jet.

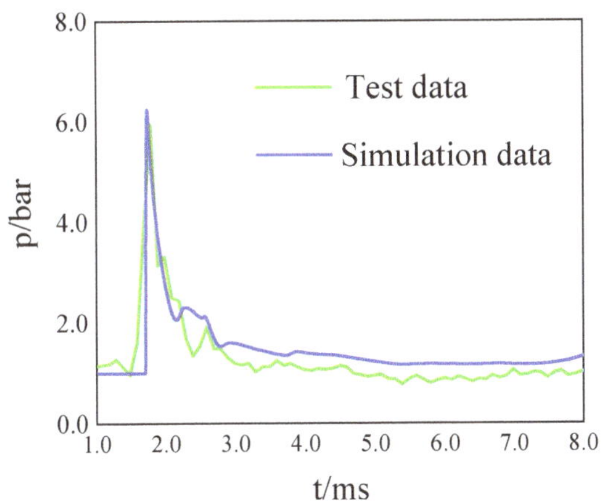

Figure 7. Comparison of the changes in pressure over time at a key point between the test and the simulation.

4. Results of Numerical Simulations

4.1. Muzzle Jet

The muzzle jet poses a series of aerodynamic problems, among which shock/shock interference, a strong shock, and a strong vortex are the main phenomena of flow in the flow field. Compared with that in the traditional muzzle jet, the moving body moves forward on the track as driven by the jet of gas in the muzzle jet, which is affected by a constrained track, and this limits the state of development of the muzzle jet in free space. Therefore, it exhibits a high degree of 3D circumferential asymmetry and generates various kinds of vortices.

Figures 8 and 9 show the contours of pressure of the muzzle jet on the plane of symmetry in the two cases. Figure 8a,b show that the moving body had just moved out of the muzzle to form the preliminary muzzle jet. However, it developed inadequately, and its range of influence was small. In addition to being hindered by the constrained track, it pushed the surrounding air to form a "semi-spherical" shock wave, in addition to a shock wave that was simultaneously formed in front of the moving body. The wave front of the shock wave was relatively regular and spherical, but it was truncated on the track.

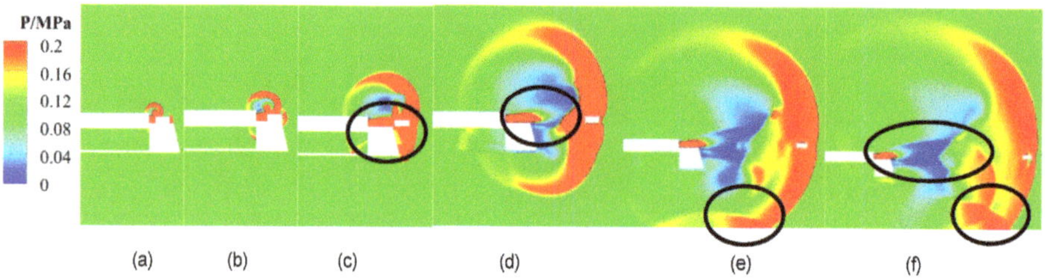

Figure 8. Contours of pressure of the muzzle jet on the plane of symmetry over time in Case 1: (**a**) t = 0.1 ms; (**b**) t = 0.2 ms; (**c**) t = 0.5 ms; (**d**) t = 1.0 ms; (**e**) t = 2.0 ms; (**f**) t = 3.0 ms.

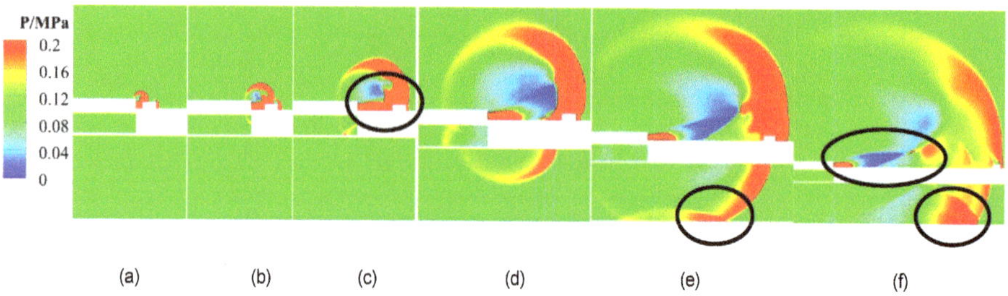

Figure 9. Contours of pressure of the muzzle jet on the plane of symmetry over time in Case 2: (**a**) t = 0.1 ms; (**b**) t = 0.2 ms; (**c**) t = 0.5 ms; (**d**) t = 1.0 ms; (**e**) t = 2.0 ms; (**f**) t = 3.0 ms.

Figure 8c,d show that the flow field of the muzzle gradually developed, and its asymmetry became more prominent. Under interference by the constrained track, the support plate, and the test platform, the jet tilted prominently toward the area above the muzzle. Although the track used in the test was not very long, the development of the shock wave was dependent on the past state of the system. The initial disturbance distorted the shock wave of the muzzle so that it changed from being circumferentially symmetric to circumferentially asymmetric. Figure 8e,f show that because the test platform was 500 mm above the ground, the shock wave of the muzzle was emitted when it impacted the ground, which caused it to propagate upward and form a new wave front. A complete wave system consisting of an asymmetric coronal shock wave, a reflected shock wave, and a multi-level shock wave composed of the Mach disk and overlapping discontinuities was formed around the muzzle at this time.

A comparison of the two cases shows that the muzzle jet, when influenced by varyingly constrained tracks, exhibited certain differences. The states of the muzzle jet shown in Figure 8a,b, and Figure 9a,b were completely consistent because they differed in time by only a few tenths of a millisecond, and the moving body was still within the range of the short track. The shock wave moved out of track and gradually developed downward, as shown in Figure 8c,d, while Figure 9c,d show that the shock wave was blocked by the "infinitely" constrained track in Case 2 and could not develop smoothly to the lower area. Figure 9f shows that the difference between the cases gradually became prominent. In Case 2, the wave reflected by the ground encountered a long track and was reflected once again. The reflected wave propagated downward to form a second area of local high pressure. This phenomenon did not occur in Case 1 and is discussed in detail below.

4.2. Evolution of Shock Waves and Vortices

The pressure induced by the reflection of the shock wave in the muzzle from the wall was consistently higher than that of the incident shock wave if the angle between the latter and the wall was non-zero. p_0, ρ_0, and $\mu_0 = 0$ represent the pressure, density, and velocity of the particles of gas at the front of the incident shock wave, respectively, and

p_1, ρ_1, D_1, and μ_1, and p_2, ρ_2, D_2, and μ_2 represent the pressure, density, and velocities of the wave and particles of the gas behind the surfaces of the incident and the reflected shock waves, respectively. The relationship given below was found to hold during the collision that led to the formation of an unsteady and strong shock wave system. The normal reflection-induced excessive pressure of the strong shock wave can be expressed as follows [21]:

$$p_2 = p_1 + \frac{2k\varphi_1(p_1 - p_0)}{(k-1)p_1 + (k+1)p_0} \tag{1}$$

where k = 1.4.

The normal reflection-induced pressure was high because the airflow behind the reflected shock wave was abruptly slowed down by the surface of the ground. Figure 10a shows a sketch of the normal reflection of the shock wave, where the subscripts 0–2 represent the corresponding states of gas. As the ground acted as a rigid body, a zone of stagnation was formed between the wall and the reflected shock wave.

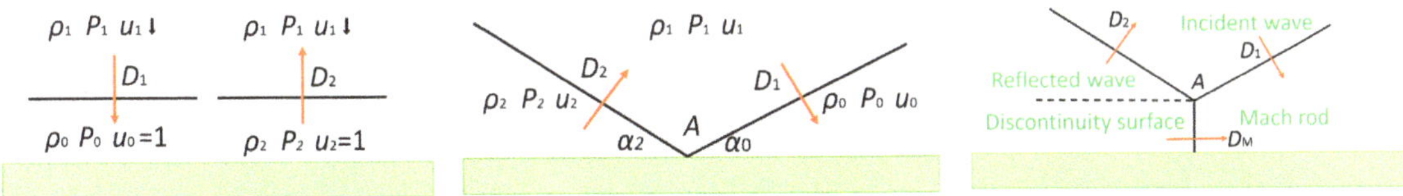

(**a**) Normal reflection (**b**) Regular reflection (**c**) Mach reflection

Figure 10. Schematic diagram of reflection of the shock wave.

A sketch of the regular reflection of the shock wave is shown in Figure 10b. Two kinds of reflections were observed: regular and Mach reflections, depending on the incident angle. The relationship between the regular oblique reflection and the pressure was as follows:

$$\frac{p_2}{p_1} = \frac{7}{6}M_1^2\sin^2\varphi_1 - \frac{1}{6} \tag{2}$$

where φ_1 is the jumping angle of densification.

When Mach reflection occurred, one wave propagated through the gas near the surface, but two waves (the incident and the reflected waves) propagated farther away through the gas.

Figures 11 and 12 show changes in the development of the vortex in the muzzle jet on the plane of symmetry over time. Under interference by the constrained track, support plate, and test platform, the shapes of the vortices in areas above and below the muzzle differed in Case 1. Both vortices were located in the low-pressure area. The streamline of the vortex in the area above the muzzle pointed outward from its center, and the spiral point was stable and difficult to dissipate. Figure 11b–d show that the development of the vortices in the flow field changed significantly. The vortex above continued to develop steadily, with its streamline pointing from its center to the outside. Its spiral point was stable and difficult to dissipate. The streamline of the vortex in the area below the muzzle pointed to its center from the outside, owing to continuous interference by the wave reflected from the ground. Its spiral point was unstable and easily dissipated.

Figure 12 shows the significant difference between the cases considered. The vortex in the upper region in Case 2 was similar to that in Case 1. It was located in a low-pressure area, and its streamline maintained its direction from its center to the outside, with a stable spiral point that was not easy to dissipate. Interestingly, however, no vortex was found in the lower region. The shock wave of the muzzle was emitted when it came into contact with the ground, and the reflected wave propagated upward. Part of it moved forward while the other part moved backward, and both formed a high-pressure area. The wave reflected by the ground was reflected once again when it met the track and subsequently developed downward to form a local high-pressure area. Figure 12e–g shows that pressures at the points of reflection were relatively high and the difference in pressure between the points was small. With the continuous development of the flow field, multiple reflections were formed between the ground and the constrained track. This state failed to yield a stable, low-pressure area such that no vortex was formed.

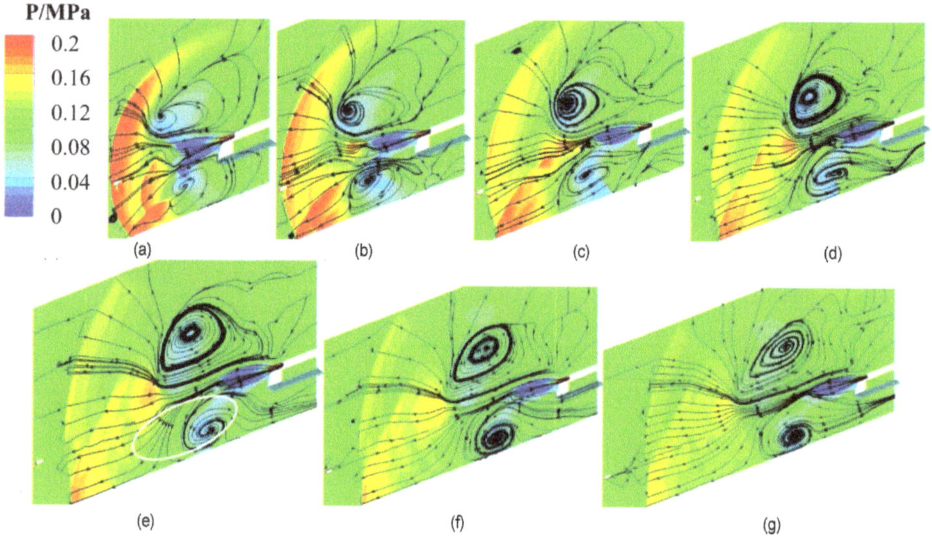

Figure 11. Evolution of vortices in a symmetric muzzle jet over time in Case 1: (**a**) t = 3.0 ms; (**b**) t = 4.0 ms; (**c**) t = 5.0 ms; (**d**) t = 6.0 ms; (**e**) t = 7.0 ms; (**f**) t = 8.0 ms; (**g**) t = 9.0 ms.

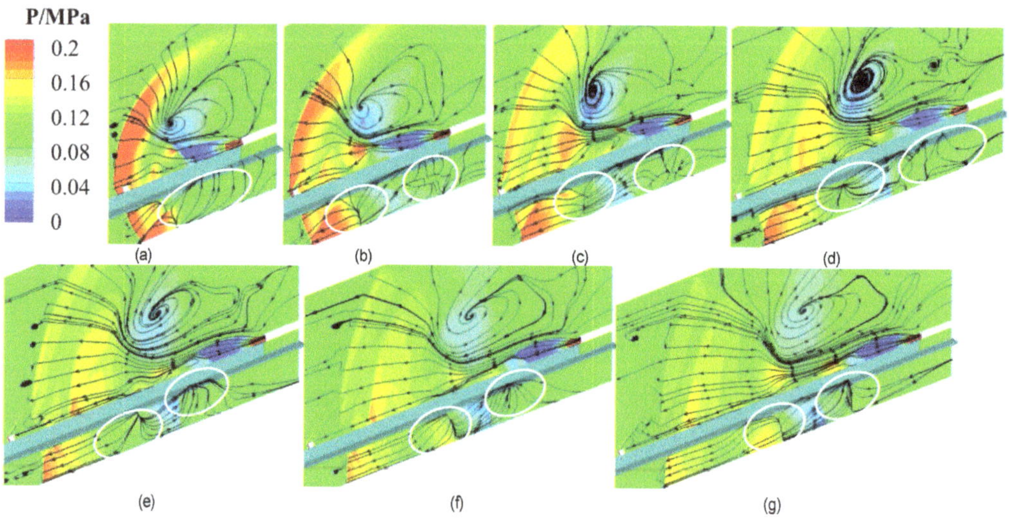

Figure 12. Evolution of vortices in a symmetric muzzle jet over time in Case 2: (**a**) t = 3.0 ms; (**b**) t = 4.0 ms; (**c**) t = 5.0 ms; (**d**) t = 6.0 ms; (**e**) t = 7.0 ms; (**f**) t = 8.0 ms; (**g**) t = 9.0 ms.

Figure 13 shows a comparison of the contours of the development and evolution of the shock waves in the area below the muzzle in the two cases. It is clear from Figure 13a that in Case 1, the shock wave of the muzzle was reflected from the ground and formed a local area of high pressure and another shock wave. The shape of the shock wave in

Case 2 was different from that in Case 1. The wave reflected from the ground was reflected once again when it encountered the track and bent downward. As shown in Figure 13b, the shock wave continues to evolve. In Case 1, as the reflection angle increased, the point of incidence was gradually lifted off the ground and formed a Mach rod. In Case 2, the shape of the second shock wave, formed by contact with the track, became increasingly prominent, and there was no significant low-pressure area in the region. This also explains why there was no vortex in Case 2. Figure 13c shows that in Case 1, the shock wave evolved continuously, and another shock wave was generated at "Position 2." The Mach rod was significantly higher than that in Case 2. The launch point on the track in the latter case shifted forward, and a prominent shock wave was generated at Position 2. Figure 13d,e show that the height of the Mach rod in Case 1 gradually increased and the reflected shock wave gradually diminished. The height of the Mach rod gradually increased in Case 2 as well, while the reflected shock gradually diminished, and the shock had a prominent shape at Position 2. On the whole, the shock wave in Case 1 was launched from the ground and formed another shock wave, while the shock wave in Case 2 was reflected multiple times between the ground and the constrained track to form a multi-channel shock wave.

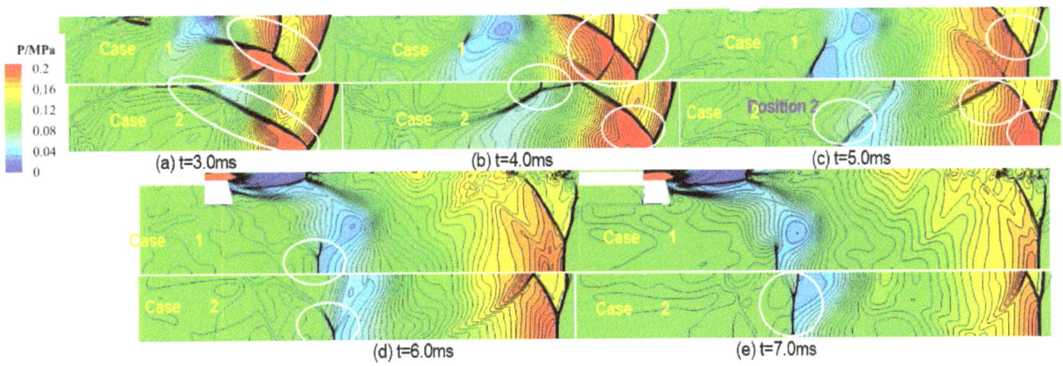

Figure 13. Comparative contours of shock waves in the area below the muzzle in the two cases: (**a**) t = 3.0 ms; (**b**) t = 4.0 ms; (**c**) t = 5.0 ms; (**d**) t = 6.0 ms; (**e**) t = 7.0 ms.

4.3. Comparison of Pressure

To further illustrate the impact of different constrained boundaries on the shock wave and the vortex, we monitored the changes in pressure at several key points over time for comparative analysis. Figure 14 shows a schematic diagram of the locations of the monitoring points, and Table 1 lists their locations.

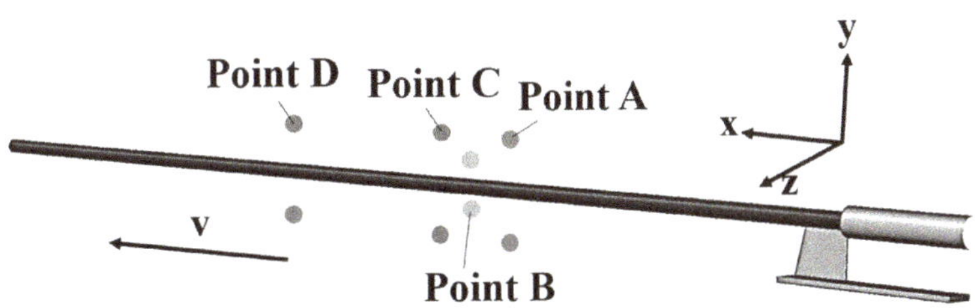

Figure 14. Schematic diagram of the locations of the monitoring points.

Table 1. Locations of the monitoring points on the track.

Point	Location (x, y, z)/mm
A-above	(1600, 60, 200)
A-below	(1600, 60, −200)
B-above	(1800, 35, 100)
B-below	(1800, 35, −100)
C-above	(2000, 60, 200)
C-below	(2000, 60, −200)
D-above	(3000, 0, 200)
D-below	(3000, 0, −200)

Figures 15a–d and 16a–d compare the changes in pressure over time at the key points in Case 1. They show that the muzzle jet exhibited an upward trend, and the monitoring points above the muzzle were disturbed earlier than those below it.

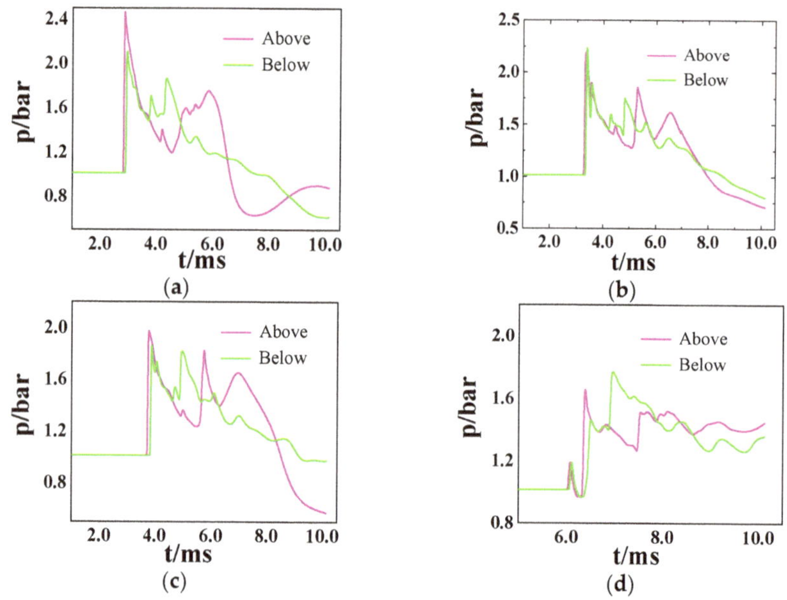

Figure 15. Comparisons of pressure at each monitoring point in Case 1 overtime: (**a**) Point A. (**b**) Point B. (**c**) Point C. (**d**) Point D.

Figure 15a shows that the body moved at a velocity of 500 m/s, and the shock wave moved to this monitoring point in 3 ms to increase the pressure. The maximum pressure at A-above should have been slightly higher than that at A-below. The pressure then decreased and slightly rose at about t = 6.0 ms. The pressure at A-below was supplemented by the reflected waves, because of which the fluctuations in it were relatively small. The shock wave gradually moved forward, and the pressure became stable. Figure 15b shows that the shock wave moved to Point B at t = 4.0 ms. Because Point B was close to the centerline of the jet, the difference in pressure at its point of symmetry was relatively small compared with those at the other monitoring points. The maximum pressures at B-above and B-below were close to each other, and the overall changes in pressure were similar. Figure 15c shows that at t = 4.0 ms, the shock wave moved to Point C, and the change in pressure at C-below was smoother than that at C-above. The change in pressure was small in the area below the muzzle because it was supplemented by the reflected wave, while the pressure in the area above the muzzle was not supplemented. Due to the distance between

Point C and the centerline of the jet, the disturbance caused by the shock wave continued over time. Figure 15d shows that the maximum pressure at D-below was greater than that at D-above. Because Point D was far from the muzzle and the muzzle jet developed more fully, it was significantly influenced by the ground, and this led to a wide range of high pressure in the lower part such that the maximum pressure at D-below was higher than that at D-above.

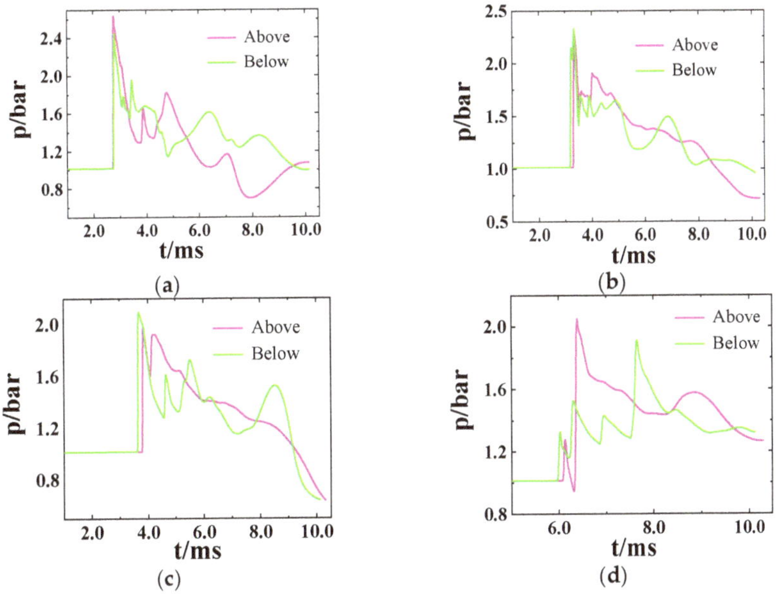

Figure 16. Comparisons of pressure at each monitoring point in Case 2 overtime: (**a**) Point A. (**b**) Point B. (**c**) Point C. (**d**) Point D.

However, the changes in pressure at the same monitoring points were significantly different in Case 2 from those in Case 1, and fluctuations in it were prominent. This also shows that the shock wave was constantly reflected between the ground and the constrained track. The disturbance due to the shock wave influenced the pressure at Point A at almost the same time in both cases, which indicates that the speed of propagation of the shock wave in the lower area of Case 2 was higher than that in Case 1. As for the pressure at Point A in Case 1, the change in the amplitude of pressure at A-below was smaller than that at A-above, but the overall fluctuations in it were prominent. It is clear from Figure 16b that the disturbance in pressure at B-below occurred earlier than that at B-above, where this is different from the situation in Case 1. Figure 16d shows that the shock wave of the muzzle had fully developed by 6.0 ms. Due to interference by multiple reflected waves, the shape of the area under the constraining rail was completely different from that of the shock wave above it. Therefore, the changes in pressure in these areas were also significantly different.

Figure 17 shows a comparison between the cases at the same monitoring point. As shown in the figure, the speed of propagation of the shock wave in Case 2 was higher than that in Case 1, which indicates that the infinitely constrained track in Case 2 accelerated its evolution. The maximum pressure in Case 2 was significantly higher than that in Case 1, which shows that the infinitely constrained track in Case 2 increased the maximum pressure of the shock wave. This is because during the rapid expansion process of the muzzle jet, the long track further limits the expansion of the shock wave, so the attenuation and decay of the shock wave will be slower, resulting in faster velocity and higher pressure. In addition, the trend of changes in pressure over time varied at all four monitoring points, which shows that the constrained track not only influenced the evolution of shock waves and vortices in the area below the muzzle but also had an impact on the area above the muzzle.

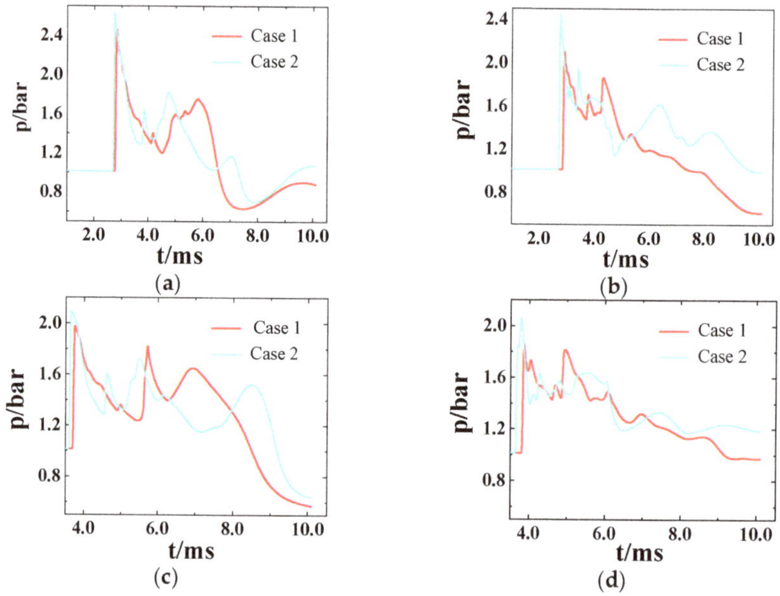

Figure 17. Comparisons of pressure at the monitoring points over time in Case 1 and Case 2: (**a**) Point A-above. (**b**) Point A-below. (**c**) Point C-above. (**d**) Point C-below.

4.4. Key Parameters of Moving Bodies

Figure 18 shows the drag forces, Mach numbers, and total thrust of the moving body in both cases. It is clear from Figure 18a that the resistance of the moving body in Case 1 was a bit larger than that in Case 2. The two cases have similar fluctuations. This also shows that the speed of the moving body in Case 2 was lower than that in Case 1. The change in the Mach number and total pressure of the moving body in Figure 18b,c verifies this. As shown in Figure 18b, the velocity of the moving body increased close to the muzzle due to the thrust of the muzzle jet. As the body moved away from the muzzle, its propulsion rapidly weakened, such that its Mach number gradually decreased under the action of resistance. The Mach number of the moving body in Case 2 was significantly higher than that in Case 1, indicating that the track restricted the development of the shock wave of the muzzle and caused it to become concentrated on the track. In Figure 18c, the maximum thrust at the bottom of the moving body is 45 Mpa, which then rapidly decreased as it moved forward. At the beginning, the trend of the two cases was basically the same because both cases had a track at the muzzle. Next, the thrust in Case 1 continued to decrease, while the thrust in Case 2 had an upward process. This is because the long rrack constrains the development of the muzzle jet, preventing it from spreading smoothly around but rather concentrating more on the orbit, which acts on the bottom of the moving body. This is also why the Mach number of the moving body in Case 2 is higher, as it receives more thrust.

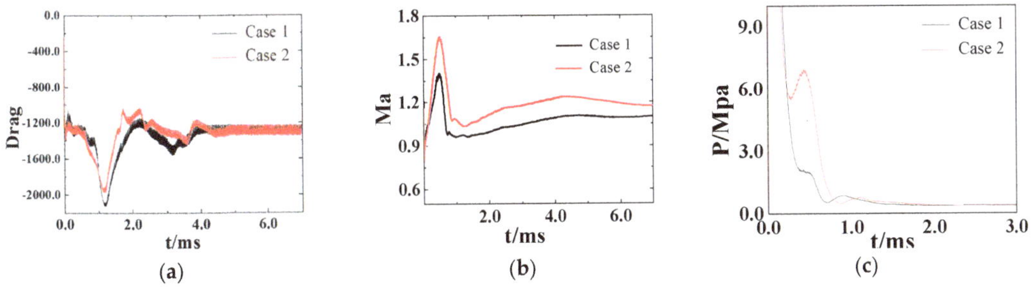

Figure 18. Drag forces, Mach numbers, and total thrust of the moving body over time in Cases 1 and 2: (**a**) Drag. (**b**) Mach number. (**c**) Total thrust.

5. Conclusions

In this study, the authors established an experimental platform and models of simulation by using the dynamic mesh method to investigate the characteristics of the evolution of shock waves and vortices in the case of a muzzle jet impinging on a moving body in the presence of varyingly constrained tracks. The conclusions are as follows:

(1) The shock wave of the spherical muzzle, which should have been circumferentially symmetrical, was distorted due to the presence of the constrained track, and the entire structure was inclined to the area above the muzzle. Because the shock wave was blocked by the track, transverse air flow was generated, which led to the formation of vortices. The characteristics of the evolution of the vortices were unique and thus different from those of the traditional model of the muzzle jet.

(2) Because the test platform was close to the ground, the shock wave of the muzzle was reflected from it, which enriched the characteristics of the evolution of the shock wave and the vortex in the asymmetric muzzle jet. This led to the formation of a Mach rod and a vortex in the area above the ground in Case 1. However, in Case 2, the shock wave was reflected multiple times between the ground and the constrained rail without forming a stable low-pressure area or a vortex.

(3) The constrained track influenced the development and evolution of the muzzle jet, where this was directly reflected by the difference in pressure and the velocity of the shock at the monitoring points between the cases considered. At the same time, the Mach number and total thrust of the moving body in Case 2 were higher than those in Case 1.

Author Contributions: Conceptualization, Z.L.; methodology, Z.L.; software, Z.L.; validation, Z.L., H.W.; formal analysis, Z.L.; investigation, Z.L.; resources, Z.L., H.W.; data curation. Z.L.; writing—original draft preparation, Z.L.; writing—review and editing, Z.L., H.W; visualization, Z.L.; supervision, Z.L.; project administration, Z.L.; funding acquisition, Z.L. All authors have read and agreed to the published version of the manuscript.

Funding: This research received no external funding.

Data Availability Statement: All data and models that support the findings of this study are available from the corresponding author upon reasonable request. All data and models generated or used during this study appear in the submitted article.

Acknowledgments: Thank you to Hexia Huang (Nanjing University of Aeronautics and Astronautics) for his opinions, guidance, and contributions to this article.

Conflicts of Interest: The authors declare that they have no conflict of interest.

References

1. Goto, K.; Kato, Y.; Ishihara, K.; Matsuoka, K.; Kasahara, J.; Matsuo, A.; Funaki, I.; Nakata, D.; Higashino, K.; Tanatsugu, N. Thrust validation of rotating detonation engine system by moving rocket track test. *J. Propuls. Power* **2020**, *37*, 419–425. [CrossRef]
2. Zhao, P.G.; Liu, Z.; Dang, T.J.; Sun, Y.; Sun, Y. Numerical and wind tunnel test research on aerodynamic characteristics of high-speed rocket sled under the influence of ground effect. *Chin. J. Comput. Mech.* **2023**, *40*, 314–322. [CrossRef]
3. Jiang, S.Y.; Wang, H.; Lin, C.J.; Wang, J.L. Establishment and simulation of two-dimensional interior ballistics model of large caliber Davis gun considering the axial movement of tubular charge clusters. *Acta Armamentarii* **2016**, *37*, 1941–1948. [CrossRef]
4. Wang, G.; Cheng, C.; Zhang, X.; Huang, X. Numerical simulation and analysis of muzzle flow during a rarefaction wave gun firing. *Propellants Explos. Pyrotech.* **2021**, *46*, 1902–1913. [CrossRef]
5. Chaturvedi, E.; Dwivedi, R.K. Computer aided design and analysis of a tunable muzzle brake. *Def. Technol.* **2019**, *15*, 89–94. [CrossRef]
6. Zhao, X.Y.; Zhou, K.D.; He, L.; Lu, Y.; Wang, J.; Zheng, Q. Numerical simulation and experiment on impulse noise in a small caliber rifle with muzzle brake. *Shock Vib.* **2019**, *2019*, 5938034. [CrossRef]
7. Debnath, P.; Pandey, K.M. Numerical Investigation on detonation combustion waves of hydrogen-air mixture in pulse detonation combustor with block-age. *Adv. Aircr. Spacecr. Sci.* **2023**, *10*, 203–222.
8. Jones, A.R.; Medina, A.; Spooner, H.; Mulleners, K. Characterizing a burst leading-edge vortex on a rotating flat plate wing. *Exp. Fluids* **2016**, *57*, 52. [CrossRef]
9. Moumen, A.; Grossen, J.; Ndindabahizi, I.; Gallant, J.; Hendrick, P. Visualization and analysis of muzzle flow fields using the background-oriented schlieren technique. *J. Vis.* **2020**, *23*, 409–423. [CrossRef]
10. Li, Z.J.; Wang, H. Effect of Precursor Flow Field of Muzzle on the Combustion Gas Jet Flow of Gun Propellant. *Chin. J. Energetic Mater.* **2017**, *25*, 282–290. [CrossRef]
11. Mizukaki, T. Visualization and force measurement of high-temperature, supersonic impulse jet impinging on baffle plate. *J. Vis.* **2007**, *10*, 227–235. [CrossRef]
12. Bin, J.; Kim, M.; Lee, S. A numerical study on the generation of impulsive noise by complex flows discharging from a muzzle. *Int. J. Numer. Methods Eng.* **2010**, *75*, 964–991. [CrossRef]
13. Florio, L.A. Effect of vent opening area and arrangement on gas flow field as gas propelled cylinder exits a flow tube. *Meccanica* **2010**, *45*, 475–501. [CrossRef]
14. Schmidt, E.M.; Shear, D. Optical measurements of muzzle blast. *AIAA J.* **2012**, *13*, 1086–1091. [CrossRef]
15. Zeqing, G.; Xiaohai, J.; Yang, W. Parallel numerical simulation of muzzle reacting flow. *Chin. J. Comput. Mech.* **2013**, *30*, 111–116+123. [CrossRef]
16. Zeqing, G.; Haitao, Q.; Xiaohai, J. Numerical simulation on the characteristics of flow field of the muzzle of embedded aircraft gun. *Acta Armamentarii* **2017**, *38*, 2373–2378. [CrossRef]
17. Zhang, H.; Chen, Z.; Guo, Z.; Zheng, C.; Xue, D. Numerical investigation on the three-dimensional flow characteristics of unsteady subsonic elliptic jet. *Comput. Fluids* **2018**, *160*, 78–92. [CrossRef]
18. Zhang, H.; Xiao, Y.; Chuan, Z. Characteristics of three-dimensional flow field structure of supersonic transverse jet. *J. Propuls. Technol.* **2023**, *44*, 30–39. [CrossRef]
19. Du, P.; Lu, J.; Feng, J.; Li, X. Numerical Simulation and Analysis of Flow Field during Dynamic Launching of Electromagnetic Rail Launcher Based on Overlap-ping Multi-block Structured Mesh. *Acta Armamentarii* **2018**, *39*, 234–244. [CrossRef]
20. Zhou, P.; Cao, C.; Dong, H. Analysis of Interior Ballistic Characteristics and Muzzle Flow Field of High-pressure Gas Launcher. *Acta Armamentarii* **2016**, *37*, 1612–1616. [CrossRef]
21. Li, Z.; Wang, H.; Chen, J. Ground effects on the hypervelocity jet flow and the stability of projectile. *Eng. Appl. Comput. Fluid Mech.* **2018**, *12*, 375–384. [CrossRef]

aerospace

MDPI

Article

Mechanism of Evolution of Shock Wave of Muzzle Jet under Initial Interference and Its Simplified Model

Zijie Li * and Hao Wang

School of Energy and Power Engineering, Nanjing University of Science and Technology, Xiaolingwei No. 200, Xuanwu, Nanjing 210016, China; wanghao1960@126.com
* Correspondence: lizijie@nujst.edu.cn

Abstract: Large-caliber and long-barrel weapons are important experimental devices for exploring the impact resistance and reliability of warheads. The force of impact of the muzzle jet has a significant influence on the overload resistance of the warhead and surrounding devices. The mechanism of motion of the body inside the tube cannot be ignored owing to the high kinetic energy at the muzzle. In this study, we designed the relevant experiment and a simulation model to analyze the structural characteristics and mechanism of evolution of the shock wave and the vortex structure in a muzzle jet. The aim was to examine the evolution of the shock wave with initial jet-induced interference. And we established three other simulation models to compare the similarities and differences between the results of the models. The results showed that, in the original complex model, the initial jet restricted the free expansion of the muzzle jet, and this led to many shock–shock collisions that retarded the development of the shock waves. Multiple reflected shock waves were thus formed under a high local pressure that distorted the shock structure, while the structure of the shock wave in the simplified models was clear and simple. The parameters of motion of the body changed by a little when the initial jet-induced interference was ignored. The difference in values of the strongest impact force measured at monitoring points far from the muzzle was small, with an error of about 2%, such that the simplified model without the initial jet could be used in place of the original complex model. The other simplified models yielded significant differences. Our results provide an important theoretical basis for further research on the muzzle jet and its applications in engineering.

Keywords: muzzle jet; initial jet; shock wave; vortex; dynamic grid; 3D

Citation: Li, Z.; Wang, H. Mechanism of Evolution of Shock Wave of Muzzle Jet under Initial Interference and Its Simplified Model. *Aerospace* **2024**, *11*, 381. https://doi.org/10.3390/aerospace11050381

Academic Editor: Bo Zhang

Received: 31 March 2024
Revised: 4 May 2024
Accepted: 6 May 2024
Published: 9 May 2024

1. Introduction

When a barrel weapon launches a projectile, the muzzle flow field will cause initial disturbance to the projectile's flight, reducing its shooting accuracy and increasing its impact point dispersion. In addition, with the development of large-caliber high-initial-velocity artillery, a series of problems such as muzzle flame, muzzle smoke, muzzle shock wave, and noise have emerged, further increasing the initial disturbance of the projectile. Therefore, studying the influence of muzzle flow field on the flight stability of projectiles is one of the important issues in weapon science research. The mechanism of operation of the muzzle jet is a challenging issue at the intersection of multiple disciplines, including gas dynamics, aerodynamic acoustics, and chemical dynamics [1,2]. The moving body accelerates and compresses the air in front of it under the propulsion of gunpowder gas in the tube to form the initial jet. The harmful disturbance caused by the muzzle jet of a large-caliber weapon launcher is extremely strong [3,4]. Due to the presence of the flame, smoke, and electromagnetic interference in the muzzle, it is difficult to measure and calculate the structure of the shock wave in a transient muzzle jet [5,6]. This makes it important to examine the muzzle jet by using numerical simulations.

Past research has provided a clear understanding of the mechanism of development of the muzzle jet as well as the law of distribution of the shock wave at the muzzle and the aftereffect of muzzle flow on moving bodies [7]. Research on the muzzle jet has benefited from advances in computational technologies and methods of numerical calculation. Merlen et al. [8] conducted theoretical analysis based on experimental observations to obtain the law of similarity of the flow field of the muzzle. Ma [9,10] considered the influence of the moving body to analyze the disturbance induced by it on the flow field in calculations of the muzzle jet. Jiang [11] used a structured grid to simulate the muzzle jet of a 122 mm caliber vehicle-mounted gun and analyzed the flow occurring during the launch of a moving body with an initial velocity of 713 m/s. Dai [12,13] numerically simulated the combustion-induced flow field of a body with an initial velocity of 735 m/s with a brake inside a muzzle with a diameter of 7.62 mm. The authors used control equations of the three-dimensional (3D) unsteady flow of the chemical reaction to describe the structure and properties of the flow field of the muzzle as well as its interactions with the moving body. Florio [14] used the 2D axisymmetric Navier–Stokes (N–S) equation and the k-ε turbulence model to simulate the flow field of a muzzle device with a side hole containing a cylindrical moving body. Jiang [15] used the ALE equation and dynamic grid technology to simulate the entire process of motion of the body from inside a 44 mm caliber bore to outside it until it flew away from the initial flow field. Guo [16,17] considered a 7.62 mm ballistic gun for an experimental analysis of the initial flow field at the muzzle and its aftereffects based on the ALE equation and a partition-structured body-fitted grid. Zhang [18] used the 3D Euler equation in combination with the Roe scheme and structured dynamic grid technology to numerically simulate the flow field at the muzzle of a 20 mm caliber gun with a brake. Schmidt et al. [19,20] measured a group of characteristic parameters of the muzzle brake structure on a 20 mm small-caliber gun using the time-accumulated shadow photography technology, obtained a clearer image of the muzzle jet, and analyzed the characteristics and flow mechanism of the muzzle jet. He observed that the coupling between the flow fields in the axial direction was very strong but the coupling along the transverse direction was very weak. Combined with the dynamic grid technology, Zhang [21] established a two-dimensional axisymmetric numerical simulation model to numerically study the explosion flow field generated by the supersonic jet. It is confirmed that the interaction between precursor shock and bow shock is closely related to the velocity of the projectile.

Previous research in the area has mostly focused on analyzing the characteristics of gas jets by using small-caliber short-barrel guns and has not adequately analyzed the interference of the initial jet. We know even less about its potential role in the characteristics of the shock wave and the vortex in the muzzle jet of a 300 mm large-diameter long-barrel launch weapon.

We use a long-barrel and large-caliber launch device as the object of research in this study. Given the structural configuration of such an experimental device, it is difficult to accurately describe the complete characteristics of its muzzle jet by using numerical simulations.

The first model of our research is the actual, whole, original complex model. This involves the movement of the body inside the tube, its movement outside the muzzle, the initial interference jet formed by the compression of gas inside the tube, and the gas muzzle jet formed by the high-temperature and high-pressure gunpowder gas injected after the moving body. All these factors also interfere with one another to form a complete muzzle jet.

The second model of our research here is that we construct simplified 3D models for analysis. It ignores the initial interference jet but includes the motion of the body outside the muzzle and the gas muzzle jet formed by the injection of gunpowder gas. This makes the physical model simpler, reduces the mesh size, and eliminates the need for coupling between dynamic meshes inside and outside the tube to significantly reduce the workload.

The third model of this study is one that neglects the model of the moving body but includes the initial interference jet formed by the gas being squeezed out of the bore as well as the gas jet formed by gunpowder gas. The lack of use of dynamic grid technology during the simulation significantly simplifies the establishment of the model.

The fourth model of our work here is a simplified model that ignores the moving body and the initial jet and considers only the free injection of the gas muzzle jet. This physical model is the simplest of the ones considered here and significantly reduces the size of the mesh.

Based on this large-barrel weapon device, the purpose of this paper is to restore its real complex three-dimensional structure characteristics of the muzzle jet by establishing a complete numerical simulation model. By comparing the simplified model, it is confirmed whether the simplified model is replaceable in some aspects. Therefore, we establish reasonable mathematical and physical models, use them to conduct experiments and numerical simulations, and provide a detailed examination of the structural characteristics of the muzzle jet, the mechanism of evolution of shock waves and vortices, changes in the parameters of the moving body, and the energy distribution of the muzzle jet.

2. Model of Numerical Simulation

2.1. Equations and Turbulence Model

Without considering the effect of external heating and body force, three-dimensional, time-dependent unsteady compressible Navier–Stokes equations are used as the governing equations:

$$\frac{\partial Q}{\partial t} + \frac{\partial f}{\partial x} + \frac{\partial g}{\partial y} + \frac{\partial h}{\partial z} = 0 \tag{1}$$

where Q is the vector of the conservative variables and f, g, and h are the vectors of the convective flux. They are expressed as:

$$Q = \begin{bmatrix} \rho \\ \rho u \\ \rho v \\ \rho w \\ \rho e \end{bmatrix} \tag{2}$$

$$f = \begin{bmatrix} \rho u \\ \rho u^2 + p - \tau_{xx} \\ \rho uv - \tau_{xy} \\ \rho uw - \tau_{xz} \\ (\rho e + p)u - u\tau_{xx} - v\tau_{xy} - w\tau_{xz} + q_x \end{bmatrix} \tag{3}$$

$$g = \begin{bmatrix} \rho v \\ \rho uv - \tau_{xy} \\ \rho v^2 + p - \tau_{yy} \\ \rho vw - \tau_{yz} \\ (\rho e + p)v - u\tau_{xy} - v\tau_{yy} - w\tau_{yz} + q_y \end{bmatrix} \tag{4}$$

$$h = \begin{bmatrix} \rho w \\ \rho uw - \tau_{xz} \\ \rho vw - \tau_{yz} \\ \rho w^2 + p - \tau_{zz} \\ (\rho e + p)w - u\tau_{xz} - v\tau_{yz} - w\tau_{zz} + q_z \end{bmatrix} \tag{5}$$

where:

$$\left.\begin{array}{l} \tau_{xx} = \frac{2}{3}\mu(2\frac{\partial u}{\partial x} - \frac{\partial v}{\partial y} - \frac{\partial w}{\partial z}) \\ \tau_{yy} = \frac{2}{3}\mu(2\frac{\partial v}{\partial y} - \frac{\partial u}{\partial x} - \frac{\partial w}{\partial z}) \\ \tau_{zz} = \frac{2}{3}\mu(2\frac{\partial w}{\partial z} - \frac{\partial u}{\partial x} - \frac{\partial v}{\partial y}) \\ \tau_{xy} = \mu(\frac{\partial u}{\partial y} + \frac{\partial v}{\partial x}) = \tau_{yx} \\ \tau_{xz} = \mu(\frac{\partial w}{\partial x} + \frac{\partial u}{\partial z}) = \tau_{zx} \\ \tau_{yz} = \mu(\frac{\partial v}{\partial z} + \frac{\partial w}{\partial y}) = \tau_{zy} \end{array}\right\} \tag{6}$$

The total energy per unit flow is expressed as:

$$e = \frac{p}{\rho(\gamma - 1)} + \frac{1}{2}\left(u^2 + v^2 + w^2\right) \quad (7)$$

where p is the pressure, ρ is the gas density, u, v, and w are the velocity components of the fluid, and γ is the gas specific heat ratio. The ideal gas state equation is $p = \rho RT$, where R is the universal gas constant and μ is the laminar viscous coefficient. τ_{xx}, τ_{xy}, τ_{xz}, τ_{yx}, τ_{yy}, τ_{yz}, τ_{zz}, τ_{zx}, and τ_{zy} are the viscous forces of different directions; k is the heat conductivity and q_x, q_y, and q_z are the volumetric heating rates in unit mass.

The realizable k-ε turbulence model is applied in the article. Compared to the stand and k-ε turbulence model, it has two main differences: 1. realizable k-ε model adds a formula for turbulent viscosity; 2. a new transfer equation has been added for the dissipation rate. Introducing Boussinesq's linear eddy viscosity assumption, the Reynolds stress expression is:

$$\begin{aligned}\tau_{ij}^{R} &= -\rho\overline{u_i'u_j'} \\ &= \mu_t\left(\frac{\partial u_i}{\partial x_j} + \frac{\partial u_j}{\partial x_i} - \frac{2}{3}\frac{\partial u_i}{\partial x_i}\right) - \frac{2}{3}\rho k\delta_{ij}\end{aligned}$$

Different eddy viscosity models have varying eddy viscosity coefficients in the model:

$$\mu_t = f\left(\frac{\rho k^2}{\varepsilon}\right)$$

where:

$$k = \frac{\overline{u_i'u_j'}}{2}$$

$$\varepsilon = v\overline{\frac{\partial u_i'}{\partial x_j}\left(\frac{\partial u_i'}{\partial x_j} + \frac{\partial u_j'}{\partial x_i}\right)}$$

The normal stress in the realizable k-ε model is $\overline{u_i'u_j'} > 0$. The Schwarz inequality for Reynolds shear stress is:

$$\left(\overline{u_i'u_j'}\right)^2 \leq \overline{u_i^2u_j^2}$$

2.2. Model of Physical Structure

We considered a 300 mm test launch device. It contained a moving body that traveled in a tube with a diameter of 300 mm at the muzzle and a wall with a thickness of 50 mm. The moving body was a simplified cylinder that weighed 160 kg and was 600 mm long. Numerically simulating such a device is a challenging problem as it needs to be simplified while ensuring a certain accuracy. We established a total of four sets of models (four cases; see Table 1) as follows:

Table 1. Characteristics of the models.

Model	Features
Case 1	Complete model; moving body; inside and outside the muzzle
Case 2	Simplified model; moving body; outside the muzzle
Case 3	Simplified model; inside and outside the muzzle
Case 4	Minimalist model; only outside the muzzle

Case 1 involved the complete structure shown in Figure 1a. It featured a moving body, a 30 m long tube, and the domain around the muzzle. Case 2 ignored the motion of the body inside the tube, as shown in Figure 1b, and considered only its motion outside the muzzle. Case 3 ignored the moving body altogether, as shown in Figure 1c, and involved only the initial jet and the muzzle jet. In Case 4, the motion of the body inside and outside the bore was ignored, as shown in Figure 1d, and only the domain outside the muzzle was considered. The flow field was 10 m long along the direction of motion of the body (X-axis), including 3 m toward the rear of the muzzle, and was 8 m long in the other two directions (Y-axis and Z-axis), as shown in Figure 1a.

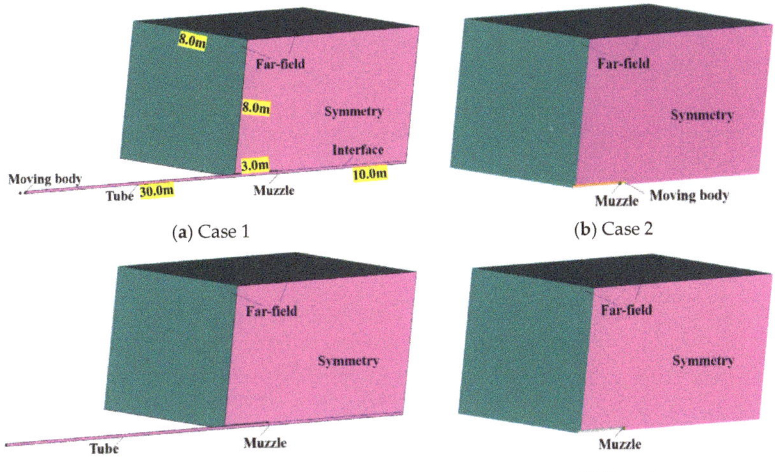

(**a**) Case 1 (**b**) Case 2

(**c**) Case 3 (**d**) Case 4

Figure 1. Schematic diagram of the model of the physical structure.

2.3. Grid Model

A widely accepted technique is to consider only a quarter of the physical model, rather than the entire model, to reduce computational time and cost. The entire computational domain was divided into structural grids, as shown in Figure 2. Case 1 featured 11 million grids, while each of the other three models had 9 million grids. Mesh refinement is applied near the muzzle and projectile, while it is relatively rough at the boundary of the computational domain. The minimum size of the mesh near the muzzle is 0.2 mm and the mesh growth rate is 1.05. When simulating complex flow fields, it is often difficult to generate a single high-quality structural mesh. In this case, the mesh can be partitioned and concatenated for processing. We thus divided the computational domain into domains with static and dynamic grids and used their interface for numerical exchange.

(**a**) Grid model of entire computational domain (**b**) Grid model of local domain

Figure 2. Schematic diagram of grid model of the muzzle jet of a large-caliber long-barrel launcher.

2.4. Boundary Conditions and Solution Methods

We used models of the interior ballistics and aftereffects as well as a model with six degrees of freedom for secondary development in FLUENT software (Fluent19.2) and numerically simulated the complete model of the large-caliber long-barrel muzzle jet.

Figure 3 shows the types of boundaries used. Specifically, the pressure outlet was used as the boundary of the calculation domain, with an environmental pressure of 101,325 Pa and a temperature of 300 K. The tube was set to be a nonslip adiabatic wall, while the moving wall was set for the moving body. Its motion was controlled by a ballistics model inside the tube and by a six-degrees-of-freedom program (six-DOF) after it had exited the muzzle. The pressure inlet was set for the muzzle and it was controlled by using a UDF program. Experimental measurements and those of the internal ballistics model yielded a muzzle pressure of 24.4 MPa and a temperature of 2000 K. According to the model of the aftereffects of the launch device, the variation in the muzzle pressure over time was $p = 24.4 \times 10^{-24.886t}$ MPa.

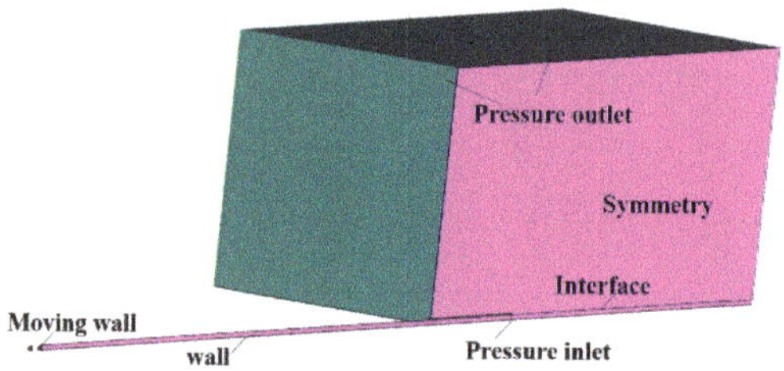

Figure 3. Schematic diagram of boundary conditions.

In addition, we processed the motichenon in the given block-based division of the mesh based on the finite volume method in combination with the structural dynamic method. The Navier–Stokes (N–S) equation solver based on density correction and the k-ε realizable turbulence model were used. The inviscid convective flux was split by using the Roe-FDS scheme, and the implicit scheme was used with the time marching method to accelerate the convergence of the numerical calculations. The equation of flow control was discretized by using the second-order upwind scheme, and the material was assumed to be an ideal gas. The time step used in this article was initially $1 * 10^{-7}$, then gradually increased to 1×10^{-6}, with 50 iterations.

2.5. Models of Interior Ballistics and Aftereffects

Specifically, the internal ballistic model is as follows:

$$\begin{cases} \frac{dz_1}{dt} = \frac{u_1}{e_1} p_1^n \\ \frac{d\psi_1}{dt} = \chi_1 \frac{dz_1}{dt} + 2\chi_1 \lambda_1 z_1 \frac{dz_1}{dt} + 3\mu_1 \chi_1 z_1^2 \frac{dz_1}{dt} \\ \frac{dv_d}{dt} = \frac{sp_d}{\varphi_d m_d} \\ \frac{dl_d}{dt} = v_d \\ \frac{dv_p}{dt} = \frac{sp_p}{\varphi_p m_p} \\ \frac{dl_p}{dt} = v_p \end{cases}$$

where:

$$\begin{aligned} & p_1(V_\psi + sl_d + sl_p) = f_b \omega_b + f_1 \omega_1 \psi_1 - \tfrac{k-1}{2} \varphi (m_d v_d^2 + m_p v_p^2) \\ & V_\psi = V_1 - \tfrac{\omega_1}{\rho_p}(1 - \psi_1) - \alpha \omega_1 \psi_1 \\ & p_1 = p_p \left(1 + \tfrac{1}{3} \tfrac{\omega_1}{\varphi_d m_d} \tfrac{\varphi_p m_p}{\varphi_d m_d + \varphi_p m_p}\right) \\ & p_1 = p_d \left(1 + \tfrac{1}{3} \tfrac{\omega_1}{\varphi_p m_p} \tfrac{\varphi_d m_d}{\varphi_d m_d + \varphi_p m_p}\right) \end{aligned}$$

where ψ_1 is the percentage of main charge burned, μ_1 is the burning rate coefficient of main charge, e_1 is the thickness of the main charge, and ρ is the density of the gas powder. f_b is the ignition powder, ρ_g is the average density of gunpowder gas in the chamber, and χ_1, λ_1, and μ_1 are the characteristics of the shape of the main charge gunpowder. k is the adiabatic index, l_d is the travel of the projectile, and l_p is the travel of the balance body. ω_b is the charge quality of ignition powder, m_1 is the mass of the projectile, m_2 is the mass of the balance body, ω_1 is the quality of the main charge, φ is the secondary power coefficient, n is the burning rate index, φ_1 is the flow correction factor, p_d is the bottom pressure of the projectile, φ_d is the drag coefficient of the bottom of the projectile, p_p is the bottom pressure of the balance body, φ_p is the drag coefficient of the bottom of the balance body, p_1 is the pressure in the tube, s is the area of the tube, Z_1 is the relative thickness of the main charge that has been ignited, v_d is the velocity of the projectile, v_p is the velocity of the balance body, V_1 is the pharmacy volume, and V_ψ is the free volume of pharmacy.

The aftereffect model is as follows:

$$
\begin{aligned}
p &= p_g e^{-At} \\
A &= \frac{s p_g}{(\beta - 0.5)\omega_q v_0} \\
\beta &= 0.5 + \frac{4-k}{2k} \frac{a_g}{v_0 \sqrt{1 + \frac{3-k}{6} k}} \\
a_g &= \sqrt{\frac{k p_g}{\rho_1}}
\end{aligned}
$$

where s is the area of the muzzle and v_0 is the velocity of the projectile when it reaches the muzzle. k is the specific heat ratio of gunpowder gas and p_g is the pressure of the muzzle.

Then, we compiled these models using C language and the language rules that UDF needs to follow, forming the UDF program. We use the UDF program to control the motion of the projectile inside the tube, as well as the initial conditions of the muzzle during the aftereffect period. The parameters corresponding to a conventional launch were obtained through an interior ballistics model as the initial boundary conditions for the numerical calculation of the muzzle jet, as shown in Table 2. The changes in pressure and velocity at the bottom of the tube are shown in Figure 4a,b. And we inferred that the variations in pressure in the muzzle over time were p = 24.4 $\times e^{-24.668t}$ MPa [22–24]. This enables us to achieve numerical simulation of the development and evolution of shock waves throughout the entire process of projectile motion.

Table 2. Results of calculation of internal ballistics.

Parameters	Value
Working time/ms	68.5
Muzzle pressure/MPa	24.4
Velocity of the moving body/(m/s)	932.0

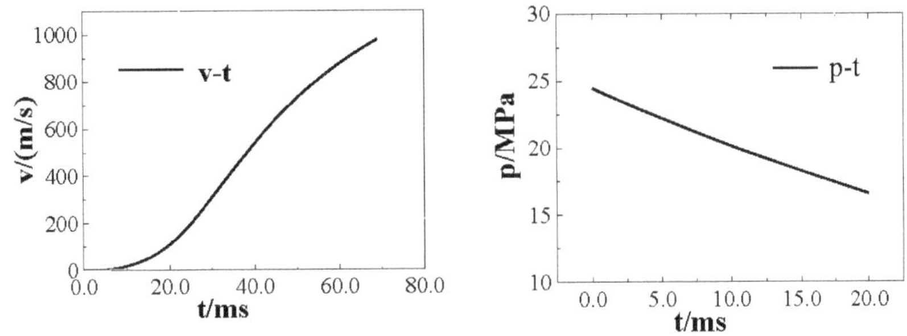

(**a**) Velocity of moving body of interior ballistics over time (**b**) Muzzle pressure over time during the aftereffect period

Figure 4. Velocity of the moving body under the effect of the muzzle pressure over time in the models of interior ballistics and aftereffects.

To achieve control of the projectile's motion throughout the entire process from its initial motion to its flight away from the muzzle, it is necessary to solve the coupled solution process of fluid control equations and rigid body dynamics equations. The motion law of a projectile is determined by the aerodynamic load it receives, and the distribution of the aerodynamic load can be determined by solving the flow field. Therefore, a rigid body six-degree-of-freedom dynamic equation system is established, including the motion equation of the rigid body's center of mass and the motion equation system of the rigid body around the center of mass. The system of six-degree-of-freedom motion differential equations for a rigid body can be expressed as:

$$\frac{dV}{dt} = \frac{1}{m}F_{x2}, \frac{d\theta_a}{dt} = \frac{1}{mV\cos\psi_2}F_{y2}, \frac{d\psi_2}{dt} = \frac{1}{mV}F_{z2}, \frac{d\omega_\xi}{dt} = \frac{1}{C}M_\xi$$

$$\frac{d\omega_\eta}{dt} = \frac{1}{A}M_\eta - \frac{C}{A}\omega_\xi\omega_\zeta + \omega_\zeta^2\tan\varphi_2, \frac{d\omega_\zeta}{dt} = \frac{1}{A}M_\zeta + \frac{C}{A}\omega_\xi\omega_\eta - \omega_\eta\omega_\zeta\tan\varphi_2$$

$$\frac{d\varphi_a}{dt} = \frac{\omega_\zeta}{\cos\varphi_2}, \frac{d\varphi_2}{dt} = -\omega_\eta, \frac{d\gamma}{dt} = \omega_\xi - \omega_\zeta\tan\varphi_2$$

$$\frac{dx}{dt} = V\cos\psi_2\cos\theta_a, \frac{dy}{dt} = V\cos\psi_2\sin\theta_a, \frac{dz}{dt} = V\sin\psi_2$$

Among them, V is the flight speed of the projectile; θ_a is the velocity elevation angle (also known as ballistic inclination angle); ψ_2 is the velocity direction angle (also known as the trajectory deviation angle); $\omega\sigma$, $\omega\eta$, and ω_ξ are the rotational angular velocity of the projectile; φ_a is the high and low angle of the projectile axis (also known as the high and low swing angle); φ_2 is the direction angle of the projectile axis (also known as the directional swing angle); γ is the roll angle; x, y, and z are the spatial co-ordinates of the projectile (horizontal distance, trajectory height, and lateral deviation); Vx, Vy, and Vz are the three components of the projectile's flight speed (horizontal velocity, vertical velocity, and lateral velocity); t is the flight time. F_{x2}, F_{y2}, and F_{z2} are the three components of all combined forces acting on the projectile in the ballistic co-ordinate system. $M\sigma$, $M\eta$, and M_ξ are the three components of all resultant moments acting on the projectile in the axis co-ordinate system; m is the mass of the projectile; A is the equatorial moment of inertia; C is the polar moment of inertia; δ_2 is the directional attack angle; δ_1 is the high and low angle of attack, which is the angle between the first axis co-ordinate system and the second axis co-ordinate system; k_{cx} is the coefficient of conformity for resistance; k_{mz} is the coefficient of conformity for static torque.

The above model is a six-degree-of-freedom rigid body trajectory equation for projectile motion, with a total of 15 variables and 15 equations. The equation system is closed and can be solved using numerical calculation methods. When the structural parameters, aerodynamic parameters, shooting conditions, meteorological conditions, and starting conditions of the projectile are known, the motion law of the projectile and the trajectory data at any time can be obtained by integration.

2.6. Dynamic Mesh Method

Dynamic meshing technology has been widely used to handle changes in the mesh caused by the motion of a rigid body during calculations, as shown in Figure 5. Commonly used dynamic grid algorithms include layering, smoothing, and local remeshing. Given that we assume here that the body moves only in one direction, the layering method can be used to simulate the motion of the rigid body. The dynamic layering method is a very effective method in dealing with the deformation of stretched mesh, which can handle changes in grid topology, add or remove nodes in the grid, and is very suitable for situations with large motion amplitude. It determines whether to split or merge the grid based on the height of the grid connected to the detection motion boundary and the combined effect of the set splitting factor (C_s), collapse factor (C_l), and grid height reference value (h_i). The dynamic layering method includes a height-based dynamic layering method and ratio-based dynamic layering method, both of which have the same principles for grid splitting and merging but have differences in the height of the generated grid. The application of the dynamic layering method is limited to three-dimensional hexahedral meshes and two-dimensional quadrilateral meshes, and it requires high quality of meshes.

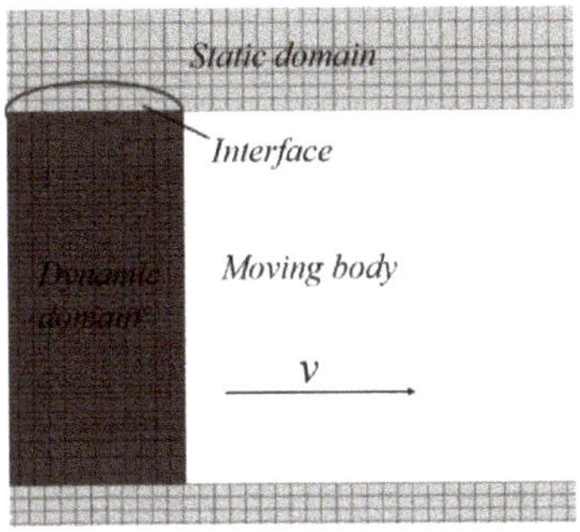

Figure 5. Schematic diagram of the boundary conditions of the model of the muzzle jet.

It requires setting the segmentation and merging factors of the grid. The grid moves forward along the *X*-axis after the moving body. A layer of the grid near the tail of the moving body is elongated, with a side length of hs. Part of the grid that exceeds the boundary of the computational domain is truncated, and its length near the boundary of the computational domain is hc. Assuming that the ideal grid size is hi, the grid is divided into two parts when the newly generated grid size satisfies $h > (1 + C_s)\ h_i$. Moreover, when the grid size satisfies $h > C_s h_i$, the grid merges into one. In this example, $h_i = 2.0$ mm, $C_s = 0.4$, and $C_l = 0.2$.

2.7. Grid Independence Verification

To verify the independence of the grid, we designed three sets of girds, namely 6.0 million, 11 million, and 14 million. They were, respectively, called Coarse grid, Fine grid, and Dense grid. Regarding the variation in a monitoring point (2000, 1000, and 0) mm in Case 4, Figure 6 demonstrates the results of numerical simulations of the different grid models considered. In this article, the region where the moving body moves is set as the moving region, while the other regions are the stationary region, meaning that the mesh is not moving. In this article, the region where the moving body moves is set as the moving region, while the other regions are the stationary region, meaning that the mesh is not moving. The dynamic region in the text is relatively small, and most of it is still a static region. For this reason, using the pressure change at a monitoring point as a grid independence verification can also demonstrate that the calculation results in this article are grid-independent. From the figure, it can be seen that the results of fine and dense grids were very close, with an error of less than 1%. But there were significant differences in the results obtained by the Coarse gird model. It can be concluded that, when using the Fine grid model, the simulation results were independent of the size of the gird.

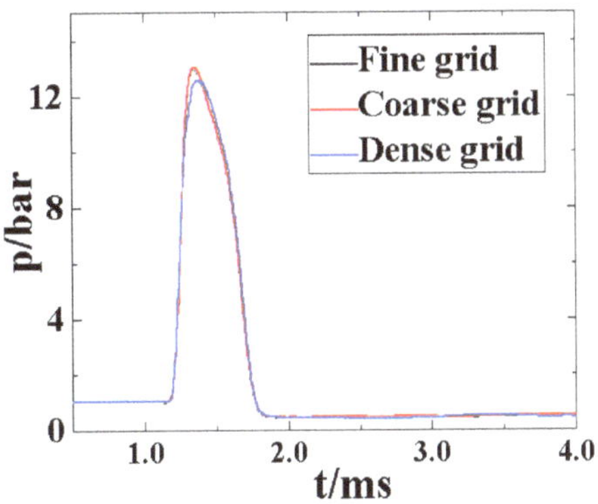

Figure 6. Pressure comparison of a monitoring point to verify grid independence.

3. Experiments

3.1. Test Equipment and Plan

The launch device used here was a test platform for weapons launch with a large caliber and a long barrel, as shown in Figure 7. The device had a diameter of 300 mm, a thickness of the wall of 50 mm, and a total distance of 30 m traveled by the moving body. A schematic diagram of the model is shown in Figure 8. The height between the ground and the muzzle is 2.2 m. Vp means the velocity of the moving body and Vb means the velocity of the balance body. The balance body is a part of the experimental apparatus and an important component for achieving the function and performance of the test, but it is not related to the model in this article. The front half of the tube is the running section of the projectile, called the projectile tube. In the middle of the tube is the chamber, and behind the chamber is the section of the balance body called the balance object tube.

Figure 7. Structure of the launch test device for large-diameter barrel weapons.

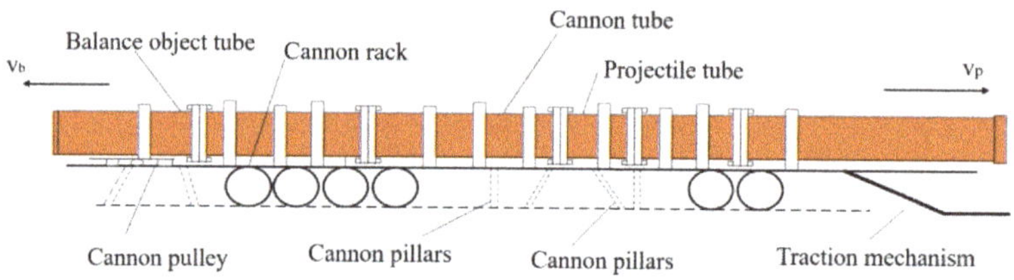

Figure 8. Schematic diagram of launch test device for a weapon with a large caliber and a long barrel.

The experimental plan considered here is shown in Figure 9. The test bench was the main launch device that pushed the moving body out of the muzzle at a speed of 932.0 m/s. The moving body flew forward after exiting the muzzle. Dynamic pressure sensors were installed at key points on both sides of its flight trajectory to monitor changes in transient pressure in the flow field. A piezoelectric dynamic pressure sensor was used to obtain the transient pressure at characteristic points in the overpressure field. All pressure signals were collected by using a DEWE2-A13 transient data recorder, produced by DEWETRON Industrial Equipment Company in Graz, Austria, with a sampling frequency of 500 kHz. We chose KISTLER-6215 and KISTLER-211B4 (Winterthur, Switzerland) pressure sensors for experimental pressure testing. Among them, the maximum range of the KISTLER-6215 sensor is 6000 bar, used to measure pressure changes in the tube and at the muzzle, while KISTLER-211B4 sensor is 500 psi, used to measure pressure changes in the flow field. And a Photon high-speed camera was used to capture images of the formation and development of the overpressure field. During the experiment, the ignition system ignited the gunpowder inside the chamber to produce high-temperature and high-pressure gas, which drove forward the moving body at a high speed. The high-speed camera was also used to capture the developmental posture of the muzzle jet and the movement-related posture of the moving body as well as to calculate its speed. We simultaneously triggered the pressure sensors to monitor the pressure at the characteristic points.

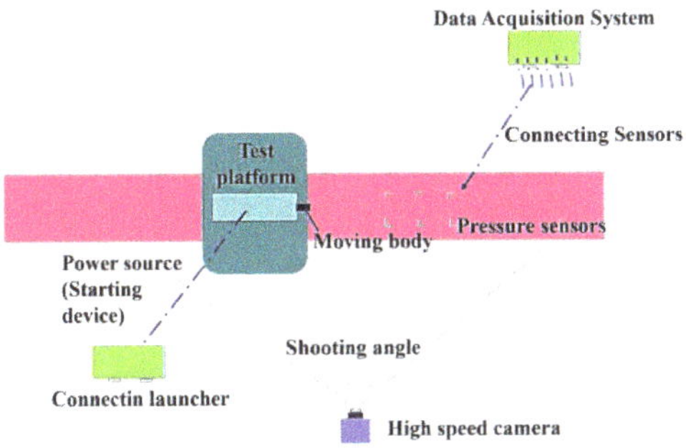

Figure 9. Schematic diagram of test scheme.

3.2. Experiment Results

The high-speed camera recorded the dynamic characteristics of the launch device, and obtained the development and structural characteristics of the supersonic muzzle jet. Owing to the large amount of charge required to launch the body and its high kinetic energy, the gunpowder burned negatively in the chamber and underwent secondary combustion after being ejected from the muzzle. The resulting muzzle flame made it difficult to determine the shape of the shock wave. As shown in Figure 10a, the moving body traveled at a high speed inside the tube, pushing the air in front of it to form compression waves. These compression waves continuously propagated toward the muzzle but did not spread to it. Figure 10b,c show that the air in the front of the moving body was squeezed out of the muzzle to form the initial jet. The sprayed gas rapidly expanded, resulting in a decrease in pressure and an increase in velocity, and diffused into the environment. The wave front of the initial jet was approximately spherical, with a weak impact force that was one order of magnitude smaller than that of the subsequent gunpowder-induced gas jet. At this point, a small amount of gas leaked from the gap between the moving body and the tube to form a small area of the flame at the muzzle.

(**a**) t = 0 ms (**b**) t = 5.0 ms (**c**) t = 10.0 ms (**d**) t = 15.0 ms

Figure 10. Development of the supersonic muzzle jet over time.

Figure 10d shows that, when the moving body left the tube, the high-temperature and high-pressure gunpowder gas inside the tube was quickly sprayed out to chase and surround the moving body. The flow field of the gunpowder gas had a spherical structure. A large amount of gunpowder and gas was sprayed out of the muzzle, causing secondary combustion and forming a large area of high-temperature turbulent combustion. As the projectile moved, the muzzle jet could not surround it, and they gradually separated. A unique shear layer was formed on the side of the moving body, starting at its head and then extending outward from its side. Because the velocity of the gas ejected from the muzzle exceeded that of the moving body, a clear shock wave was formed at its bottom.

Changes in the pressure at key points were also monitored during the test. Figure 11 compares the curves of pressure versus time between the test and the simulation at the point (3000, 500, and 200 mm), and it is clear that they were consistent. Specifically, the center point of the muzzle is defined as the origin. The *X*-axis represents the direction of projectile motion, with the Y-axis perpendicular to the X-axis pointing upwards and the Z-axis perpendicular to the XY plane pointing sideways. Therefore, the method of simulation used in this paper could accurately simulate the characteristics of the muzzle jet of the models.

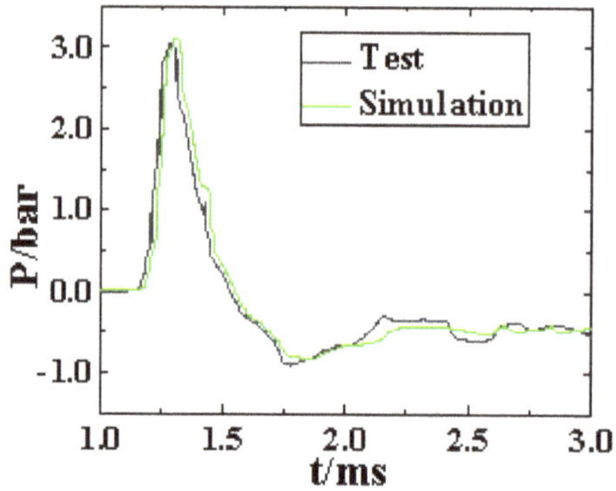

Figure 11. Comparison of the changes in pressure over time at a key point between the test and the simulation.

4. Results of Numerical Simulations

We analyze the evolution of the morphology of the shock wave in the gas jet and its role in enhancing the mixing process in order to describe the strong unsteady characteristics of 3D flow and mechanism of mixing of the gas jet with interference by the initial jet. The muzzle jet with initial interference was complicated as it involved such phenomena as collisions between shock waves, those between shock waves and vortices, and collisions between vortices. This led to the development of reflected and focused shock waves. The complexity of the structural features of the shock waves decreased significantly in three of the simplified models considered here. We focus here on (1) the analysis and comparison on the characteristics of the muzzle jet in the four models on the plane of symmetry, including the mechanism of evolution of the morphology of the shock wave and changes in the morphology of the vortex, (2) analysis and comparison of the 3D morphology of the shock waves, (3) comparison of changes in pressure and velocity over time at the key feature points, and (4) comparison of the parameters of motion of the body.

4.1. Structure of Flow Field

We used the physical model and the model of interior ballistics to determine that, driven by the propellant gas in the tube, the moving body moved for 68.5 ms within the tube before reaching the muzzle. Figure 12 shows the pressure of the initial interference jet formed by compressed air in front of the moving body in the inner tube over time (from 63 ms to 68 ms). As shown in Figure 12a, the initial jet was about to form at 63 ms and the velocity of the moving body was 913.6 m/s at this time. At 63 ms, the moving body is moving inside the tube and the shock wave before it has reached the muzzle. At this point, the initial flow field is about to form and the moving body is 6.5 m away from the muzzle, with a velocity of 901.8 m/s. The pressure at which the shock wave before the projectile reaches the muzzle is approximately 1.1 MPa. Figure 12b–e show that, as the moving body continued to accelerate, the air in front of it was compressed out of the muzzle. The air rapidly expanded, the pressure decreased, and the velocity of the body increased to form the initial jet, including bow-induced shock waves (BS), Mach disks (MD), and bottle shock waves (BoS) (shown in Figure 12f). The body had moved out of the muzzle by 68.5 ms, at which point its velocity was 932 m/s.

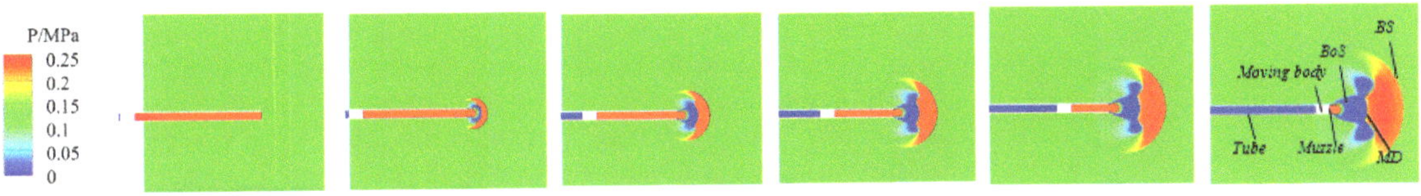

(a) t = 63.0 ms (b) t = 64.0 ms (c) t = 65.0 ms (d) t = 66.0 ms (e) t = 67.0 ms (f) t = 68.0 ms

Figure 12. Changes in the pressure of the initial jet over time.

Figure 13 shows the pressure (0.5 ms~10.0 ms) over time in Case 1 (the model with interference by the initial jet) on the plane of symmetry. As shown in Figure 13a, the initial jet had already influenced the area around the muzzle at t = 0.5 ms. High-temperature and high-pressure gas was sprayed out of the muzzle with the movement of the body, expanded rapidly with a reduction in pressure and an increase in velocity, and continuously diffused toward the surrounding area. The coronal shock wave (CS) in front of the moving body encountered the bottle-shaped shock wave (BoS) of the initial jet, leading to a shock–shock collision that led to the disappearance of the CS of the moving body. The structure of the BoS of the initial jet was also destroyed, and it was replaced by two reflected shock waves (RS). The MD of the initial jet was compressed and deformed and led to the formation of the spherical shock wave (SS) of the muzzle jet.

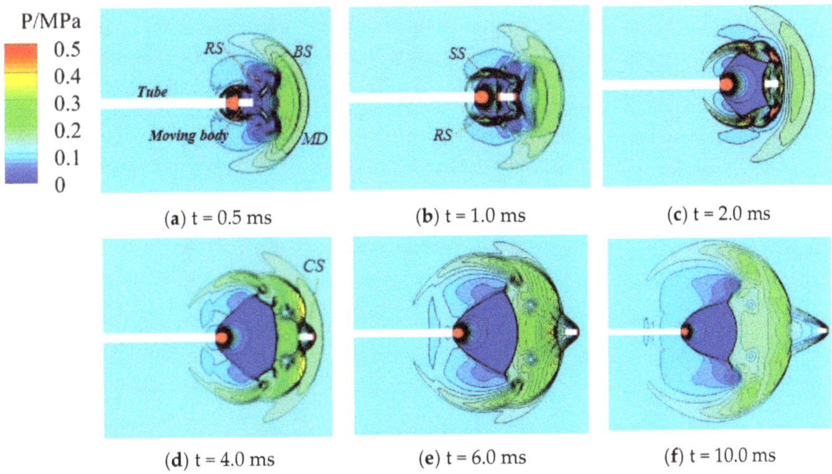

(a) t = 0.5 ms (b) t = 1.0 ms (c) t = 2.0 ms

(d) t = 4.0 ms (e) t = 6.0 ms (f) t = 10.0 ms

Figure 13. Changes in the pressure of the muzzle jet over time in Case 1.

As shown in Figure 13b, the body continued to move forward and the muzzle jet continued to eject gas at t = 1.0 ms. The deformed CS in front of the moving body encountered the MD of the initial jet, causing a shock–shock collision. The MD of the initial jet disappeared and the CS did not form. Instead, two RSs were generated, and the SS of the muzzle jet was clearly visible. It was slightly deformed due to interference from the initial jet. We also observed that a clear shock wave appeared at the bottom of the moving body because the high-temperature and high-pressure gas jet expanded at a much higher rate than the velocity of the moving body. It expanded more quickly under the influence of the initial jet. The muzzle jet encountered the moving body such that this suppressed its development and led to a reflection in the shock wave and a change in its direction of propagation. The muzzle jet continuously sprayed out gas, and the forward-moving shock wave collided with the reflected shock wave to suppress its development. The reflected shock wave was thus stopped and formed a "horizontal" shock wave at the bottom of the moving body.

Figure 13c shows that the body continued to move forward while the muzzle jet continued to spray gas. All shock waves except for the BoS of the initial jet had been quelled by this time, and the BoS of the muzzle jet was formed. Due to the suppression of the development of the SS in the muzzle jet by the surface of the outermost SS of the initial jet, the BoS of the muzzle jet caught up with the SS to yield a shock–shock collision. This led to the formation of multiple RSs that were sufficient to disrupt the shape of the SS and distort the BoS.

Figure 13d shows that the structure of the shock wave of the muzzle jet gradually became clear and the moving body was about to cross the area surrounding the initial jet. A CS gradually formed in front of the moving body, and the BoS and MD gradually became clear. The SS of the muzzle jet was still twisted and deformed due to the limitation of the initial jet and the impact of the BoS. As is shown in Figure 13e, the moving body had completely broken through the hood of the initial jet at t = 6.0 ms, but the initial jet still had a slight impact on the muzzle jet. Although the development of the shock waves was dependent on their history, they formed a complete muzzle jet including the CS, BoS, MD, bottom shock waves of the moving body, and SS.

Figure 13f shows that the shadow of the initial jet was not visible at t = 10.0 ms, and the entire area was filled with jets from the high-temperature and high-pressure gunpowder gas. A complete muzzle jet was thus formed. Although there were some differences between this state and that without interference from the initial jet, the influence of the latter gradually disappeared.

Figure 14 shows changes in the pressure (0.5 ms~10.0 ms) in Case 2 (the model without interference by the initial jet) on the plane of symmetry over time. Due to the absence of interference by the initial jet, the structure of the muzzle jet was more regular than that in Case 1. The body moved forward at a high speed to form a CS at the front. The high-temperature and high-pressure gas was sprayed behind the moving body, rapidly expanded at a speed above 1900 m/s, and chased and impacted the body to form shock waves at its bottom and on both sides of it. The moving body gradually freed itself from the impact of the muzzle jet and flew forward at a high speed. At this point, a complete muzzle jet was formed containing a CS, BoS, MD, bottom shock waves of the moving body, and an SS.

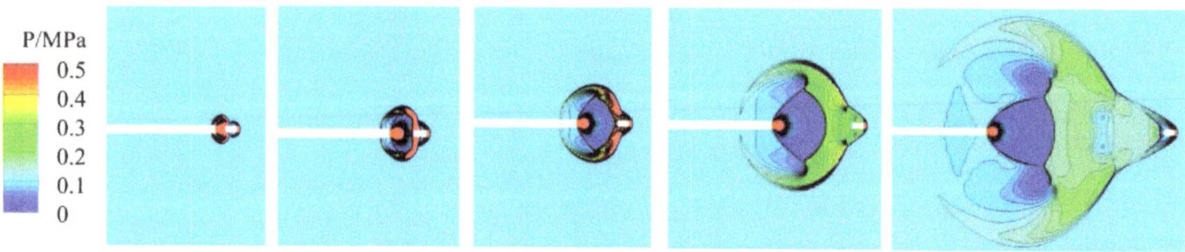

(**a**) t = 0.5 ms (**b**) t = 1.0 ms (**c**) t = 2.0 ms (**d**) t = 4.0 ms (**e**) t = 10.0 ms

Figure 14. Changes in the pressure of the muzzle jet over time in Case 2.

The above results show that the main difference between Cases 1 and 2 was that the speed of expansion of the muzzle jet was lower in the latter case due to the absence of interference by the initial jet, and the degree of enclosure of the moving body weakened to cover only half of its length. The structure of the muzzle jet was clear and distinct, without complex shock–shock collisions or the formation of an additional reflected shock wave. The shape of the muzzle jet was smooth and followed a regular pattern over time. However, under interference from the initial jet, there was no consistent pattern of development of the shock wave in the first 4 ms. The structure of the muzzle jet followed a certain pattern of development until the interference by the initial jet had been completely eliminated.

Figure 15 shows changes in the pressure (0.5 ms~10.0 ms) in Case 3 (the model without the projectile but with interference by the initial jet) on the plane of symmetry. The development of the shock wave was similar to that in Case 1. Under interference by the initial jet, the high-temperature and high-pressure gunpowder gas was sprayed out and encountered the BoS of the initial jet for a shock–shock collision to form two reflected RSs. Following this, the muzzle jet encountered the MD of the initial jet to form a local area of high pressure and two RSs. The muzzle jet continued to develop, and a complete jet was subsequently formed that contained a CS, BoS, MD, bottom shock waves of the moving body, and an SS.

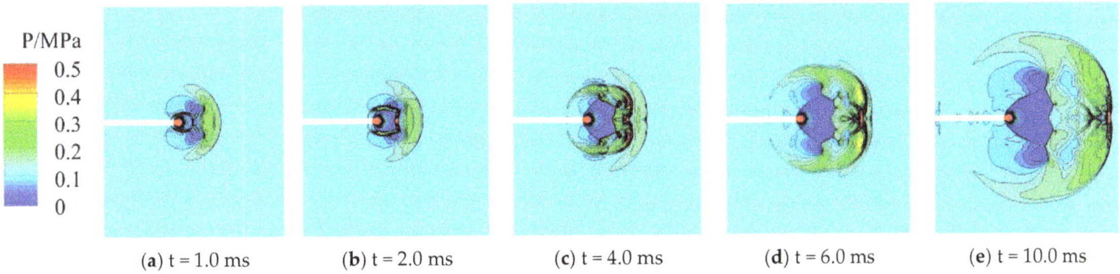

Figure 15. Changes in the pressure of the muzzle jet over time in Case 3.

Figure 16 shows changes in pressure in Case 4 (the model without either the projectile or interference by the initial jet) over time on the plane of symmetry. It clearly shows the typical shock wave of the muzzle jet. High-temperature and high-pressure gas was sprayed out of the nozzle, expanded rapidly to reduce the pressure and increase the velocity, and diffused in a 3D symmetrical manner in the surroundings. The phenomenon of a complete wave system consisting of shock waves, including SSs, RSs, and an MD with overlapping discontinuities, has been widely observed.

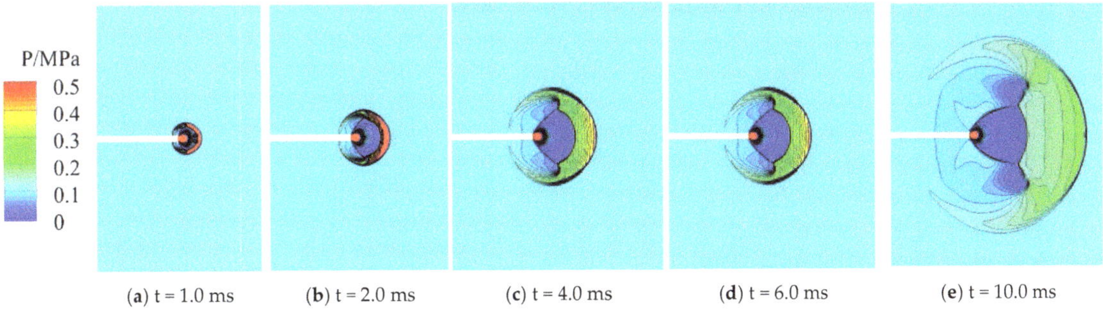

Figure 16. Changes in the pressure of the muzzle jet over time in Case 4.

4.2. Structure of the Vortex

The muzzle jet disturbed by the initial jet featured not only shock–shock collisions but also shock–vortex and vortex–vortex interactions. To comprehensively analyze the characteristics of flow in a muzzle jet with interference by the initial jet, we monitored the temporal changes in the vortices as shown in Figure 17a, which shows that the vortex of the initial jet still existed at t = 1.0 ms and was symmetrical. The vortex was located in the low-pressure region, with the streamline pointing from the center to the outside, while the spiral point was stable. As shown in Figure 17c, the strong shock wave of the muzzle jet encountered the weak vortex of the initial jet at t = 2.0 ms and caused it to disappear. A vortex behind the flow field was formed and gradually became clear, as shown in Figure 17d,e.

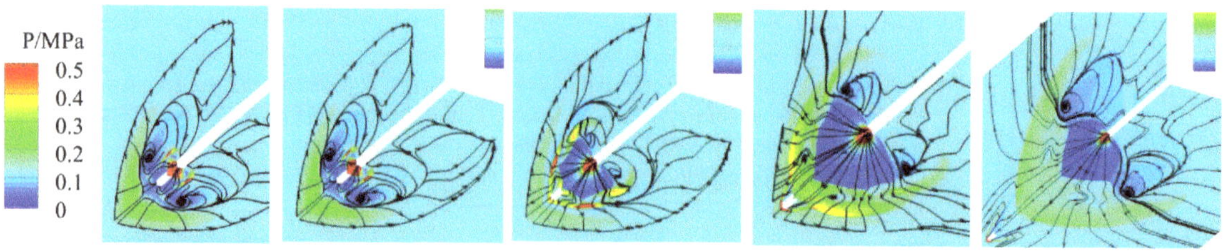

(**a**) t = 0.5 ms (**b**) t = 1.0 ms (**c**) t = 2.0 ms (**d**) t = 4.0 ms (**e**) t = 10.0 ms

Figure 17. Changes in the streamlines of the muzzle jet over time in Case 1.

The vortex in Case 2 was relatively simple and clear (as shown in Figure 18) and was formed at around t = 2.0 ms. Due to the absence of interference from the initial jet, the vortex persisted in the low-pressure area, with its streamline pointing from the center to the outside and the spiral point being stable. Compared with Case 1, Case 3 lacked a coupling of the moving body, while its impact on the collision between shock waves and vortices was minor. We can conclude that the development and evolution of vortices in Case 3 were similar to those in Case 1 (as shown in Figure 19).

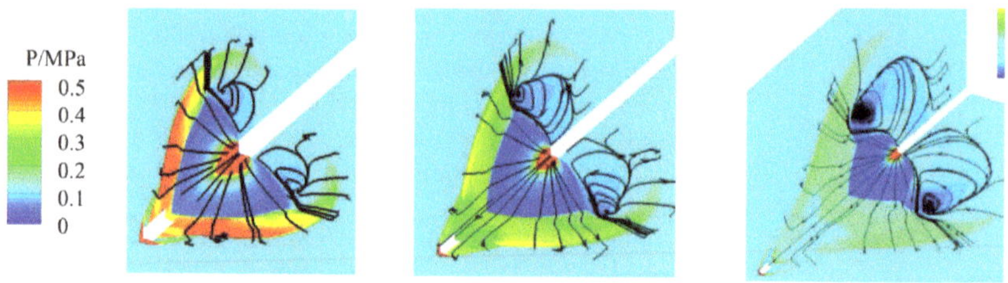

(**a**) t = 2.0 ms (**b**) t = 4.0 ms (**c**) t = 10.0 ms

Figure 18. Changes in the streamlines of the muzzle jet over time in Case 2.

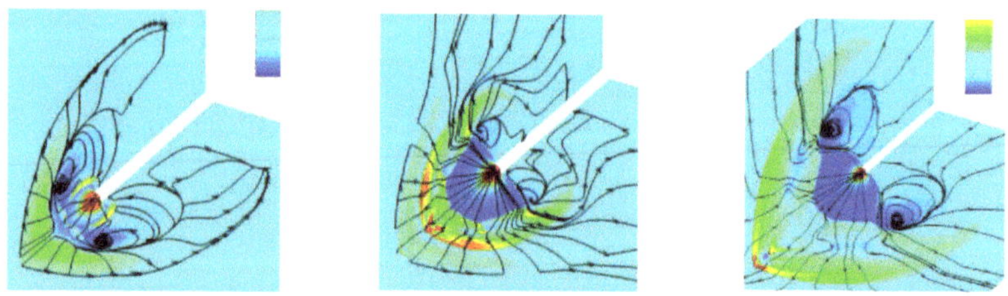

(**a**) t = 0.5 ms (**b**) t = 3.5 ms (**c**) t = 10.0 ms

Figure 19. Changes in the streamlines of the muzzle jet over time in Case 3.

4.3. Structure of the Shock Wave

To fully capture the interference by the initial jet in the shock wave of the muzzle jet, Figure 20 shows a schematic diagram of the shock wave in Case 1 at different times. Figure 20a shows that the outermost spherical shock wave of the initial jet surrounded the complex shock wave inside. The muzzle jet was ejected and encountered the shock wave of the initial jet to yield strong shock–shock collisions, shock–vortex collisions, and vortex–vortex collisions that hindered the expansion of the shock wave and formed an area of local high pressure. As shown in Figure 20b, the shock wave of the muzzle jet, which was disturbed by the initial jet, remained complex over time. The spherical shock wave of the muzzle jet broke through the wave front of the spherical shock of the initial jet, and the two wave fronts collided to arrest the expansion of the shock wave and form an area of local high pressure. The bow shock in front of the moving body had already appeared at this time. Figure 20c shows that, as the muzzle jet developed at t = 8.0 ms, the influence of the initial jet almost disappeared and the shock wave gradually became clear and simple.

(**a**) t = 20 ms (**b**) t = 4.0 ms (**c**) t = 8.0 ms

Figure 20. Development of the shock wave in Case 1.

Figure 21 shows a schematic diagram of the structure of the shock wave at different times in Case 2. The structure of the shock wave of the muzzle jet was clearer and simpler than in Case 1. The bow shock bent on both sides around the shear layer of the jet in 3D space, forming a spherical shock wave front that enveloped the other shock wave structures inside. A flow field composed of cylindrical shock waves, MDs, and RSs was formed in a highly under-expanded jet basin.

(**a**) t = 2.0 ms (**b**) t = 4.0 ms

Figure 21. Development of the shock wave in Case 2.

4.4. Parameters of the Moving Body

Figures 22 and 23 show a comparison of the parameters of the moving body (such as its resistance and lift) over time between Cases 1 and 2. The impact force and the velocity of the moving body were different between the models. Because the initial jet had already disturbed the area in front of the muzzle in Case 1, the pressure in this area was lower than the ambient pressure. When the high-temperature and high-pressure gas was sprayed out, it led to a higher ratio of a drop in pressure compared with the model without disturbance from the initial jet. This accelerated the expansion of the gas. The high-speed gas impacted the bottom of the moving body, causing it to be subject to a greater impact force (as shown in Figure 22a). The velocity of the moving body was also slightly higher than that in Case 2 (as shown in Figure 22b), but this difference was not large enough to have a significant impact on a moving body with an initial velocity of 1000 m/s or higher. We can infer that the resistance experienced by the moving body was also higher than in Case 2 (as shown in Figure 23a). The muzzle jet disturbed by the initial jet involved several shock–shock and shock–vortex collisions that added to the complexity of the flow field. The high-pressure and low-pressure regions alternated, resulting in an unstable lift on the moving body compared with that in Case 2.

(**a**) Impact force (**b**) Velocity

Figure 22. Comparison of the impact force and velocity of the moving body.

(**a**) Drag (**b**) Lift

Figure 23. Comparison of the resistance and lift of the moving body.

Table 3 shows errors in the parameters of the moving body in the two models (Cases 1 and 2). When a large-caliber weapon and a projectile with a high kinetic energy were used, errors in the drag and lift of both models were about 2.5%, the error in the thrust was about 9%, while the error in velocity was the smallest at 0.01%. These results provide a reference for the analysis of numerical simulation for engineering applications.

Table 3. Errors in the parameters of motion of the body in Cases 1 and 2.

Parameters	**Error (Case 2–Case 1)/Case 1**
Drag	−2.055%
Lift	−2.705%
P	−8.938%
v	−0.0074%

4.5. Pressure Comparison of Key Points

To better understand the pressure distribution in the muzzle jet when it was disturbed by the initial jet, we monitored the changes in pressure over time at several key points. Figure 24 is a schematic diagram of the distribution of pressure at these points and Table 4 shows their locations (with the muzzle as the center).

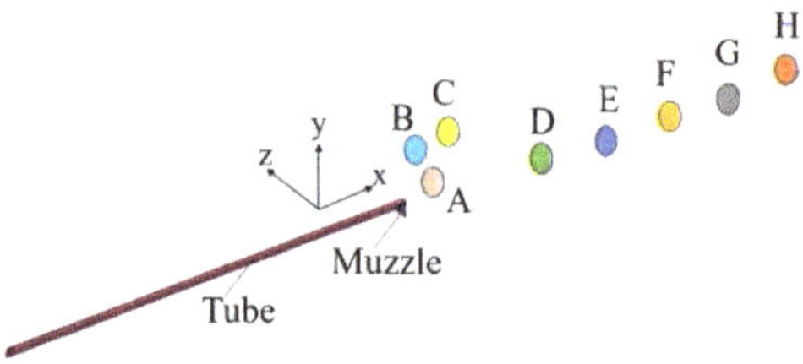

Figure 24. Distribution of feature points.

Table 4. Positional parameters of the feature points.

Picture Title	Points	Location (mm, mm, mm)
(a)	Point A	(500, 200, 200)
(b)	Point B	(500, 500, 500)
(c)	Point C	(1000, 500, 500)
(d)	Point D	(2000, 0, 0)
(e)	Point E	(3000, 0, 0)
(f)	Point F	(4000, 0, 0)
(g)	Point G	(5000, 0, 0)
(h)	Point H	(6000, 0, 0)

Figure 25 shows a comparison of the changes in static pressure over time at various monitoring points. We sought to determine whether the simplified models (Cases 2–4) could attain the same accuracy as the original model (Case 1) for engineering applications. It is clear from the figure that the changes in pressure in Cases 2 and 4 were the simplest and clearest. That is, when the shock wave passed a given monitoring point, the pressure at it increased and then decreased. Slightly more complex scenarios were encountered in Cases 1 and 3 due to interference from the initial jet. The fluctuations in pressure at the monitoring points near the muzzle were significant and similar. Figure 25a,c show that, due to the proximity of the monitoring point to the muzzle, the latter was located in the region of the expansion wave of the initial jet, and the pressure here was lower than the ambient pressure. Therefore, the pressure of the muzzle jet at the initial moment in Cases 1 and 3 was lower than the ambient pressure (i.e., less than zero relative to the ambient pressure). The monitoring points shown in Figure 25b were not disturbed by the initial jet, because of which the initial pressure at this point in all four models was identical to the environmental pressure. The two monitoring points in Figure 25d,e were located in the high-pressure zone of the initial jet, because of which the pressure at these points was higher than the ambient pressure. The three monitoring points shown in Figure 25f–h were located far from the muzzle such that they were unaffected by the initial jet, and thus their initial pressure was the ambient pressure. The changes in pressure at monitoring points near the muzzle in Cases 1 and 3 were similar because they were severely affected by the initial jet and the effect of coupling of the moving body was not strong. At several monitoring points far from the muzzle, the changes in pressure in Cases 1 and 2 were similar, as the muzzle jet had already separated from interference by the initial jet and was influenced only by its coupling with the moving body.

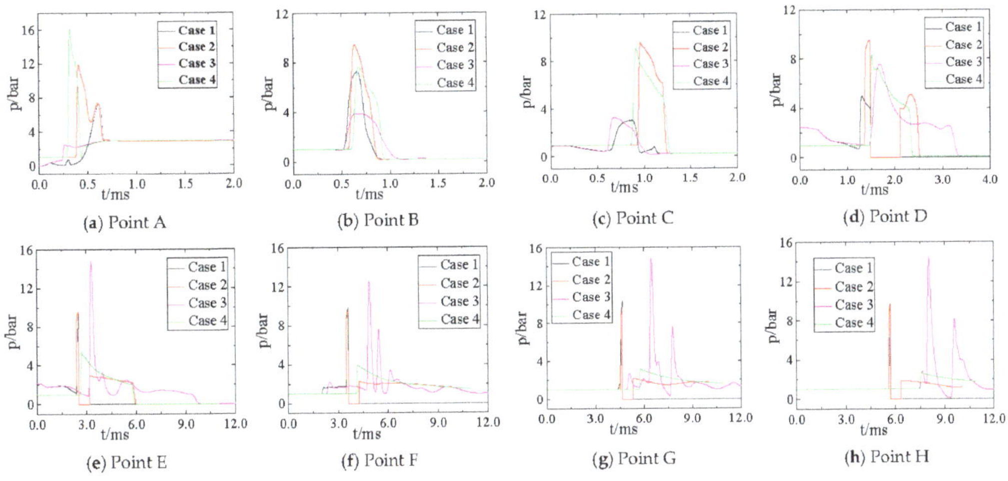

(**a**) Point A (**b**) Point B (**c**) Point C (**d**) Point D

(**e**) Point E (**f**) Point F (**g**) Point G (**h**) Point H

Figure 25. Comparison of pressure at the monitoring points in the four cases.

Figure 26 shows a comparison of the changes in the velocity of the moving body over time at each monitoring point. It is clear that the velocity exhibited the same trend as the static pressure over time at each monitoring point. At monitoring points near the muzzle, the changes in velocity in Cases 1 and 3 over time were similar, as both cases involved the initial jet and the effect of coupling of the moving body was not strong. Changes in velocity in Cases 1 and 2 at several monitoring points far from the muzzle were similar because the flow had already separated from interference by the initial jet and was influenced only by the coupling between the muzzle jet and the moving body.

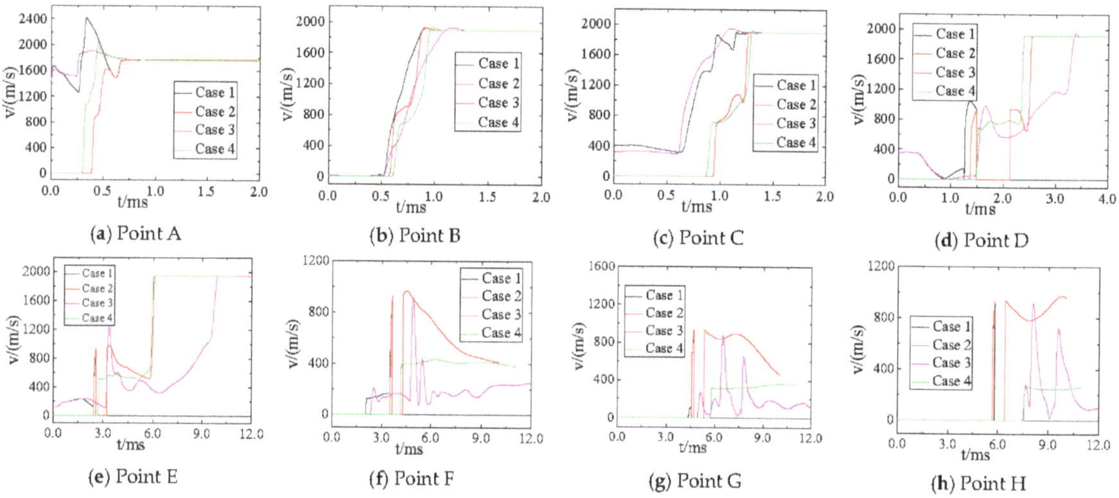

Figure 26. Comparison of velocity at the monitoring points in the four cases.

Taken together, our findings indicate that, under the working conditions considered here—that is, a muzzle pressure of 25 MPa, an initial velocity of the moving body of 900–1000 m/s, a static pressure near the muzzle of approximately to 10–20 atm, and a dynamic pressure of around 20 atm—the muzzle jet without interference from the initial jet (Case 2) was a good choice as a simplified model of calculation beyond a distance of 5 m. When considering a location close to the muzzle, the muzzle jet subjected to interference by the initial jet but without a moving body (Case 3) is suitable as a simplified model of calculation. This solves the problem of the difficulty of simulating the muzzle jet of a large-caliber and long-barrel launcher.

5. Conclusions

In this study, we examined the characteristics of development of shock waves and vortices in a muzzle jet in case of interference by the initial jet for a launch device with a large diameter and a long barrel with a high kinetic energy of the muzzle. The conclusions are as follows:

(1) The shock wave of the muzzle jet under interference by the initial jet was complex and featured shock–shock collisions and shock–vortex collisions in the flow field that suppressed the expansion of the muzzle jet and led to the formation of multiple reflected shock waves and high-pressure zones. The strong shock wave of the muzzle jet collided with the weak vortex of the initial jet, causing it to disappear and leading to the formation of a stable vortex of the muzzle jet in the flow field. The muzzle jet without interference by the initial jet had a clear and simple structure.

(2) Owing to the low energy of the initial jet—one order of magnitude lower than the energy of the muzzle jet—the force of its impact on the high-speed moving body was relatively small, while the impact on its lifting force was relatively large but could be ignored.

(3) If the distance to the muzzle is not considered under the operating conditions considered here, the muzzle jet without the initial jet-induced interference can be used as a simplified model for calculation beyond a distance of 5 m. When considering a location close to the muzzle and ignoring the moving body, the muzzle jet under interference by the initial jet can be used as a simplified model for calculation.

Author Contributions: Conceptualization, Z.L. and H.W.; methodology, Z.L.; software, Z.L.; validation, Z.L.; formal analysis, Z.L.; investigation Z.L.; resources, Z.L.; data curation, Z.L.; writing—original draft preparation, Z.L.; writing—review and editing, Z.L.; visualization, Z.L.; supervision, H.W.; project administration, H.W.; funding acquisition, H.W. All authors have read and agreed to the published version of the manuscript.

Funding: This research received no external funding.

Data Availability Statement: All data and models that support the findings of this study are available from the corresponding author upon reasonable request. All data and models generated or used during the study appear in the submitted article.

Acknowledgments: Thank you to Hexia Huang (Nanjing University of Aeronautics and Astronautics) for his opinions, guidance, and contributions to this article.

Conflicts of Interest: The authors declare that they have no conflicts of interest.

References

1. Li, Z.; Wang, H.; Chen, J. Ground effects on the hypervelocity jet flow and the stability of projectile. *Eng. Appl. Comput. Fluid Mech.* **2018**, *12*, 375–384. [CrossRef]
2. Chaturvedi, E.; Dwivedi, R.K. Computer aided design and analysis of a tunable muzzle brake. *Acta Armamentarii* **2019**, *015*, 89–94. [CrossRef]
3. Jiang, S.Y.; Wang, H.; Lin, C.J.; Wang, J.L. Establishment and simulation of two-dimensional interior ballistics model of large caliber Davis gun considering the axial movement of tubular charge clusters. *Acta Armamentarii* **2016**, *37*, 1941–1948.
4. Schmidt, E.M.; Shear, D. Optical measurements of muzzle blast. *AIAA J.* **2012**, *13*, 1086–1091. [CrossRef]
5. Wang, G.; Cheng, C.; Zhang, X.; Huang, X. Numerical simulation and analysis of muzzle flow during a rarefaction wave gun firing. *Propellants Explos. Pyrotech.* **2021**, *46*, 45–46. [CrossRef]
6. Zhao, X.Y.; Zhou, K.D.; He, L.; Lu, Y.; Wang, J.; Zheng, Q. Numerical simulation and experiment on impulse noise in a small caliber rifle with muzzle brake. *Shock Vibr.* **2019**, *12*, 1–12. [CrossRef]
7. Moumen, A.; Grossen, J.; Ndindabahizi, I.; Gallant, J.; Hendrick, P. Visualization and analysis of muzzle flow fields using the background-oriented schlieren technique. *J. Vis.* **2020**, *23*, 409–423. [CrossRef]
8. Merlen, A.; Dyment, A. Similarity and asymptotic analysis for gun-firing aerodynamics. *J. Fluid Mech.* **1991**, *225*, 497–528. [CrossRef]
9. Ma, D.W. Calculation of unsteady flow field containing moving objects. *Explos. Shock* **1991**, *11*, 31–36.
10. Ma, D.W.; Le, G.G.; Yu, S.H. Numerical simulation of supersonic free jet with insufficient expansion. *J. Comput. Phys.* **1994**, *11*, 362–366.
11. Jiang, K.; Wang, H.; Huang, M. Numerical simulation of the firing flow field of a gun with muzzle brake. *Int. Symp. Ballist.* **2010**, *22*, 51–53.
12. Dai, S.L.; Xu, H.Q. Numerical simulation of three-dimensional detonation waves induced by high-speed flying projectiles. *Propuls. Technol.* **2007**, *28*, 3.
13. Dai, S.L.; He, Z.D.; Xiao, Z.L. Numerical simulation of combustion flow field at the muzzle of negative oxygen balanced propellant. *J. Explos. Explos.* **2011**, *34*, 3.
14. Florio, L.A. Effect of vent opening area and arrangement on gas flow field as gas propelled cylinder exits a flow tube. *Meccanica* **2010**, *45*, 475–501. [CrossRef]
15. Jiang, X.; Chen, Z.; Fan, B.; Li, H. Numerical simulation of blast flow fields induced by a high-speed projectile. *Shock Waves* **2008**, *18*, 205–212. [CrossRef]
16. Guo, Z.; Jiang, X.; Wang, Y. Parallel numerical simulation of muzzle reacting flow. *Chin. J. Comput. Mech.* **2013**, *30*, 111–116+123.
17. Guo, Z.; Qiao, H.; Jiang, X. Numerical simulation on the characteristics of flow field of the muzzle of embedded aircraft gun. *Acta Armamentarii* **2017**, *38*, 2373–2378.
18. Zhang, H.; Xiao, Y.; Zheng, C.; Chen, Z.; Xue, D.; Zhu, S.; Song, W. Characteristics of three-dimensional flow field structure of supersonic transverse jet. *J. Propul. Technol.* **2023**, *44*, 30–39.
19. Schmidt, E.M.; Fansler, K.S.; Shear, D.D. Trajectory Perturbations of Fin-Stabilized Projectiles Due to Muzzle Blast. *J. Spacecr. Rocket.* **2015**, *14*, 339–344. [CrossRef]
20. Schmidt, E.M.; Gordnier, R.E.; Fanslert, K.S. Interaction of Gun Exhaust Flowfields. *AIAA J.* **1984**, *22*, 516–517. [CrossRef]

21. Zhang, X.B. *Gun Ballistics*; Beijing Institute of Technology Press: Beijing, China, 2014.
22. Jiang, Z.; Takayama, K.; Skews, B.W. Numerical study on blast flow fields induced by supersonic projectiles discharged from shock tubes. *Phys. Fluids* **1998**, *10*, 277–288. [CrossRef]
23. Li, Z.J.; Wang, H. Effect of Precursor Flow Field of Muzzle on the Combustion Gas Jet Flow of Gun Propellant. *Chin. J. Energetic Mater.* **2017**, *25*, 282–290.
24. Jin, Z.M. *Advanced Internal Ballistics*; Higher Education Press: Beijing, China, 2003.

MDPI AG
Grosspeteranlage 5
4052 Basel
Switzerland
Tel.: +41 61 683 77 34

Aerospace Editorial Office
E-mail: aerospace@mdpi.com
www.mdpi.com/journal/aerospace

www.ingramcontent.com/pod-product-compliance
Lightning Source LLC
Chambersburg PA
CBHW061455210926
51850CB00008B/1723
9783725820115